INFORMATION, LANGUAGE, AND COGNITION
Edited by Philip P. Hanson

Vancouver Studies in Cognitive Science is a series of volumes in cognitive science. The volumes will appear annually and cover topics relevant to the nature of the higher cognitive faculties as they appear in cognitive systems, either human or machine. These will include such topics as natural language processing, modularity, the language faculty, perception, logical reasoning, scientific reasoning, and social interaction. The topics and authors are to be drawn from philosophy, linguistics, artificial intelligence, and psychology. Each volume will contain original articles by scholars from two or more of these disciplines. The core of the volumes will be six articles and comments on these articles to be delivered at a conference held in Vancouver. The volumes will be supplemented by about nine articles especially solicited for each volume and will undergo peer review. The volumes should be of interest to those in philosophy working in philosophy of mind and philosophy of language; to those in linguistics in psycholinguistics, syntax, language acquisition and semantics; to those in psychology in psycholinguistics, cognition, perception, and learning; and to those in computer science in artificial intelligence, computer vision, robotics, natural language processing, and scientific reasoning.

VANCOUVER STUDIES IN COGNITIVE SCIENCE
forthcoming volumes

VOLUME 2 *Formal Grammar: Theory and Implementation*
Editor, Robert Levine, Linguistics,
Ohio State University

VOLUME 3 *Connectionism: Theory and Practice*
Editor, Steven Davis, Philosophy,
Simon Fraser University

The articles in Vancouver Studies in Cognitive Science are indexed in *The Philosopher's Index*.

information, language, and cognition

edited by Philip P. Hanson

University of British Columbia Press
Vancouver 1990

Printed in Canada

ISBN 0-7748-0327-4

ISSN 0847-0502

Printed on acid-free paper ∞

Canadian Cataloguing in Publication Data
Main entry under title:

Information, language, and cognition

(Vancouver studies in cognitive science, ISSN 0847-0502 ; 1)
Includes bibliographical references.
ISBN 0-7748-0327-4

1. Information theory. 2. Language and languages.
3. Cognition. I. Hanson, Phillip P. II. Series.
Q360.I53 1990 001.539 C89-091616-0

UBC Press
6344 Memorial Rd.
Vancouver, B.C. V6T 1W5

This book has been published with the financial support of
the Advanced Systems Institute of British Columbia
and with the assistance of a grant from the Social Sciences and
Humanities Research Council of Canada, using funds provided by
the Canada Council.

Contents

PREFACE

vii

CHAPTER 1

What Is Information? David Israel and John Perry 1
Comment John W. Heintz 20

CHAPTER 2

Verbal Information, Interpretation, and Attitudes
Nicholas Asher 29
Comment Edward P. Stabler, Jr. 57

CHAPTER 3

Truth Conditions and Procedural Semantics Robert F. Hadley 73
Comment Zenon W. Pylyshyn 101

CHAPTER 4

Putting Information to Work Fred Dretske 112
Comment Brian Cantwell Smith 125

CHAPTER 5

Concept Formation and Particularizing Learning
Lee R. Brooks 141
Comment Paul Thagard 168

Contents

CHAPTER 6

Information and Representation Jerry Fodor 175
Comment Ali Akhtar Kazmi 191

CHAPTER 7

Rountable Discussion
Nicholas Asher, Lee R. Brooks, Fred Dretske, Jerry Fodor, David Israel, John Perry, Zenon W. Pylyshyn, and Brian Cantwell Smith 198

CHAPTER 8

Belief and Mental Representation Scott Soames 217

CHAPTER 9

Partial Information, Modality, and Intentionality
Fred Landman 247

CHAPTER 10

Unifying Partial Descriptions of Sets Carl J. Pollard and M. Andrew Moshier 285

CHAPTER 11

Animals and Individualism Kim Sterelny 323

CHAPTER 12

When Is Information Explicitly Represented? David Kirsh 340

CHAPTER 13

Psychological Inference, Constitutive Rationality, and Logical Closure Ian Pratt 366

CHAPTER 14

Intrinsic Information John D. Collier 390

CONTRIBUTORS 410

Preface

In February 1988, Simon Fraser University inaugurated its Cognitive Science Programme by hosting the first in a projected annual series of interdisciplinary conferences. This volume — also the first in a projected annual series — contains the proceedings of that conference, much of it revised, together with additional solicited papers of related interest. The principal speakers at the conference are identifiable from the table of contents as those whose papers are followed by a 'Comment.' These latter also contributed much to the occasion. Their authors could themselves readily have served as principal speakers, as, of course, could the authors, not present at the conference, of the additional solicited papers! The conference closed with a 'Roundtable Discussion' of the proceedings involving principal speakers and commentators and a substantially edited transcript of that lively conversation has also been included here.

Our aim has been to give leading exponents and critics of 'information-based' theories of cognition and language a forum for reflection on the philosophical underpinnings of their respective approaches, an opportunity to identify and explore points of convergence and divergence with respect to such questions as the following. What is information? In what does the informational content per se of an event or structure consist? What is it for an event or structure to *have* such content and what is it for a cognitive agent to be *attuned* to it? Where, if at all, do we locate informational content in the causal nexus? What are its materialistic credentials? More immediately, where, if at all, do we locate informational content being stored or processed in some information processing system? What explanatory efficacy can the appeal to informational content have in theories of language and of cognition, in particular in semantics, learning theory, communication theory, and theories of perception, inference, and belief? Does such explanatory efficacy require the causal efficacy of informational content per se? What, if anything, does the semantics of sentences reporting beliefs and other propositional attitudes suggest about the nature of those attitudes? Conversely, how if at all does our theory of propositional attitudes constrain our semantics of natural language attributions of them? Must information processing be 'computational' in some interesting sense that constrains the architecture of implementing processors? Is the possibility of falsehood and misrepresentation, in language and in thought, compatible with information-theoretic accounts of them? These are just some of the issues explored, explicitly or implicitly, in this collection.

The contributors also represent a broad spectrum of background theoretical commitment. For instance, we hear from situation semanticists, discourse representational theorists, head-driven phrase structure grammarians, exponents of the semantics of direct reference, and a procedural semanticist. We hear from exponents of the view that information is mentally encoded in 'local' explicit sentence tokens of a 'language of thought' and that cognitive activity is computation on those tokens; from those who instead promote the explanatory virtues of postulating 'distributed' mental representations; and from those who suggest that information may be encoded only implicitly, cognitive activity in part involving its 'recovery' via appropriate procedures, as context may demand.

Perhaps not surprisingly, some points of divergence turn out to hinge on discrepancies in the use of the term "information." For example Perry, Israel, Dretske, Fodor, and Collier tend to use it in a 'truth value biased' way, such that, very roughly (ignoring such nuances as the *de dicto/de re* distinction, relativity to constraints, and different types of declarative sentence), if a signal or whatever carries the information that p, where 'p' stands in place of some declarative English sentence, then it is true that p. Others, such as Asher, Hadley, Landman, and Kirsh, tend to use it in a 'truth value neutral' way, as akin to something like propositional content or meaning, thus, trivially, permitting false information. It is arguable that this latter, truth value neutral, use is the one more natural to semantic theory, theories of linguistic competence, and theories of computation. But interest in the former, truth value biased, use has recently increased in philosophical circles, due in large measure to two important books: Fred Dretske's *Knowledge and the Flow of Information* (1981) and John Perry's and Jon Barwise's *Situations and Attitudes* (1983). For instance Dretske, in the work just cited, attempts to explain the truth value neutral intentional contents of beliefs and other cognitive states as arising out of truth value biased informational properties. This may strike one as a priori implausible. But if intentional contents are to be explained in a materialistically acceptable, naturalistic way as Dretske wants, then they must ultimately be explained in terms of phenomena that are themselves 'truth value resistant': that admit of *neither* truth *nor* falsity. If truth value biased information, admitting just of truth, can be explained in a materialistically acceptable way, then perhaps it can, as Dretske hopes, provide a crucial causal-explanatory link between truth value neutral intentional contents and their truth value resistant naturalistic base. If in addition the particular semantic contents of natural language sentences can be explained, as projected by Paul Grice's well-known semantic program, in terms of the intentional contents of cognitive states of natural language users, then a truth value biased notion of information can turn out to be causal-explanatorily prior to a truth value neutral notion not only in the theory of intentionality but in semantic theory as well.

Obviously these are all big if's. And the problems arising that are specific to information-based approaches are inextricably tied to other problems arising within other approaches as well. Some problems of the latter sort that are also at least broached in this collection include: the semantics of propositional attitude attributions; the issue of 'individualism'; the tenability of realism about the propositional attitudes of 'folk psychology' both as to their existence and as to their causal efficacy in the production of behaviour; and the psychology of concept acquisition, storage, and use.

I thank the members of our Board of Readers for their reading and recommendations, and our Editorial Advisory Board for its advice and support. I especially thank Steven Davis, series editor, for his organizational initiatives, energy, and vision. But there are many other individuals and institutions whose participation and support must also be acknowledged. I am grateful for editorial assistance from Robert Hadley, Ali Kazmi, Jeffry Pelletier, Alice ter Muelen, William Rounds, and Steven Savitt. I would like to thank John Madden, the then acting president of the Advanced Systems Institute of British Columbia, my colleague Raymond Jennings, Cognitive Science secretary Mary Bruegeman, and students of the Cognitive Science Programme for their participation and help with the conference. Special thanks is due to Nick Cercone, director of Simon Fraser University's Centre for Systems Science, for his support of our endeavours. The conference would not have been possible without the generous financial support of the following institutions: Simon Fraser University, the University of British Columbia, the University of Victoria, the Advanced Systems Institute of British Columbia, and the Social Sciences and Humanities Research Council of Canada (grant no. 443-87-

0196). Additional thanks are owed to the latter two institutions, ASI and SSHRC, for their further generous support of this publication.

Philip P. Hanson

CHAPTER ONE

What Is Information?

David Israel and John Perry

INTRODUCTION

In this paper we provide an account of information and informational content, show how it accords with certain intuitive principles of information, and use it to resolve an apparent tension among those principles.[1]

Our aim is not to provide a semantics for talk about information, but to provide an account of information itself. Still, it will be helpful to begin with some observations about the structure and logical properties of *information reports*.

(1) The x-ray indicates that Jackie has a broken leg.
(2) The acoustic waves from the speaker carry the information that the announcer said, "Nancy Reagan is irritated."
(3) The fact that the x-ray has such and such a pattern indicates that Jackie has a broken leg.

(1) and (2) have a structure similar to reports of propositional attitudes. We call an information verb or verb phrase ('shows,' 'indicates,' 'carries the information'), together with the preceding noun phrase, an *informational context*. We shall call the proposition designated by the 'that'-clause the *informational content*. The object designated by the initial noun phrase of a report like (1) or (2) we shall call the *carrier* of the information; the fact designated by the initial noun phrase of a report like (3) we shall call the *indicating fact*.

Both styles of information reports are *factive*. That is, if the report is true, the informational content is true too. If the x-ray indicates that Jackie has a broken leg, then she does. In particular, if the fact that the x-ray has such and such a pattern indicates that her left hind leg is broken, then it is. In this way information reports differ from reports of some cognitive attitudes and linguistic acts, but are similar to others. What is believed or conjectured or asserted or denied need not be true, although what is seen or known must be. And in this way information reports differ from reports of what is possible, and are similar to reports of what is necessary.

Information contexts, like modal contexts and propositional attitude contexts, are clearly not truth-functional. Reports of cognitive attitudes and linguistic acts differ from modal statements, in that substitution of necessarily equivalent statements in the latter preserve truth value, while this is not so with the former. On this issue, information reports are like reports of cognitive attitudes and linguistic acts. For example, it does not follow from (1) that the x-ray shows that Jackie has a broken leg and $7 + 5 = 12$.

Like 'believes,' and 'is necessary,' 'indicates' distributes across *and* but not across *or*.

> If x indicates that P and Q, then x indicates that P and x indicates that Q.
> x may indicate that P *or* Q, even though x neither indicates that P nor indicates that Q.

Reports of linguistic acts and cognitive attitudes are notoriously opaque — substitution of codesignative terms in the content sentences does not always preserve the truth value of the whole report. As we might expect given the emphasis in recent philosophy of language on the different semantic properties of names and definite descriptions, it is important to distinguish two kinds of opacity. Opacity with respect to definite descriptions is relatively noncontroversial, common, and well understood. Modal, cognitive, linguistic, and informational reports all exhibit such opacity, when read in a certain way. Even though Jackie is Jonny's dog, we cannot infer from (1) that the x-ray indicates that Jonny's dog has a broken leg, *if* we take this to mean that Jonny's ownership of the dog is part of what is indicated.

Opacity with respect to proper names is less common, more controversial, and less well understood. It seems that if Tully was necessarily human, so was Cicero; that if Cicero was possibly the best philosopher of his age, so was Tully. But it seems that someone might well say or believe that Cicero was the best Roman philosopher, while not believing or even explicitly denying that Tully was. Should we say that such a person said or believed that Tully was the best Roman philosopher, simply because he said or believed that Cicero was? This would be misleading. Would it be false? Philosophers differ over whether the opacity is real and semantic or apparent and pragmatic. If the former, then it seems that on this score, information reports are closer to modal statements than to reports of cognitive attitudes and linguistic acts. If the entrails of some animal showed that Tully had a broken leg, then they showed that Cicero did.

Information contexts, then, are factive and not truth-functional; substitution of necessarily equivalent content statements does not preserve truth; they distribute across *and* but not across *or*; they may support opaque readings with definite descriptions in the content sentences, but not with proper names.

With our intuitions thus sharpened, let us turn from information reports to information itself.

THE PRINCIPLES

We now turn to stating some intuitive principles of information. We use such terms as 'fact' and 'situation' in their ordinary senses; in the next section we give explications of them within situation theory.

We take the second sort of information report, exemplified by (3), as canonical. Where *X* is a noun phrase designating the carrier of the information, '*X* indicates that such and such' is elliptical for '*X*'s being so and so indicates that such and such.' For example, we bring Jackie, who has been limping badly, to the vet, who takes an x-ray of Jackie's left hind leg, the one she's been favouring. The x-ray is developed. At this point the vet might say something fully explicit like (3).

So our first principles are as follows:

(A) Facts carry information.[2]
(B) The informational content of a fact is a true proposition.

What underlies the phenomenon of information is the fact that reality is lawlike; that what is going in one part of reality is related to what is going on in some other part of reality, by laws, nomic regularities, or as we shall say, *constraints*. Our point of view may be taken as a generalization of Hume's. He took constant conjunctions to be contingent matters of fact, that one type of event was *constantly conjoined* with another. We take constraints to be contingent matters of fact, that one type of situation *involves* another. Involving implies constant conjunction: if one type of situation involves another, then if there is a situation of the first type, there is one of the second type. But we leave open the question of whether constant conjunction implies involvement.

In a world knitted together by constraints — whether these be constant conjunctions or some more metaphysically potent connections — situations carry information. The fact that there is a situation of one type carries the information that there are situations of the types that one involves. If it is a constraint that objects left unsupported near the surface of the earth fall, then the fact that a certain apple near the surface of the earth is left unsupported carries the information that it will fall.

This conception licenses the notion of the information carried by a fact relative to a constraint. It is this relative notion of informational content that we think is implicit in our actual thinking about information and important for theoretical purposes. From it one *might* derive an absolute notion of the information carried by a situation, as that information carried by the situation relative to some constraint or other. We do not think this is a useful notion.

(C) The information a fact carries is relative to a constraint.

Hume saw constant conjunction as supplying the world with enough structure so that events contained information, which experience enabled us to recognize. But this structure did not require that there be any intrinsic connections between events; no event contained information in virtue of its intrinsic properties. If the event were embedded in a different sort of world, where different constraints held sway, it would carry quite different information than it actually does.

(D) The information a fact carries is not an intrinsic property of it.

The informational content of a fact can involve objects quite remote from those involved in the fact. Jackie is not a part or aspect of the x-ray mentioned in (3), nor is she a constituent of the fact that the x-ray has such and such a pattern, but something remote from it. The x-ray is not broken, and does not have bones. Information typically involves a fact indicating something about the way things are elsewhere and elsewhen, and this is what makes information useful and interesting.

(E) The informational content of a fact can concern remote things and situations.

This conception of information can explain how an x-ray could carry the information that some dog had been x-rayed and had a broken leg. But it is not clear how it can account for the specific information the x-ray carries about Jackie that is reported in (3). As we noted above, Jackie is not a part of the x-ray, and it does not seem that her having a broken leg could be constantly conjoined with x-rays exhibiting the pattern that the vet recognizes. So how can the informational content of the x-ray have her as a constituent?

We shall call the sort of information reported, e.g., in (3), *incremen-*

tal information. The conception is most easily grasped if we think of what the x-ray indicates *to* the vet. If she does not know which dog the x-ray is of, it simply indicates that some dog has been x-rayed that has a broken leg. We call this *pure* information. But if she knows that Jackie was x-rayed, then the pattern on the x-ray indicates to her the additional or incremental information that Jackie has a broken leg. This is the incremental information carried by the x-ray, given the fact that the x-ray is of Jackie. The fact that is given *connects* the indicating situation and the specific objects the information is about, so we shall call it the *connecting fact*. We must be careful that this example does not mislead as to our intentions, however. Incremental information is important in understanding the use humans make of information, but humans and mental states need not be brought into its analysis. Incremental information about specific objects is an objective feature of the world that is there for us to use.

(F) Informational content can be specific; the propositions that are informational contents can be about objects that are not part of the indicating fact.
(G) Indicating facts contain such information only relative to connecting facts; the information is incremental, given those facts.

If we put the x-ray in a drawer for a day or a month, it will still indicate that Jackie had a broken leg. After a month, of course, it will not indicate that Jackie has a broken leg *then*, for the leg might have mended. It will still indicate that Jackie had a broken leg at the time the x-ray was taken. This illustrates two important points about information.

First, different facts can carry the same information. Suppose that t is the time the x-ray was taken, and t' is a month later. The fact that the x-ray exhibits a certain pattern at t and the fact that it exhibits that pattern at t' are different, yet they carry the same information. And of course many facts, exhibiting more radical differences from the original state of the x-ray, could carry the same information about Jackie's leg — the way she limped, the vet's remarks after feeling the leg, and the notations in Jackie's file years later.

In the case of the stored x-ray, though, the later fact contains the information that Jackie had a broken leg *because* the earlier one did. And this is the second point illustrated by the example. For this is a (very) simple case of the storage of information. Note that what goes on in this case is that the carrier of the information is itself stored, in the drawer say, whence it and the information it carries can be retrieved. This storage system works so long as the manner in which the indicating object is stored preserves the indicating property. The world is to be arranged in such a way that the carrier has the indicating property over a usefully long stretch of time. No storage system works forever.

Now imagine that a xerox is made of the x-ray and the copy sent to a vet in another city. It, too, will indicate that Jackie has a broken leg. That is, it's having such and such a pattern will indicate that Jackie has a broken leg. This is a simple case of the *flow of information*. Notice here that it's crucial that the copy be a copy, that is, that it be related in a certain way to the original carrier and its indicating property. But things can be otherwise. In some cases storage and transmission of information involves varying both the carrier and the indicating properties. Thus, the x-ray's indicating pattern could be digitized and sound waves produced by scanning the binary array. (This process could even be reversible, up to the stipulated quantization.)

Later we look closely at a more complex case of the flow of information due to Dretske.[3] An announcer speaks into a microphone, the microphone is connected through a transmitter to a transmitting antenna. The modulation of the electromagnetic signal transmitted by the antenna contains information about what the announcer says because it contains information about the way the microphone diaphragm vibrates, and that contains information about the voice.

(H) Many different facts, involving variations in objects, properties, relations and spatiotemporal locations, can indicate one and the same informational content — relative to the same or different constraints.

(I) Information can be stored and transmitted in a variety of forms.

The x-ray's being such and such at *t* carries the information that Jackie's left hind leg is broken; but what good does this do the x-ray? None. What good does it do Jackie? Perhaps a lot. If, that is, the vet has the information that Jackie has a broken leg, the chances of her doing Jackie some good increase dramatically. It has often been noted that books contain a lot of information, too, yet that fact seems not to be of any use to books. There is a distinction between carrying or containing information and *having* information. We shall suggest that an agent or device has the information that P just in case it is in a state that both carries the information that P, and controls the behaviour of the device in a way appropriate to the truth of P.

(J) Having information is good; creatures whose behaviour is guided or controlled by information (by their information carrying states) are more likely to succeed than those which are not so guided.

There is a certain tension between (J) and the rest of our principles. They all emphasize that the information carried by an agent or device being in a certain state is a contingent, extrinsic fact about that agent or device. Given different constraints or specific facts, the information carried by the agent or device being in the state would be quite different. Yet the effect that being in the state has on the device cannot depend on these remote contingencies. How then can such states control behaviour in ways that are appropriate to this information? We shall see, however, that when we think through our principles in a careful way, this tension is only apparent.

THE FRAMEWORK

We have noted that indication is a relation between facts and propositions. We have also said that what underlies information are laws or constraints involving types of situations. So what are all these things? And what precisely are we saying about them?

Situations

A basic idea of situation theory is that there is a concrete reality, which has concrete parts but not concrete alternatives. This reality can be thought about, perceived, studied, and analysed in a variety of different ways, from a variety of different perspectives, for a variety of different purposes. But ultimately everything that exists, everything that happens, everything that is true, has its status because of the nature of this reality. The parts of reality are what we call *situations*. Situation theory is committed to there being situations, but not to there being a largest total situation of which the rest are parts.

Relations, argument roles, locations, individuals, issues, positive and negative states of affairs

When we think or talk about reality, we need some way of analysing it. This we call a *system of classification and individuation*. Such a system consists of domains of situations, relations, locations, and individuals. The commonplace that different schemes can be used to study the same reality is one to which situation theory subscribes. But this fact should not be thought of as showing that situations are structureless, with their properties projected onto them by language or thought. Rather, situations are rich in structure, and support a variety of schemes, suited (or unsuited) to various needs.

Each relation R comes with a set of argument roles. For example, the relation of *eating* comes with the roles of *eater*, *eaten*, and *location of eating*. Objects of appropriate sorts play these roles. The eater must be some sort or organism. The eaten must be a physical object or quantity of stuff. The location of eating must be a spatio-temporal location.

A relation, together with appropriate objects assigned to its roles, gives rise to an *issue*, namely, the issue of whether or not the objects stand in the relation. There are two possibilities, and each of these we call a *state of affairs*.

Example. If *eating* is the relation, Reagan is the eater, a certain quantity of succotash is the eaten and the White House at a certain time is the location (call it *l*), then there are the following two states of affairs:

$\ll$ *Eats, Loc*: *l*, *Eater*: *Reagan*, *Eaten*: *the succotash*;1 $\gg$
$\ll$ *Eats, Loc*: *l*, *Eater*: *Reagan*, *Eaten*: *the succotash*;0 $\gg$

The first state of affairs resolves the issue positively, the second, negatively. We say the first has a positive and the second a negative polarity.

The relation of *eating* is the major constituent of these states of affairs; Reagan, the location *l* and the quantity of succotash are the minor constituents. The polarities should not be thought of as constituents at all.

Officially, we don't assume that the argument roles of a relation have a natural order — that is, an order independent of the order in which they are expressed in a given language or in a given construction in a language. But in this paper we shall often use the order suggested by English to identify argument roles, without explicitly mentioning them.[4] For the first state of affairs above we write:

$\ll$ *Eats, l, Reagan, the succotash*; 1 $\gg$

Facts and other states of affairs; makes-factual; the partiality of situations

Situations determine whether a given state of affairs or its dual (the one with the same relation, and same assignment of constituents to roles but the other polarity) is a fact. This primitive relation we call *making factual* or *supporting*.

$s \models \sigma$ means that *s makes σ factual*.

We will also make use of the property of *being factual*. A state of affairs is factual iff some real situation supports it.

$\models \sigma$ means that *σ is factual*.

The following are uncontroversial theses about the $\models$ relation:

Given a state of affairs and its dual,

- some situation will make one of them factual;
- no situation will make the other one factual;
- some situations will leave the issue unresolved, i.e., will make neither of them factual.

The following is a controversial thesis about this important relation:

- some situation resolves all issues.

This, of course, is the thesis that there is a largest total situation.

The third thesis tells us that situations are partial. They do not resolve all the issues (except, perhaps, for the total situation called for in the fouth thesis). Because of the partiality of situations, we must distinguish between two ways a situation s can fail to make a given state of affairs σ factual:

— s may make the dual of σ factual
— s may fail to resolve the σ issue one way or the other.

Parameters and anchors

For theoretical purposes, it is useful, though not strictly necessary, to have a domain of *parameters* corresponding to individuals and locations.[5] Where $\ll \ldots, a, \ldots \gg$, is a state of affairs with a as a minor constituent, and $\mathbf{a}$ is a parameter, $\ll \ldots, \mathbf{a}, \ldots \gg$ is a *parametric state of affairs* or *infon*.[6]

The step from states of affairs to infons is a sort of abstraction.[7] To get from the infons back to states of affairs, we need *anchors*. An anchor is a partial function from the domain of parameters to appropriate objects. Where f is an anchor, $> \ \ll \ldots, \mathbf{a}, \ldots \gg [f] = \ll \ldots, f[\mathbf{a}], \ldots \gg . <$

An anchor f *satisfies* an infon i relative to a situation s iff $s \models i[f]$. An anchor f satisfies an infon i *simpliciter* iff $\models i[f]$, i.e., if there is a situation s such that $s \models i[f]$.

Compound infons

We need to characterize two sorts of compound infon, the *meet* of a set of infons, $\wedge I$, and the *existentialization* of an infon with respect to parameter x, $\exists x(i)$. We characterize the conditions under which an anchor satisfies each:

- f satisfies $\wedge I$ iff $i[f]$ is factual for each $i \in I$.[8]
- f satisfies $\exists x(i)$ iff for some object a, $i[f_{x/a}]$ is factual.

Types, constraints, and involvement

Now we need to define the notions which are at the heart of our account of information.

Where σ is a state of affairs, $[s \mid s \models \sigma]$ is the *type of situation* that supports σ. Where i is an infon (i.e., a parametric state of affairs), $[s \mid s \models i]$ is a *parametric type*, and i is the *conditioning infon* of T ($cond(T)$). A situation s is *of* parametric type T relative to f if $s \models i[f]$, where i is the conditioning infon of T and f is defined on all of the parameters of i.

Since infons and parametric types are the entities most used from now on, we shall mean parametric types when we say "types"; non-parametric types may be thought of as the special case.

We take constraints to be states of affairs with types of situations as constituents. *Simple involvement* is a binary relation. If T involves T', then for every situation of type T, there is one of type T'.[9] We write:

$\ll Involves, T, T', 1 \gg$

Relative involvement is a ternary relation. If T involves T' relative to T'', then, for any pair of situations of the first and third types, there is a situation of the second type. We write:

$\ll Involves_R, T, T', T'' 1 \gg$

Propositions

We take propositions to be nonlinguistic abstract objects that have absolute truth values. From the perspective of situation theory, this means that a proposition requires not only a type — that which corresponds or doesn't correspond to the way things are — but also a situation for the type to correspond to. Two basic kinds of propositions are recognized in situation theory. An *Austinian proposition* is determined by a situation and a type and is true if the situation is of the type. A *Russellian proposition* is determined by a type alone, and is true if some situation or other is of that type. If we adopted the fourth thesis, that there is a total situation, Russellian propositions could be taken as Austinian propositions determined by this total situation.[10]

Propositions are not infons. Infons characterize situations; propositions are truth bearers. We shall assume that for each type of situation and each situation there is an Austinian proposition that is true just in case that situation is of that type. With respect to Russellian propositions, we shall assume that for each type, there is a proposition that is true just in case some situation is of that type. This last is a strong assumption, that can lead to paradox. We shall not concern ourselves with such issues in the present paper; instead we urge all interested parties to consult the treatment in Barwise and Etchemendy's *The Liar*.[11]

Infons may have individuals and locations as constituents. When an infon with an individual or location as a constituent is the conditioning infon of a type, then we also say that the type has that individual or location as a constituent, as does the proposition determined by that type. A proposition whose type contains no such constituents, because each argument role has been quantified over, is *general*, in Kaplan's terminology; others are *singular*. We shall say that a singular proposition is *about* its constituents.

INFORMATION

We now turn to constructing our theory of information within the version of situation theory just sketched.

Let C be some constraint. The fact *σ carries the pure information that P relative to* C iff

(1) $C = \ll Involves, T, T'; 1 \gg$.[12]
(2) For any anchor f such that $\sigma = cond(T)[f]$, $P =$ the proposition that $\exists s'(s' \models \exists a_1, \ldots a_n(cond(T')[f]))$.

Informally, we would have the following in the case of the x-ray:

The x-ray's being Φ-ish indicates that there is a dog, of whom this is an x-ray, and that dog has a broken leg.[13]

We are often interested in more specific information. For instance, to guide her action appropriately, the vet has to know which dog. It is not enough for her to be acquainted with the indicating fact and aware of the constraint. She must know that the x-ray was of Jackie's left hind leg; she must know that the information it carries is *about* Jackie — what we have called *incremental* information. To capture the notion of incremental information, we need a more complex constraint, one of relative involvement, the third type being the *connecting type*, the type of the *connecting situation*. We also call such constraints, *relative constraints*.

Let C be some relative constraint, then the fact *σ carries the incremental information that P relative to C and the fact σ′* iff

(1) $C = \ll Involves_R, T, T', T''; 1 \gg$.
(2) For any anchor f such that $\sigma = cond(T)[f] \wedge \sigma' = cond(T'')[f]$, $P =$ the proposition that $\exists s' (s' \models \exists a_1, \ldots a_n(cond(T')[f]))$.

Again, informally, in our case, the connecting fact is that the x-ray in question is of Jackie's left hind leg, and it is in virtue of Jackie's being a constituent of this fact that she is a constituent of the indicated proposition, the proposition that Jackie's left hind leg is broken.[14]

An application of the theory

Let's now apply the theory more formally and fully to our example involving Jackie's leg and the x-ray. We can consider this as a case of *pure information* or of *incremental information*.

In both cases, the indicating fact σ is the x-ray's being of a certain type at t. When we consider the pure information, we have in mind the following simple constraint: whenever there is a state of affairs consisting of some x-ray's having such and such a pattern at some time t, then there is a state of affairs involving a dog's leg having been the object of that x-ray and that leg's being broken at t.[15] So the indicated proposition is that there is a dog of which this is the x-ray, and it has a broken leg. The pure information is about the x-ray, but not about Jackie, or her leg.

Using the resources of situation theory, we represent the simple constraint as follows:

$$T = [s \mid s \models \ll \textit{X-ray}, \mathbf{x}, \mathbf{t}; 1 \gg \wedge \ll \textit{Has–pattern–}\Phi, \mathbf{x}, \mathbf{t}; 1 \gg]$$
$$T' = [s \mid s \models \ll \textit{Is–x–ray–of}, \mathbf{x}, \mathbf{y}, \mathbf{t}; 1 \gg \wedge \ll \textit{Has Broken Leg}, \mathbf{y}, \mathbf{t}; 1 \gg]$$
$$C = \ll \textit{Involves}, T, T'; 1 \gg$$

The indicating situation, σ, is

$$\ll \textit{X-ray}, a, t'; 1 \gg \wedge \ll \textit{Has – pattern} – \Phi, a, t'; 1 \gg$$

where a is the x-ray and t' the time. We assume that σ is factual, that is that $\exists s(s \models \sigma)$. Now let f be any anchor defined on $\mathbf{x}$ and $\mathbf{t}$ (at least) such that

$$\sigma = \textit{cond}(T)[f] = \ll \textit{X-ray}, \mathbf{x}, \mathbf{t}; 1 \gg \wedge \ll \textit{Has–pattern–}\Phi, \mathbf{x}, \mathbf{t}, 1 \gg [f]$$

(Thus, $f(\mathbf{x}) = a$ and $f(\mathbf{t}) = t'$.) Then P = the proposition that

$$\exists s'(s' \models \exists y(\ll \textit{Is–x–ray–of}, x, y, \mathbf{t}; 1 \gg \wedge \ll \textit{Has Broken Leg}, y, \mathbf{t}; 1 \gg)[f])$$

Thus P is the proposition that the state of affairs which consists of some dog being the object of a, the x-ray in question (at t', the time in question) and that dog's having a broken leg (at the time in question) is factual. Or, more simply, it is the proposition that there is some dog whose leg is depicted by a at t' and whose leg is broken at t'.

When we consider this as a case of incremental information, we have in mind the relative constraint that if an x-ray is of this type, and it is the x-ray of a dog, then that dog had a broken leg at the time the x-ray was taken. The fact that the x-ray was of Jackie is the connecting fact, and the incremental informational content is the proposition that Jackie has a broken leg. This proposition is about Jackie, but not about the x-ray.

The relevant relative constraint is:

$$C' = \ll Involves_{R}, T, T', T'';1 \gg$$

where T, the indicating type is as before. T', the indicated type is

$$[s \mid s \models \ll \textit{Has Broken Leg}, \mathbf{y}, \mathbf{t}; 1 \gg]$$

and T'', the connecting type is:

$$[s \mid s \models \ll \textit{Is–x-ray–of}, \mathbf{x}, \mathbf{y}, \mathbf{t}; 1 \gg].$$

As before, σ is:

$$\ll X-ray, a, t'; 1 \gg \wedge \ll Has-pattern-\Phi, a, t'; 1 \gg.$$

Again, we assume that σ is factual. Further, we assume that the connecting state of affairs, σ' is factual. Where b is Jackie, $\sigma'/\mathbf{y}$-*is*

$$\ll Is-x\text{-}ray-of, a, b, t'; 1 \gg.$$

Any anchor f, such that $\sigma = cond(T)[f]$ and $\sigma' = cond(T')$, must be defined on the parameter $\mathbf{y}$ of the connecting type, in particular, it must anchor $\mathbf{y}$ to Jackie. Thus, for any such anchor f, the proposition carried incrementally by σ relative to C and σ' is the proposition that

$$\exists s''(s'' \models \ll \textit{Has Broken Leg}, b, t'; 1 \gg).$$

This is a singular proposition about Jackie, and not at all about the x-ray. And it is, after all, Jackie that we're concerned about.

The flow of information

Now consider a case in which it is natural to speak of "information flow." The manner in which the diaphragm of a certain microphone is vibrating carries information about what a certain announcer is saying. The modulation of the electromagnetic signal arriving at a certain antenna carries information about the way in which that microphone diaphragm is vibrating. And finally, the modulation of the elecromagnetic signal arriving at the antenna carries information about what the announcer is saying, for instance, "Nancy Reagan is irritated."

How does the modulation of the electromagnetic signal at the antenna carry information about the words the announcer spoke? Let's look at the constraints and connecting facts that are involved.

The first constraint we call $C_{voice\text{-}info}$: If the diaphragm of a microphone vibrates in a certain way (T_{mike}), then the announcer's voice produced certain sounds (T_{voice}), given that the announcer was speaking into the mike ($T_{mike\text{-}voice}$).

$$C_{voice\text{-}info} = \ll Involves_R, T_{mike}, T_{voice}, T_{mike\text{-}voice}; 1 \gg$$

The connecting fact, that the announcer was speaking into the microphone we call $\sigma_{\text{mike-voice}}$.

The second constraint we call $C_{mike\text{-}info}$: if the electromagnetic signal reaching the antenna is of a certain type ($T_{antenna}$), then the diaphragm of the microphone vibrates in a certain way (T_{mike}), given that the antenna and the microphone are connected in a certain way ($T_{antenna\text{-}mike}$).

$$(C_{mike\text{-}info}), = \ll Involves_R, T_{antenna}, T_{mike}, T_{antenna\text{-}mike}; 1 \gg$$

The connecting fact, that the antenna and the mike are connected in this way, we call $\sigma_{\text{antenna-mike}}$.

The third constraint we call $C_{info\text{-}flow}$: if the electromagnetic signal reaching the antenna is of a certain type ($T_{antenna}$), then the announcer's voice produced certain sounds (T_{voice}), given that the announcer was speaking into the mike and the antenna and the microphone are connected in a certain way ($T_{antenna\text{-}mike} \wedge T_{mike\text{-}voice}$).[16]

$$C_{info\text{-}flow}) = \ll Involves_R, T_{antenna}, T_{voice}, (T_{antenna\text{-}mike} \wedge (T_{mike\text{-}voice}); 1 \gg$$

The connecting fact, that the antenna, mike and voice are connected in this way, we call ($\sigma_{\text{antenna-mike}} \wedge \sigma_{\text{mike-voice}}$).

Let *P* be the proposition that the announcer said "Nancy Reagan is irritated." Let *Q* be the proposition that the diaphragm of the mike in question is vibrating in such and such a way — the way that is the major constituent of type T_{mike}. Then

- σ_{antenna} carries the information that *Q* relative to $C_{mike\text{-}info}$, given the connecting fact $\sigma_{\text{antenna-mike}}$.

Moreover, we say that

- σ_{antenna} carries the information that *P* relative to $C_{info\text{-}flow}$, given ($\sigma_{\text{antenna-mike}} \wedge \sigma_{\text{mike-voice}}$).

In such a case, we also say information of type T_{voice} flows along channels of type ($T_{antenna\text{-}mike} \wedge T_{mike\text{-}voice}$). We can derive something more like the normal use of the term if we move to the individual objects, the carriers, that are constitutents of the connecting types.[17]

We might as well generalize, and state the following principle of information:

(K) There are laws of information flow.

Laws of information flow involve compound infons and relations among the parameters of those infons. In plain(er) English: laws of information flow involve relations among states of various components of information systems, states of various objects remote from those systems and connections between these. These laws are useful or exploitable to the extent that these relations and connections are controllable or at least knowable.[18]

The Helpfulness of Information

Now we must face the question of how information, as we conceive it, can be helpful. Clearly, for information to be of use to some agent, or to enable some device to do what it is supposed to, it is not enough that the states of the agent or device *carry* the information. The agent or device must in some sense *have* the information.

We want to develop an account of having information as being in a state that plays two roles. First, the agent's being in the state carries certain information relative to a constraint. Second, an agent's being in that state has an effect (relative to some other constraint) that is appropriate given the information. In that case, we want to say that the agent not only carries but has the information.

But there seems to be a large problem standing in the way of developing such an account on the basis of an approach to information of the sort we have been putting forward. On our account, the information that an object carries in virtue of being in a certain state is not *intrinsically* connected to the object's being in that state. The x-ray's being in state Φ carries the information that Jackie's leg is broken *only* relative to a constraint and a fact. Relative to other constraints and facts, that very same state will carry different information.

But the effect of that state on other parts of the system of which it is a part will not depend on these constraints and facts. How then can the resulting response be appropriate to the information? Our strategy for analysing the having of information, given our account of information, seems to require something like action at a distance.

These problems, however, are merely apparent.

Let's consider a simple example. You stick a pencil in an electric pencil sharpener; a lever is depressed; a circuit is closed; the motor turns on, the blade spins; the pencil is sharpened. In this case, the insertion of the pencil caused the pencil sharpener to be in a certain state, having a lever depressed, that carried information. Under normal usage, this state only occurs when a pencil is inserted, and so carries the information that this is so. This state causes things to happen inside the pencil sharpener: the circuit closes, the motor starts, the blade spins. So, the state of having the lever depressed plays two roles. It carries information, relative to constraints, about the wider circumstances in which the system finds itself — that a pencil has been inserted. And it causes things to happen in the system.

Note that we say that the electric pencil sharpener worked, or did what it was supposed to do, or responded appropriately because we have in mind the goal of sharpening pencils. Relative to another goal, say frustrating people who want their pencils sharpened, it did not work. One might convert a device that works, relative to the first goal, to one that works, relative to the second, by putting in a different kind of blade — say one that leaves the end of the inserted pencil blunt. Whenever we talk about success or failure, or the appropriateness of an action, we have in the background some goal or measure of success.

Let us suppose:

(1) G is a goal, say of sharpening pencils.
(2) *Lever-depressed*, *circuit-closed*, and *blades-spinning*, etc. are states of systems of a certain kind K.
(3) There is a constraint $C_{pure\text{-}info}$: if a system a of kind K is in *lever-depressed* at location l, then there is a pencil inserted in a at l.
(4) There is a constraint C_K that governs the internal workings of the system: if a is in state *lever-depressed* at l, a will go into state *circuit-closed*, then state *blades-spinning*.
(5) There is a constraint $C_{pure\text{-}result}$: if a is in state *blades-spinning* then if there is a pencil in contact with the blades of a, that pencil will be sharpened.

The pencil sharpener of kind K has been designed so that the state that carries the information that a pencil has been inserted sets in motion a chain of events that promotes the goal for which it was designed, sharpening pencils. The design will be successful only in an environment in which the depressing of the lever will carry the information that a pencil has been inserted, and the motion of a blade against a pencil will leave it sharpened. We say that the system is *attuned* to these constraints, relative to the goal of sharpening pencils.

Thus there is no particular problem about how an agent or device may be caused, by the state that carries remote information, to respond in ways appropriate to that information, and so be said not merely to carry but to have that information. The problem may still seem to apply to the case of incremental information, however. What can be the point of saying that the state of a device or an agent has the information that . . . *b* . . . ? In such a case, the state of the agent or device will *carry* the information that . . . *b* . . . relative to some fact that connects the agent or device with *b*. It could have been connected with some other object, *c*, in which case it would carry the information that . . . *c* . . . . But it could be in exactly the same state *S* in the two cases. In this case the effects would be the same, so in what sense can the presence of *b* rather than *c* be relevant?

This argument is fallacious, however, for the effects need not be the same. The immediate effects of being in state *S* will be the same, no matter what the remote cause of being in state *S* might be. But the remote effects of being in state *S* may depend on facts that vary with the remote causes. These remote effects may be the ones that are relevant to the success or failure of the system's response, relative to a given goal.

In our example, the insertion of a pencil *p* into our sharpener depresses a lever, which closes a circuit, causing a motor to impel the blades. The lever would have been depressed in the same way, had another pencil *p′* been inserted rather than *p*. But given that *p* is inserted, *p* gets sharpened; had *p′* been inserted, *p′* would have been sharpened. The remote effects of inserting the different pencils differ, even though the local effects are the same.

The following factors are involved:

(1) *G* is a goal.
(2) *lever-depressed, circuit-closed, blades-spinning,* etc. are states of systems of a certain kind *K*.
(3) There is a relative constraint $C_{inc\text{-}info}$: *if a system a* kind *K* is in *lever-depressed* at location *l*, then pencil *c* is inserted in *a*, given that *c* is depressing the lever of *a*.
(4) There is a constraint C_K that governs the internal workings of the system: if *a* is in state *lever-depressed* at *l*, *a* will go into state *circuit-closed*, then into state *blades-spinning*.
(5) There is a relative constraint $C_{inc\text{-}result}$: if *a* is in state *blades-spinning*, pencil *c* will be sharpened, given that *c* is in contact with *a*'s blade.
(6) There is a constraint, $C_{how\text{-}things\text{-}are}$: if a pencil *c* is depressing the lever of a device of kind *K*, it will be in contact with the blades of that device.

Note the last constraint, that connects the fact of a pencil depressing the lever of the sharpener and the pencil being in contact with the blades of the sharpener. These facts are rather intimately connected, given the construction of an electric pencil sharpener.[19] In cases involving information flowing to an agent from more remote events, which then performs actions whose appropriateness moreover depends on remote effects, the analogous connection between facts that accounts for the agent's success, may be quite a bit more complex and fragile. When such contingencies relate the objects an agent has information about with the objects its actions need to affect to promote its goals, mere *attunement* may not suffice. In such cases, having information may require a system of representation to keep track of these contingencies.[20]

NOTES

1 The work presented here was supported by the System Development Foundation under a grant to the Center for the Study of Language and Information, Stanford University, and SRI International. Many of the ideas reported here reflect the work of the STASS (Situation Theory and Situation Semantics) group at CSLI; we are particularly indebted to Jon Barwise.

2 As we shall see, this is really shorthand for the following. Situations carry information in virtue of making certain states of affairs factual.

3 *Knowledge and the Flow of Information* (Cambridge, MA: MIT Press 1981), 58.

4 The argument role for spatiotemporal locations will always be displayed last.

5 In our paper *What Are Parameters?* (in preparation), we address the issue of what parameters are, and whether they are a necessary part of situation theory.

6 The term 'infon' is to suggest that parametric states of affairs are theoretical entities that are the basic units of information. It is due to Keith Devlin. 'Infon' and 'state of affairs' are close in meaning: an infon is a parametric state of affairs; a state of affairs is a nonparametric infon.

7 This is a bit misleading; see *What Are Parameters?*.

8 This characterization partially reflects the postulation of a complete lattice of infons. For the meet of two infons σ, σ', we use the notation $(\sigma \wedge \sigma')$.

9 Note that the definition does not require that when s is of type T that *it* also be of type T'.

10 See Jon Barwise and John Perry, *Situations and Attitudes* 139–40. The distinction is further clarified and the present terminology introduced in Jon Barwise and John Etchemendy, *The Liar* (Oxford: Oxford University Press 1987) where it plays a key role in the treatment of semantic paradox. See also the postscript to the 2nd edition, forthcoming.

11 In what follows, we will avail ourselves of propositions in which the situation parameter is existentially quantified. The reader should not take this as the expression of a substantive commitment on our part.

12 We simplify by treating the *involves* relation, as well as the relation of *relative involvement* as not having an argument role for locations. We should also note that relative constraints are not to be confused with what in *Situations and Attitudes* were called conditional constraints.

13 In what follows, we shall simply assume that the x-ray's being Φ-ish indicates that it is of a dog's leg.

14 As we shall see, this reflects the fact that the anchor for the connecting type, that is for T'', must assign Jackie to the role of being the object whose leg is x-rayed and thus the indicated type, T' — and the indicated proposition — will be about her.

15 For the sake of simplicity, we shall assume that the x-ray is developed essentially instantaneously.

16 The complete lattice of infons induces a complete lattice of types. A situation s is of $(T_1 \wedge T_2)$ iff s is of T_1 and s is of T_2.

17 Strictly speaking, such objects figure as the values of anchors for the parameters of the conditioning infons.

18 The reader should compare the above with what Dretske, *Knowledge and the Flow of Information*, calls the *Xerox Principle*.

19 Given, that is, *how-things-are*.

20 For more on these issues, the reader may consult Israel's *The Role of Propositional Objects of Belief in Action*, CSLI Report No. 72 and Perry's *Circumstantial Attitudes and Benevolent Cognition*, CSLI Report NO. 53, reprinted from J. Butterfield, ed., *Language, Mind and Logic* (Cambridge, Eng.: Cambridge University Press 1986), 123–33. We hope to address these issues further in a series of papers.

Comment
What can a pencil sharpener know?

John W. Heintz

In their paper, David Israel and John Perry give an account of two important features of information-carrying facts: that they carry information about remote situations — situations outside of themselves — and that they can contain information about specific objects which may also be remote from the information-carrying fact.

Israel and Perry develop some useful machinery for displaying their account. My main purpose in these remarks is to develop, by way of some suggestions, the part of their account that addresses the goal-orientation of some information-carrying systems, the difference between *carrying* and *having* information.

Israel and Perry begin by enunciating a number of principles, such as:

(*A*) facts carry information;
(*B*) the informational content of a fact is a proposition; and
(*C*) the information a factual state of affairs carries is relative to a constraint.

An x-ray provides an example:

(3) The fact that the x-ray has such and such a pattern indicates that Jackie has a broken leg.

The x-ray itself is the *carrier* of the information. That Jackie has a broken leg is the *content*. The fact that the x-ray has such and such a pattern is the *indicating fact*. The *constraint* is a further fact, that any x-ray's having such and such a pattern *involves* there being (or having been) a dog with a broken leg. That the content is specific (specifically about Jackie) is due to a further connecting fact, that the x-ray is of Jackie.

An obvious question, one raised in print and discussion by some at the conference, is how to select among the very many (true) constraints which cluster around any given indicating fact. For example, this x-ray's having its particular pattern also involves there having been an x-ray technician taking the x-ray. Indeed, in conjunction with the fact that Mickie Williams was on duty that day, the x-ray carries the information that Mickie Williams took the x-ray. Israel and Perry accept, I trust, examples of this sort, because they are analysing one familiar use of the word "information." It is the notion we use when we say that the ice falling off the edge of Antarctica carries information about when it first fell on Antarctica as snow, and when we say that light reaching earth from a star carries information about that star's chemical composition. We tend to use the word "information" once we (by which I mean *them* — the relevant scientists) figure out how to infer the date of snowfall from some measurements taken from the air trapped in the ice. This is a sense of "information" which leaves no room for misinformation. If we get it wrong it is because the constraint statements we believed were not true. It is in this sense of information that we select from many indicating facts a few which we count as meaningful.

In jargon of an earlier day Israel and Perry's indicating facts were called *signs,* or *natural signs,* and the information they carried was referred to as the *signification* of the respective signs. Ryle used this language when he said that a sign is evidence of what it signifies.[1] It is this use that Reginald Jackson elaborated upon in distinguishing signs from symbols:

> A sign is a fact or, if not a fact, it is at least factual. . . . The knowledge of what a sign signifies is the basis of a valid inference. For it justifies an assertion other than the assertion that the sign signifies this, namely, the assertion that what the sign signifies is the case. E.g., the knowledge that this peal of thunder signifies that a flash of lightning has occurred justifies the assertion that a flash of lightning has occurred.[2]

Grice's notion of *natural sign* is much the same.[3]

What do Israel and Perry have to add to these earlier accounts? They go beyond Ryle, Jackson, and Grice in two ways. They employ the machinery of situation theory[4] to give a semantics for those facts which are signs, the propositions which are their contents, and the various relations among these. Second, they are interested in giving an account of when an agent or artifact that is in an information carrying state actually has the information carried by that state. Charles W. Morris was concerned with something like this when he imposed a further constraint on the use of "sign": "something is a sign only because it is interpreted as a sign of something by some interpreter."[5]

For Israel and Perry, too, it is a matter of principle: "(J) Having information is good; creatures whose behaviour is guided or controlled by information (by their information carrying states) are more likely to succeed than those which are not so guided."[6]

Three problems suggest themselves. First, it appears that there is altogether too much information. Recall principle (C):

> (C) The information a factual state of affairs carries is relative to a constraint.

Different constraints yield different information. So relative to one constraint, the fact that the x-ray has such and such a pattern carries the information that Jackie's leg is broken. Relative to another it carries the information that Mickie Williams took the x-ray. Relative to other constraints it carries the information that its film base was manufactured by Kodak, its emulsion contains silver, it was developed in Rodinal, that it was exposed by a certain intensity of radiation, and so on indefinitely. According to Israel and Perry, what whittles down the number of constraints, and hence the variety of information, is that the indicating fact *has an effect that is appropriate given the information.*[7] An agent may carry lots of information, but will only be said to have it where the indicating fact has an appropriate effect. You may be in a state which carries the information that you are in the presence of food, but if that state does not have an effect appropriate to your being in the presence of food, you do not have the information; the state is a sign you fail to *interpret*.

The second problem is that there is too little information. Constraints are constant conjunctions: whenever there is a fact of type *IF* (for indicating fact) there is a situation of type C (for content).[8] The trouble is that neither art nor nature supplies us with very many constant conjunctions. Thunder is not quite always accompanied by lightning. X-rays sometimes have misleading patterns. Litmus paper usually, but not always, turns blue in a base.

The third problem was raised by J.A. Fodor in discussion at the conference: the content is underdetermined. It will generally be the case that for a particular indicating fact *IF* and content C such that *IF* produces an effect appropriate to C against a background *B*, there will be another content *C′* for which *IF* produces an appropriate effect relative to background *B′*. Nothing in nature, Fodor claimed, determines whether *IF* carries the information C or *C′*. Consider this example: suppose that, whenever a fly flies within a certain range (C), a frog's visual system goes into a certain state (*IF*) which in turn has the effect of the frog's tongue snaking out to snatch the fly. This is an *appropriate* effect, given that flies are good frog food (*B*). Suppose, however, that the frog's visual system goes into the very same kind of state (*IF*) when something within a certain range of size and colour moves within a certain range of the frog (*C'*). Suppose that *IF* has the effect of the frog's tongue snaking out to snatch the object (fly or not). This effect is appropriate given that *most of the time*, whenever something within that range of size and colour moves within that range of the frog (*C'*) it is good frog food. What is there to choose between saying that the frog, in virtue of *IF*, has the information *C* and saying that is has the information *C'*? Fodor says nothing can.

Mohan Matthen has recently offered an account that attempts to deal with the first two problems together. Where Israel and Perry spell out an effect's being appropriate in terms of a *goal* or *measure of success*[9] (their example is a mechanical device), Matthen (thinking of biological systems) talks of *functions*.[10] Matthen's definition works as follows:

IF has the function *E* in circumstance *B* iff
(1) *IF* has the consequence *E* in circumstance *B*.
(2) *IF* is there because it does *E* in circumstance *B*.[11]

Matthen then defines a *quasi-perceptual state*: "A quasi-perceptual state *IF* has content *C* iff the function of *IF*-states is to bring about the functionally appropriate response *E* whenever *C* is the case."[12] *IF* may be constantly conjoined with any number of situations, but *IF*'s effect *E* will, very likely, only be appropriate to one type of situation *C*, in which case *C* will be the content. Indeed, *IF* will itself have any number of consequences in circumstances *B*, but of only one will it be the case that *IF* is there *because IF* does it. That one will be *E*. Matthen looks to evolution to underpin the "because." The state of the frog's visual system will have all sorts of consequences, including effects on itself, but it is the fly-catching (*E*), in the presence of flies (C), which evolution has selected.

Israel and Perry's example is an electric pencil sharpener. Its *IF* is having a pedal depressed, *C* is there being a pencil inserted, and *E* is the blades spinning away, sharpening the pencil.[13] The sharpener's internal state of having a little pedal depressed thus carries the information that a pencil is inserted. Not evolution, but the device's designers and users, determine that the blades' spinning is the appropriate response to a pencil's being inserted.

Matthen's account, unlike Israel and Perry's, does not in fact require a constant conjunction between the indicating fact and the content. It may be the function of the indicating fact to bring about the functionally appropriate response whenever *C* is the case, even though *C* is not always the case when *IF* occurs.

Matthen's account is thus ready to deal with the second problem. Indeed, he considers three types of case where the conjunction between *IF* and *C* is not constant. These types of case arise both in creatures and in artifacts.

(1) Malfunction: If a system is not working normally, because of genetic damage or injury, then misrepresentations may result.[14] We may say that normally *IF* states involve *C* states. "Normally" covers cases where the system functions so that *C* states produce *IF* states and *IF* states produce *E* states. If the system is broken they may not.

(2) Maladaptation: If a system is exposed to environmental circumstances to which it is not adapted, it may be "fooled."[15]

(3) Normal misperceptions: These result from the use of indicators that are imperfect but the best available, where this use confers an overall advantage upon the organism, despite the occasional occurrence of error.[16] Normal misperceptions occur because the indicators nature (or humankind) has to choose from are often imperfect. The constraint, in Israel and Perry's terms becomes "*normally mostly* when *IF* states occur, so do *C* states."

It turns out that Israel and Perry's pencil sharpener fits many features of Matthen's functional account including, surprisingly, the three types of error. The pencil sharpener operates under a constraint:

> There is a constraint $C_{inc\text{-}info}$: if a system *a* of kind *K* is in pedal-depressed at location *l*, then there is a pencil inserted in *a* at *l*.[17]

I bought one of the pencil sharpeners yesterday and played with it. I found that if I jammed a pencil in hard enough and often enough to bend the pedal against the contact, it produced a *malfunction*. As a result of the damage, the system was in pedal-depressed state at a time and place but there was no pencil inserted in it at the time and place. So I took the thing apart and bent the pedal back and put it back together, to correct the malfunction.

I then produced a *maladaptation*. I put 150 pencils through one after the other, grinding each one down to the eraser. This produced so many shaving that they escaped from the reservoir, squeezed into all the empty spaces, and depressed the pedal. The machine was not intended to be used without ever emptying the reservoir. The pedal was depressed, the blades whirled merrily away, but again no pencil was inserted. The "environmental" circumstances I exposed the machine to were not ones it was designed for.

I then unplugged the machine, took it apart, cleaned it out, put it together, and tried to produce a normal misperception. I was successful. That is why my little finger is bandaged today. The pedal was depressed, the circuit closed, the blades whirled, but no pencil was inserted. The system was designed to fit pencils and hold them against the blades so the pencil would get sharpened. Unfortunately, other slender cylindrical objects produce the same effects. The pencil detector is (by economic necessity) an imperfect indicator.

The crude and unhelpful thing to say would be that $C_{inc\text{-}info}$ is false. But I do think that the machine's being in pedal-depressed state *does* carry the information that a pencil is inserted. So I want some version of the constraint to be true, something such as

> *IF intentionally* carries the information *C* in circumstances *B* iff *B* have been so constructed that
> (1) situations of type *IF* involve situations of type *C*,
> (2) situations of type *IF* bring about a desired response to situations of type *C*.

Here again "involves" means "normally, mostly involves," and "normally" and "mostly" are relativized to clause (2). Situations of type pedal-depressed may involve situations of type finger-inserted, but not normally, because the machine was not made for sharpening fingers. What nature does for selecting constraints in cases of functional information, design does for selecting constraints in intentional information. Unless some such modification is made to Israel and Perry's simple constraint, it will be false, and the pedal-depressed state will not carry the information a pencil is inserted. So this kind of mechanical information flow depends (unlike the biological cases) on the intentions of human agents and is inconceivable without it. Evolution and design not only make information useful, they make it information.

This leaves us with the third problem. Fodor's problem is rather like Quine's argument for the indeterminacy of translation. One's first natural response to it is the same: scientific theories are always underdetermined by the evidence. Whatever additional constraints (simplicity, explanatory power, etc.) are used to select among observationally equivalent scientific theories elsewhere will do as well to select among the alternative candidates for content of an indicating fact. This natural response has not always satisfied followers of Quine, and it may not satisfy Fodor. A little more can be said for it, however. Someone who puts forward seriously an account of the information carried in a perceptual state of an organism will already have done (or relied upon) a good deal of investigation into the physiology and behaviour of the organism. That person will therefore accept a large number of generalizations about the organism, and some of these will be organized into a theoretical or quasi-theoretical framework. Identifying the perceptual state as having the (imperfectly carried out) function of fly-catching may well fit more smoothly with these generalizations than identifying it as having the function of C′-thing catching. The predicate "fly" may appear frequently in these generalizations, for example, whereas "C′" may not appear at all. There may be other states, for example, which appear to have as their function catching flies: they produce reactions to situations of which flies are part, but which are not very like C′. They may, for example, be auditory states, constraints for which do not involve size and colour (remember that C′ was a kind of state in which something was within a certain range of size, colour, and distance). So the framework of accepted generalizations may make "fly" just the right predicate to use here. None of this implies that "C′" could not be used in another observationally equivalent framework, given suitable adjustment to many of the generalizations — hence the comparison with Quine's indeterminacy thesis. Nonetheless, given what the scientist has already accepted, for *this* framework "fly" is the word and C is the content of the visual state.

In a talk in the spring of 1987 Ruth Millikan offered what may be a variation or refinement of the above response to Fodor's argument.[18] Her idea is to focus on representation *consumption*, rather than representation production. The devices that *use* representations determine certain states to be representations and at the same time determine their content. Something is a representation only by virtue of having a use within the system, a use whereby it functions as a representation. Later in her talk Millikan suggested that a representation may not have a single function, but may serve many different functions, all of which require that the representation (what we have been calling the "indicating fact") have the same content vis-à-vis the same (that is "normal") conditions. This suggestion may help with a final apparent problem.

Consider the x-ray example with which we began. Matthen's account of content rests on the x-ray's having a function, as Israel and Perry's account rested on there being some appropriate effect. There is, however, no specific appropriate effect for an x-ray to have. All we can say is that x-rays are meant to be interpreted for diagnosis. Whatever specific effects x-rays have as they perform their general function of being used to diagnose, each of these effects is appropriate only if the x-ray accurately represents what it is an x-ray of. Whether guiding the veterinarian in applying a cast, in performing surgery, or recommending euthanasia, none of these effects will be appropriate if Jackie does not have a broken leg. So for each of these effects, that Jackie has a broken leg counts as the content of the x-ray.

True perceptual states relate to the quasi-perceptual states of Matthen's account rather in the way that x-rays relate to pencil sharpeners.

What Matthen calls quasi-perceptual states have specific appropriate effects. Normal human perceptual states do not; they produce belief states which in turn are employed in all sorts of ways in our mental lives. These uses (many or most of them, anyway) require for their appropriateness that the content of perceptual states be constant and accurate. Matthen put forward a definition of "perceptual state" which in our jargon reads:

> A perceptual state *IF* has content C iff the function of *IF* states is to *detect* when C is the case.

Here "detect" covers the range of effects (or uses) which the perceptual state may have. In a recent reply to a published criticism,[19] Matthen comments: "Perceptual systems must surely have evolved from . . . automatic control systems, but they do not serve directly to initiate action; rather they interact with memory, reason, desire and so on — in short, with the faculty of judgement. If action results, it comes out of judgement, not directly out of perception."[20] Following Millikan's suggestion we may say that it is the function of the perceptual state to engage in these interactions when its content is the case.

A final point: information reports are not extensional; they do not, in general, retain their truth through substitution of co-extensive expressions. Israel and Perry make this point at the outset, and most participants in the discussion concur.[21] Alexander Rosenberg, in the aforementioned comment on Matthen, challenges the kind of functional account of information or perception, on the ground that any functional claim retains its truth value through substitution of co-extensive expressions. Clause (2) in Matthen's definition of function says

> *IF* is there *because* it does *E* in circumstances *B*.

As everyone knows, "because" is not extensional. Among Baby David's blocks just one block is square and that one is blue. That block fails to fit through the triangular slot in David's mailbox *because* it is square. It is not true that it fails to fit through the triangular slot *because* it is blue.[22]

Israel and Perry have refined a constant conjunction account of information. They have given it point by adding the notion of something's having information, and analyzing that notion in terms of appropriate effects. The problems their account faces are eased by adopting a functional account which weakens the requirement of constant conjunction and by recognizing that an information carrier may have, not a single appropriate effect, but a range of effects, all relying on the same content. The resulting account is messier than theirs, with a lot of loose ends, but it seems at this point a promising development of their ideas.

NOTES

1 Gilbert Ryle, "Are There Propositions?," *Proceedings of the Aristotelian Society* (1930): 95.
2 Reginald Jackson, "The Conventional Basis of Meaning," *Proceedings of the Aristotelian Society* (1933): 207, 210–11.
3 H.P. Grice, "Meaning," *Philosophical Review* (1957): 377–88.
4 Jon Barwise and John Perry, *Situations and Attitudes* (Cambridge, MA: Bradford MIT Press 1983).
5 "Foundations of the Theory of Signs," *The International Encyclopedia of Unified Science*, I, 2 (1939; reprint, University of Chicago Press 1955), 82.
6 "What Is Information?," 6.
7 Ibid., 7.
8 See "What Is Information?," 4, for the formal semantics.
9 Ibid., 16.
10 Mohan Matthen, "Biological Functions and Perceptual Content," *Journal of Philosophy* (1988): 5–27.
11 I have changed Matthen's variables to my mnemonic ones.
12 Again, I have modified Matthen's definition to fit the terminology of this paper.
13 "What Is Information?," 17–18.
14 Matthen, "Biological Functions and Perceptual Content," 13.
15 Ibid., 13.
16 Ibid., 15.
17 "What Is Information?," 17.
18 In a paper entitled "Biosemantics" at the conference of the Three Alberta Universities, April 1988. I paraphrase. I do not do justice to the richness and variety of Millikan's examples. See her *Language, Thought and Other Biological Categories* (MIT Press 1984).
19 Alexander Rosenberg, "Perceptual Presentations and Biological Function: A Comment on Matthen," *Journal of Philosophy* 86:1 (1989): 38–44.
20 Mohan Matthen, "Intentionally and Perception: A Reply to Rosenberg," manuscript.
21 "What Is Information?," 1–3.
22 Matthen makes similar points, more directly about the character of perceptual content, in his "Intentionality and Perception: A Reply to Rosenberg."

CHAPTER TWO

Verbal Information, Interpretation, and Attitudes

Nicholas Asher

In his book *Knowledge and the Flow of Information*, Fred Dretske tells the following story to distinguish verbal information from conventional meaning. Dretske is asked by an opponent at a duplicate bridge tournament what his partner's bid means. As his partner has bid 5 clubs to Dretske's 4 no trump using the Blackwood convention, Dretske replies that his partner's bid means that he had 0 or 4 aces. If the opponent knows anything about bridge, however, she already knows the conventional meaning of the bid. According to Dretske, she does not want to know what the bid means but what "information" Dretske received from his partner's signal. That information is that the partner has no aces. Dretske writes: "this information was communicated to me because I had three aces in my own hand; hence he could not have all four aces. My opponent (having no aces in her hand) did not get this piece of information. All she could tell (given my partner's bid) was the *he had either 0 or 4 aces*."

Dretske's story suggests a concept of verbal information that I would like to explore in this paper. It suggests that verbal information is a function of the conventional, linguistic meaning of the verbal message and what the recipient knows about the situation which the message describes. More specifically, the information contained in a message for a recipient is what the linguistic meaning of the message would contribute to the recipient's overall picture of the situation or beliefs in general were he to accept the message. Thus, verbal information might be thought of as a mapping from one set of beliefs or one overall cognitive state to another set of beliefs or another cognitive state. The analysis of this notion depends not only on a theory of belief but also on an analysis of acceptance and verbal communication.

In other papers I have tried to sketch out a theory of attitudes and attitude reports within the framework of Discourse Representation Theory (DRT).[1] I will draw on that work here to offer an account of verbal information. My account is largely orthogonal to a materialistic notion of information like the one that Dretske advances. The materialist notion of information develops a particular account of the causal connections between an agent and his environment to produce a materialistically acceptable reduction of the content of mental states. I will make use of certain causal connections between real world properties and mental concepts (or representations thereof) in my account of verbal information, but I will suggest, contra Dretske and other materialistically inclined informationalists, that such causal connections constitute only part of the story concerning information. My purposes are also rather different from the materialist's; I seek no reduction of the content of mental states. In fact, I suspect that it is not possible. I am interested in the concept of information and its interaction with other mentalistic notions like that of belief and acceptance, and it seems to me that this interaction merits a closer look than it has received to date.

I

Verbal information is, I claim, functionally dependent on an account of linguistic meaning. So let me begin by sketching an account of linguistic meaning. What should semantical rules for a natural language give us? They should determine for a given assertoric sentence in isolation its content or truth conditions, and they should determine the semantical contribution a sentence makes to the content of a discourse of which it is part. This last task is important for my purposes because the sorts of considerations and problems that arise in the integration of the content of a sentence with the content of a preceding discourse resemble the more complex task of integrating the content of a message with one's antecedent beliefs. Moreover, it is more complex than simply "adding" the semantic content of the sentence to the content of the sentences previously used in the discourse. Consider for example the following two-sentence discourse:

(1.1) The meltdown at Chernobyl has ended.
(1.2) Every West European will remember it.

The second sentence in (1) makes a specific contribution to the overall content of the discourse that depends upon the content of the first sentence in a very specific way. Its truth conditional meaning depends on the first sentence too, for without it the pronoun *it* is an unbound variable and the second sentence an open sentence *without* any determinate truth conditions. An adequate semantic theory should give some indication as to how the sort of integration of content present in intersentential anaphora might proceed. The account of NPs in DRT is the best developed account in semantics that has tackled these questions in a systematic and formal way. I turn now briefly to an account of meaning in DRT.

DRT has two basic components: the so called **construction algorithm** and the **correctness definition**. The construction algorithm maps a natural language discourse (or rather the sequence of its constituent sentences' syntactic structures) to a *discourse representation structure* (DRS), the semantic representation that the theory posits. The correctness definition yields a set of *embedding* functions from DRSs into models that (1) provide the truth conditions of a sentence in isolation and (2) show how the content of a sentence contributes to the content of the antecedent discourse it follows.[2] A DRS is itself an ordered pair, consisting of a set of *discourse referents* and a set of *conditions*. Discourse referents come in various types. I will denote discourse referents by letters 'u,' 'v,' 'w,' 'x,' 'y,' and 'z' with or without subscripts for individual discourse referents, by the letter e with or without subscripts for event discourse referents. I won't use any other types of discourse referents in the construction of DRSs in this paper.[3] Conditions are property ascriptions to discourse referents; they are defined recursively using a list of basic *predicates* (many of which are derived from English nouns, verbs, intersective adjectives) and logical operations. The syntax of the DRS language used to express these conditions has become quite complicated — in part because of the amalgamation of several different versions of DR theory, partly for more theoretically principled reasons which I cannot go into here. I will stick with the diverse ways of expressing conditions to maintain a connection with how DRT has been so far formulated, but the reader unfamiliar with DRT should not worry too much about the details of the syntax of the DRS language.

DRS construction proceeds in a stepwise fashion through the syntactic parses of the discourse's constituent structures. As an example of a DRS, consider the one for (1.1) in (2.1) on the following page.

The construction of a DRS for the whole discourse in (1) proceeds by adding the material from subsequent sentences to the DRS for the first sentence. When applied to a parse of the second sentence of (1), the construction rules yield the following DRS with just one condition.

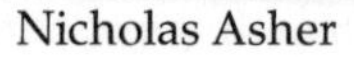

(2.1)

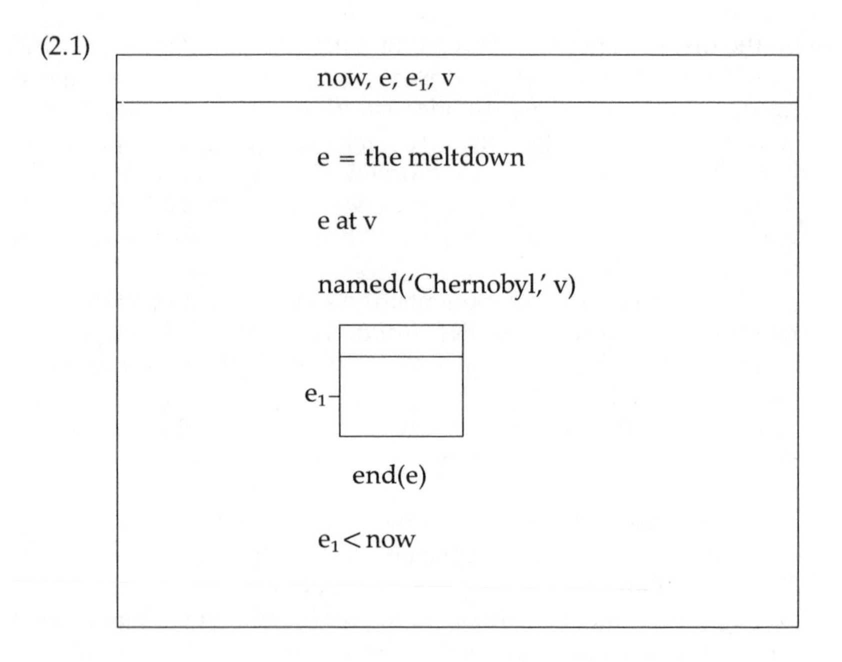

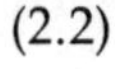
(2.2)

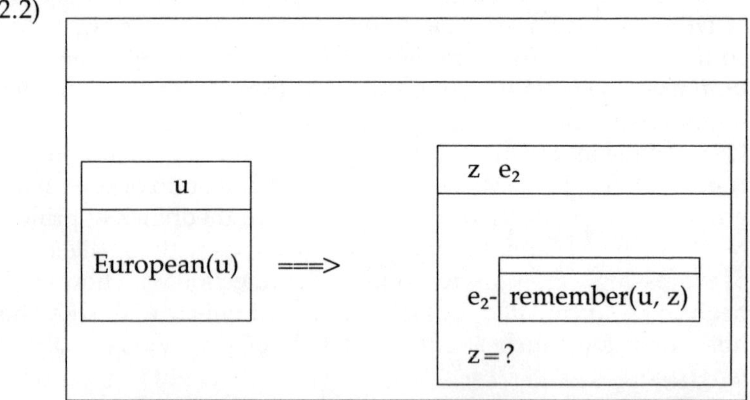

Note that a peculiar condition obtains in (2.2); z is a discourse referent introduced by the anaphoric pronoun and the rules of DRS construction require that it be identified with some antecedently introduced discourse referent. However, at the present stage of construction, the anaphora is unresolved, because there is no discourse referent with the appropriate features to be identified with z.

The final step of DRS construction is to add the discourse referents and conditions introduced in (2.2) to those declared in (2.1). Since 2.1 contains a discourse referent with the appropriate features, the unresolved equation in (2.2) can now be resolved with (2) as the final result:

(2)

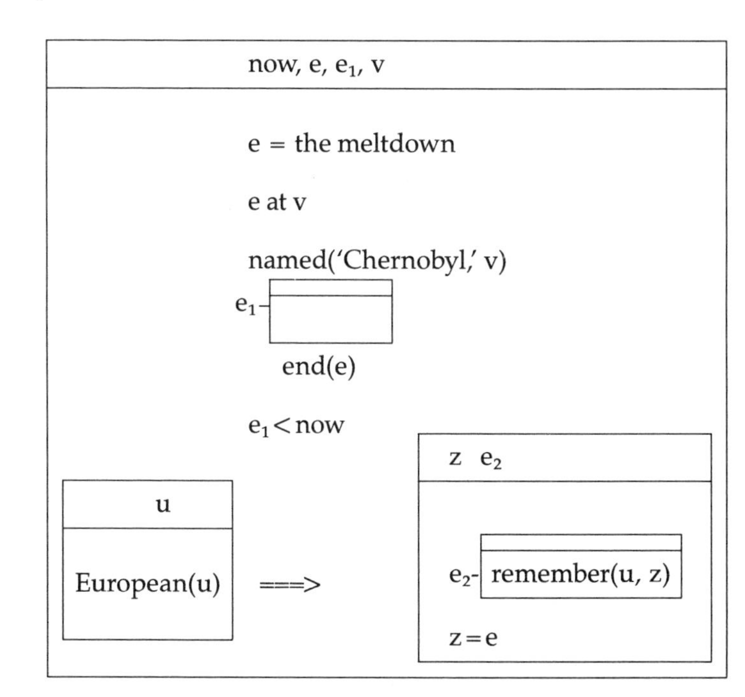

The interpretation of DRSs uses the model-theoretic idiom of possible worlds semantics. For our purposes here we may simplify matters somewhat from the full definition of the interpretation of DRSs as it is found in Asher (1986, 1987). Define a DRS model M to be a quadruple $\langle D, W, T, [\![\,]\!] \rangle$, where W and T are the set of worlds and times respectively and where D is a function from W $\times$ T onto a family of non-empty sets ($D_{\langle w,t \rangle}$ is the "domain of w at t") and where $[\![\,]\!]$ is an intensional interpretation function. The truth conditions of a DRS are given by a *proper embedding*. Roughly, f is a *proper embedding* of a DRS K at a world w and a time t in a model M just in case f maps the elements of the universe of K onto objects in the domain of w and t in M such that all the conditions in the condition list of K are satisfied in M relative to $[\![\,]\!]$ and that assignment at w and t. (For more details, see the appendix.) I will define the *content* of a DRS K to be the set of indices at which K has a proper embedding. So, for example, f is a proper embedding of (2) in w and at t in M iff f maps the discourse referents in the universe of (2) onto elements of D(w) so that (1) there is an object f(u) in D(w) such that f(u) is an object named 'Chernobyl' in w; (2) there are two events, f(e) and $f(e_1)$ in D(w) such that: (a) f(e) is the meltdown in w; (b) $f(e_1)$ consists in the end of f(e) in w; (c) $f(e_1)$ occurs before the time of utterance of the discourse. Further (turning now to the last complex condition in the condition list of (2)), any extension g of the function f that maps the discourse referents declared in the universe of the DRS on the left side of the conditional onto objects in D(w) such that g(u) is a European in w in M can be extended to a function h that maps z and e_2 onto two events h(z) and $h(e_2)$ such that $h(e_2)$ is a remembering by g(u) of h(z) and h(z) = f(e).

There are a couple of remarks that should be made concerning the construction procedure and interpretation of DRSs. I have exhibited only one DRS for the discourse, and, presumably, it is the one that most speakers would choose as conveying the intended meaning of the discourse. However, there are other possible DRSs that the construction procedure would yield for (1). There are other discourse referents beside e, for instance u, that have the grammatically appropriate properties to be identified with z. So another DRS that could result from the processing of (1) is a DRS just like that in (2) except that it contains the condition z = u instead of z = e. DRS construction typically provides a set of possible DRSs for a single discourse. Moreover, the full construction procedure requires many different sources of information to furnish a variety of constraints on anaphoric coreference.[4] But even these constraints do not always determine a single antecedent for every occurrence of an anaphoric pronoun. Some of these constraints only prefer some antecedents over others. Another difficulty that may prevent the construction procedure from determining a unique DRS for a particular discourse concerns the scope of quantifiers and other operators in the DRS representation. So in the general case, the construction algorithm yields a set of DRSs for a particular discourse.[5]

There is also an unsatisfactory feature of (2) concerning the treatment of definite, singular terms. It is worth discussing the complications required for a proper treatment, however, since these have, I think, implications for the notion of information. Proper names, indexicals and demonstratives, as the direct reference theorists have persuasively argued, contribute the same denotation in every possible world, no matter in what nonquotational context the name occurs. But as I have stated the semantic contribution of proper names, this is by no means assured. One means for remedying this difficulty would be to mimic within the DR-theoretic framework the direct reference proposal that such expressions contribute the individuals they denote to the propositions expressed by sentences containing them; such a proposal would involve using a device separate from the DRS in which scope distinctions may occur to fix the assignment to the discourse referent. Such a device is easy to construct in DRT; it is a partial function from discourse referents in the universe of a DRS to objects and times in the domain of the model. I'll call such a function an *external anchor* for the DRS. A DRS with an external anchor is said to be *externally anchored*. I'll also call a real individual that is the value of this function with the input of a discourse referent **x** in a DRS K the *external anchor* for **x**. A DRS like (2) should thus be externally anchored; viz. the discourse referent introduced by 'Chernobyl' should have as external anchor the object, Chernobyl. I'll indicate such an external anchor by adding under the diagram the pair <u, **Chernobyl**>. External anchors constrain the interpretation of the DRS, so that if K is a DRS containing a discourse referent **x** and **x** has an external anchor **b**, then f will be a proper embedding of K (at a world w) only if f(**x**) = **b**.

While the device of external anchoring does appear to accommodate within the DR-theoretic framework some of the insights of the doctrine of direct reference, it does not address another problem noted and emphasized by direct reference theorists. This problem has to do with the conditions that singular terms like proper names introduce in a DRS. Suppose we try to process the sentence "Kripke is necessarily identical with Kripke." If the condition "named (x, 'Kripke')" is left inside the DRS as part of its satisfaction conditions, then the account appears to predict that this obviously true statement expresses the falsehood that the individual named Kripke necessarily has the property of being named "Kripke."[6] Since a name gives rise to an externally anchored discourse referent, the condition it contributes is otiose for determining the real individual that is the value of the discourse referent under a proper embedding of the DRS. Further, the presence of such a condition threatens to make the DRS construction procedure and interpretation yield incorrect truth conditions in certain cases.

Such conditions do serve a purpose, however. The conditions associated with a proper name or other "directly referential expression" supply the agent with the means for connecting up the content of the message with his antecedent beliefs in the right way. The construction of such "internal connections" is an essential part of the task of "interpretation," in which the recipient of a verbal message integrates the content of the message with his antecedent beliefs. These internal connections also end up playing a crucial role in gauging the behaviour of singular terms within belief contexts. The contribution (for the agent) of a definite NP like a definite description towards the information expressed by the statement in which the definite occurs is often simply to single out some individual for which a discourse referent has already been introduced in the processing of the discourse or which is otherwise available from context, not to contribute to the set of property ascriptions attributed to the individual picked out. This way of characterizing the conditions introduced by names (but also other definites) pertains to DR theory's formalization of the familiarity theory of definites discussed at length in Heim's dissertation.[7] Thus, I think it is essential to a proper account of names, indexicals, demonstratives, and other directly referential expressions that they do introduce some conditions on discourse referents — including such metalinguistic information as is suggested by the use of the condition "named (,)."

An acceptable account of the contribution of proper names and other definites to DRS construction exploits the distinction between two sorts of conditions: those that directly contribute to the truth conditions of a report, and those conditions that are relevant to the interpretation of the report by the agent. I'll call the two sorts of conditions *attributive* and *referential* respectively, in honour of Donellan's similar distinction for definite descriptions.[8] Conditions produced by names are almost always referential, while conditions derived from other definites may be referential and attributive. By treating the conditions of definite descriptions as either attributive or referential, one in effect reproduces the familiar scope distinctions for definite descriptions and operators like those introduced by modalities. But to really fill out this account, I need to say something about belief and interpretation, and that is what I turn to now.

II

Suppose that an agent A receives a verbal message ϕ. Suppose further that a DRS is at least a plausible semantic representation of the content of ϕ. Since the complement clauses of attitude reports have the same linguistic meaning within such contexts as they do when standing alone, it follows that the DRS for ϕ, K_ϕ, might also serve to characterize the object of the belief that accepting ϕ might lead A to have. Kamp (1985) has called such a correspondence between meaning and belief the *Unity of Thought and Information* (UTI).

As it stands, UTI is a very weak claim about the correspondence of meanings and beliefs. It applies to the semantic representations posited by any reasonable theory, regardless of one's metaphysical analysis of belief. But it is also incorrect as it stands because it fails to take into account how A's other beliefs form a context for interpretation that affects what belief ϕ might lead A to have. If one holds a representational theory of belief, however, then UTI can be corrected, as well as strengthened. Suppose that belief is a matter of standing in some relation to a representation (and perhaps some other things like being situated in one's environment in certain ways so that certain causal connections obtain between elements of these representations and the environment). These representational objects of belief, which I'll call *belief objects*, are complexes of individuals and property ascriptions to those individuals. They involve, not real individuals, but rather what I shall call *conceptual individuals*. These conceptual individuals serve as "pegs" for property ascriptions.[9] Suppose further that as DRT hypothesizes, DRSs are an abstract, partial description of the way speakers encode some of the semantic aspects of verbally transmitted information. Then a stronger and more interesting version of UTI would assert that the structure posited by DRT for DRSs and their interactions in the process of interpretation and belief formation also constitute a partial description of the structure of attitude objects. DRSs will describe belief objects, insofar as discourse referents will serve as the theoretical reconstruction of conceptual individuals, and conditions the theory's reconstruction of an agent's property ascriptions to those conceptual individuals. The notion of an embedding function makes more precise the notion of ''standing proxy."[10]

Despite appearances to the contrary, UTI is *not* a hypothesis about psychological processes. Both the applications of DRT to verbal processing and to belief and interpretation are highly abstract descriptions. The DRS construction procedure is not detailed enough to serve as a psychological model of verbal processing of belief formation, although, in the same way as some syntactic theories do, the rules for processing that it proposes might suggest psychological models of certain kinds.[11] A thesis like UTI, together with the DRS construction procedure, purports to explain semantic data and conceptual difficulties with our "ordinary" notions of the attitudes and language understanding. As such, of course, they are compatible with a wide number of psychological hypotheses about how the general structures for verbal processing and belief formation they posit are actually realized. UTI is also an empirical thesis of sorts. To the extent that positing something like DRS structures for belief objects is successful in addressing puzzles about belief and belief reports, to that extent UTI will be successful.

Let's now see how UTI might lead to an account of an agent's interpretation of a verbal message and belief formation relative to a context of antecedent beliefs. The process of interpretation is a process of integration and it proceeds in several steps. The first step in forming a new belief is to construct an attitude Object for the content of a verbal message ϕ which is just the DRS K constructed from it. The second step will be to integrate and to co-ordinate this attitude Object with antecedently held attitude Objects. The final step prior to evaluation and acceptance or rejection will be to delete those conditions in the DRS K whose purpose is only to establish the right internal connections for the new belief once their purpose has been served. Conditions introduced by proper names and definite descriptions (under at least one reading) fall into this category.[12] This last step will serve to distinguish within the theory those conditions that are part of the descriptive content of a belief or a message from those that are not. Finally, if the result of interpretation is accepted, then a new belief is formed.

Let's look at the second stage of interpretation in a little more detail. One sort of integration required at this stage is that the new attitude object must be co-ordinated with antecedently acquired beliefs about the same subject-matter. When everything goes right, the way that such a potential belief, for example, gets incorporated is that the agent's beliefs about the same subject-matter share one or more common conceptual individuals as constituents. Let's go back to the example of (1.1). Assume that I have other beliefs about Chernobyl and the event of the meltdown. But not among them, since I am supposing that the content of the first statement (1) is novel to me, is the belief about the event of the meltdown's ending. In that case, then my beliefs will not only share a conceptual individual that stands proxy in the mind of the agent for Chernobyl but also one that stands proxy for the event of the meltdown.

Translating this suggestion into the DR-theoretic framework, the various DRSs that characterize beliefs about the same conceptual individual must share at least one discourse referent. I need some way of talking about collections of attitudes or, if you will, portions of an agent's total cognitive state. I shall describe an agent's total cognitive state by means of a *delineated DRS* — a structure composed of a set of DRSs each characterizing some object of one of the agent's attitudes, a set of discourse referents common to these DRSs (the delineated DRS's *universe*), and a set of conditions on those DRSs that I'll ignore here. An example of such a structure concerning some portion of my beliefs about Chernobyl is given in (3) on the following page.

Now suppose that an agent with a (portion of) a delineated DRS such as that in (3) processes (1) into the putative belief Object described by (2) as the result. An acceptance of (2) should yield the structure in (4), where the discourse referents standing proxy for Chernobyl and the meltdown in (2) have been identified with the ones that are common for Chernobyl and the meltdown in the agent's delineated DRS in (3).

(3)

e_2, v_1, u_1	
v_1 in u_1	e_2 at v_1
u_1=the Ukraine	e_2=the meltdown
named(v_1, 'Chernobyl')	accident(e_2)
	dangerous(e_2)

(4)

v_1, u_1, e_1, e_2, e, u		
v_1 in u_1	e_2 at v_1	e=the meltdown
u_1=the Ukraine	e_2=the meltdown	e at u
named (v_1, 'Chernobyl')	accident(e_2)	named (u, 'Chernobyl')
	dangerous(e_2)	e_1 end(e)
		e_1 now
		e_2=e
		u=v_1

The rule for identifying discourse referents in this way must, of course, be a rule based on the agent's beliefs themselves — namely, the agent must believe that the two occurrences of the name are *intended* to be coreferential. Without evidence to warrant such a belief, the agent would perhaps not identify discourse referents in the way exhibited in (4). Another element of the integration exhibited in (4) is that the discourse referents in the topmost position of (2) are now declared in the universe of the delineated DRS. This aspect of acceptance and belief formation has an import even for those discourse referents not identified with already introduced discourse referents in the universe of the delineated DRS. For instance, in accepting (2), the agent comes to believe that there is an event of the ending of the meltdown; to declare that discourse referent in the universe of the delineated DRS is one way of representing this existential commitment.

Now for the final step of our account of interpretation which distinguishes between two sorts of conditions. The internal connections relevant to the new belief have been established but the information used to make those connections has not yet been eliminated. In choosing to eliminate all the conditions introduced by definites in the DRS 2.1, the following picture of the relevant portion of the agent's total cognitive state emerges:

(5)

v_1, u_1, e_1, e_2, e, u		
v_1 in u_1	e_2 at v_1	e_1 end(e)
u_1 = the Ukraine	e_2 = the meltdown	e at u
named (v_1, 'Chernobyl')	accident(e_2)	e_1 now
	dangerous(e_2)	$u = v_1$
		$e_2 = e$

Note that the conditions introduced by definite NPs such as proper names are never entirely deleted from a delineated DRS. They are not deleted, for instance, when they are introduced by a proper name or a definite description for the first time. Nevertheless, such conditions seem to have a status different from other descriptive conditions and so should be treated as a special case. This information serves to correlate a discourse referent with the linguistic item and is needed to establish the right match in establishing internal connections for a new belief.

(5) is the last stage in this simple story about interpretation. To recapitulate, the interpretation of a discourse about familiar subject-matter leads to an expanded cognitive state with shared discourse referents. The process of interpretation generates these "internal" connections between the new content and antecedently given attitudes when intuitively different attitude Objects are about the same conceptual individual (or the same subject-matter). If the evaluation by the subject of (5) is positive, then the acceptance of this new content results in the formation of a new belief, a new "belief-component" of the agent's delineated DRS. That is, the agent judges what would happen to his overall cognitive state if he were to accept the belief. On

the basis of reviewing the proposed modification to his beliefs, he may reject the statement, accept it, and form a belief or remain undecided about its truth.

Besides internal connections, the interpretation of a verbal message may lead to a belief with connections to the agent's environment. When the processing of a verbal (or other) input yields an externally anchored DRS, the agent's acceptance of this content will add the external anchors provided by this new information to the external anchors for the agent's antecedent beliefs. Such externally anchored DRSs characterize *de re* beliefs, in which the content of the agent's belief depends on the agent's relations to his environment.

This picture of belief formation leads to an important consequence concerning the nature of belief: the content of a belief is often only fully determined when it is interpreted relative to its place in an agent's total cognitive state. That is, the content of the belief may be completely specified only by specifying the contents of connected attitudes in such a way as to make clear whether the attitudes are about the same "conceptual individual" or not. I have tried to show how this is possible in Asher (1986, 1987a, 1987b), but I won't go into those technical details here.

III

With this view of belief in place, I can now characterize information semantically by looking to the interpretation of the delineated DRS that results from the amalgamation of the antecedent attitudes and the newly formed belief. Dretske suggests himself that information should be characterized in terms of what possibilities it eliminates for the agent. So suppose that an agent A processes a message ϕ to yield a DRS $K\phi$. Suppose then that A has an antecedently given delineated DRS K and that in the interpretation of ϕ, he adds $K\phi$ to K to yield a new delineated DRS K', which contains $K\phi$ as a component. (I'll write 'K is a component of K'' as '$K < \mathbf{K}'$'.) Suppose finally that A accepts ϕ so that K' is the resulting delineated DRS of which $K\phi$ is a belief component. Then, the *information conveyed by ϕ to* A is $\bigcap_{K < K}$ Content(K) $-$ $\bigcap_{K < K'}$ Content(K). In other words, the information conveyed by ϕ consists in that additional content that differentiates the content of **K** from that of K'. ϕ conveys some information to A just in case $\bigcap_{K < K}$ Content(K) $-$ $\bigcap_{K < K'}$ Content(K) $\neq \emptyset$. If the veridicality for information is required, we must simply stipulate that ϕ conveys information only if K_{ϕ} has a proper embedding in the actual world. One may also want to relativize this definition of information to a "relevant" portion of the agent's delineated DRS — consisting perhaps of all those portions of *K* that are internally connected to K_ϕ. This formal characterization of information using possible worlds semantics for DRT is a simple consequence of the notion that a belief has a definite content only relative to an interpretation of other attitude objects to which it is internally connected. It suggests obvious parallels to Dretske's notion of information as an elimination of possibilities.[13] Stalnaker's work on a contextual theory of propositions should be mentioned as a forerunner of this notion of belief and information.[14]

The DRT framework, however, does add something to the Stalnaker-elimination semantics view of information. The DR-theoretic approach to belief exploits the representational format of the DRS delineated DRS; particularly important are the discourse referents which are taken to represent the conceptual individual constituents of an agent's beliefs. This formalism naturally yields a definition of *information for* an agent *about a conceptual individual* he has in mind. The delineated DRS format invites a rearrangement of the conditions contained in all its component DRSs along the following lines: for each discourse referent **x** declared in the universe of a delineated DRS *K*, there is a collection of conditions in the various components K of *K* in which **x** appears as an argument. Let's call this subset of $\cup_{K < \mathbf{K}} \mathrm{Con}(K)$ *the file for* **x** *in K*.[15] There is a natural DRS associated with the file for **x** in **K**:K(file(**x**,**K**)) = $<U, \mathrm{file}(\mathbf{x},K)>$, where $U = \{\mathbf{v} \in U_{\mathbf{K}} : \mathbf{v}$ occurs as an argument of $\phi \in \mathrm{file}(\mathbf{x},K)\}$. Now define the information for an agent A about a conceptual individual **x** delivered by a verbal input ϕ relative to the agent's background cognitive state **K** as: Content(K(file(**x**,*K*))) – Content (K(file(**x**,*K'*))), where *K'* is the result of updating *K* with K_ϕ.

Information about a conceptual individual is an important type of information. Growth of an agent's self-knowledge is a special case of an agent's acquiring more information about a particular conceptual individual. There are many examples that show that the particular kind of information about a conceptual individual that marks self-knowledge is important. Moreover, the information that typifies self-knowledge cannot be expressed in terms of simple possible worlds semantics.[16] One needs a theory of belief that distinguishes property ascriptions and the objects to which these properties are ascribed. Information cannot be captured in a simpler theory of information that does not exploit a representational theory of belief of the sort that I have sketched. For instance, consider the story told by John Perry.[17] While shopping at a supermarket, Perry notices on the floor a trail of sugar. The information Perry gets from the scene in the supermarket is that someone is carrying a bag of sugar with a hole in it. In fact he is that person, but he doesn't know it. What the agent is missing is that connection between the victim in the belief and the agent himself; were the agent to make the connection that it is he who is spilling the sugar, he would act quite differently than he does without this information. It is this connection that makes the information take on a wholly different character.

As he acquires information about the person spilling the sugar on the floor, Perry forms beliefs about some conceptual individual. This individual is represented by a discourse referent **x** in the delineated DRS for Perry's total cognitive state. But until Perry learns that it is he who has been spilling the sugar on the floor of the grocery, he does connect **x** with the special discourse referent **i** that represents him to himself. He acquires self-knowledge when he makes that identification. Such an identification adds to the file about **i**. Given the representational nature of my theory of belief, it is possible to single out one of these conceptual individuals as a special one that stands proxy in the mind of the agent for himself.

Such ascriptions of properties *de se* can also be handled to some extent in other frameworks (for instance in a theory of belief like Lewis's).[18] But the DR-theoretic view of belief and belief formation provides us with a way to track the development of the various beliefs in Perry's puzzle. Presumably, if Lewis's view of belief were to furnish a view of belief formation to describe this case, it would follow the DRT formulation sketched above. But, further, Lewis countenances ascriptions of properties only to objects that one knows by acquaintance or by a description that essentially individuates them, and this appears to fail to take account of the following sort of case. Suppose a member of the local police force around Chernobyl finds evidence of sabotage at the plant. He forms the idea that there was a saboteur, although in fact there was none. In this situation he does not have an acquaintance with the saboteur nor does he have a description that essentially individuates the saboteur.[19] Gradually the policeman builds a case, gathering more and more information about the "the saboteur," until his evidence is sufficient, he thinks, for him to discover and to apprehend the saboteur. He doesn't find anyone, however, who could meet all the qualifications for being the "saboteur." The policeman then uses this information to conclude that perhaps there was no saboteur after all. A reconstruction of the formation and transformation of the policeman's beliefs is quite straightforward if one ascribes, as I do in my account, properties to conceptual individuals; without positing conceptual individuals, however, it seems difficult to tell a plausible story about the detective's beliefs. Another area in which the DR-theoretic construal of belief and information goes beyond a theory like Lewis's is that DRT offers a way of translating verbal messages into objects of belief, an account of which I sketched very briefly in the last section. These verbal inputs may involve discourses with pronominal anaphora. These translate into discourse referent identifications within the beliefs of the agent, and these identifications again are an integral part of the information that the agent receives from the input.

If I am right in thinking that the notion of information about a particular conceptual individual is an important one, then we can no longer characterize information solely relative to the content of our beliefs. We must also characterize information in terms of their structure. This modification does *not* deny that truth conditional content has a role to play in the notion of information. I propose only to refine the notion of information so as to characterize it in terms of the structure, as DRT sees it, as well as the content of an agent's beliefs. A first step in that direction is to exploit the structure accorded to beliefs insofar as they contain certain discourse referents as constituents. This yields the notion of information about a conceptual individual.

The DRT framework, however, also invites us to exploit the structure of conditions in DRSs in a theory of information. By exploiting this structure, we can make the definition of information much more sophisticated. While information as I have defined it so far is always strictly cumulative, it might appear that in some cases the acceptance of the result of interpreting some verbal message is not simply a matter of accumulating another belief but rather of acquiring one belief but losing a number of others.[20] By exploiting the structure of beliefs in a delineated DRS, however, it is possible to distinguish further stages of belief integration beyond those discussed in the previous section. For instance, the acceptance of the result of interpretating a verbal message may lead to a delineated DRS K that is inconsistent, in the sense that there is no proper embedding of K at any world w in the intended model. A further stage of belief integration would consist in a process of "revision" or "contraction" to yield a new, consistent delineated DRS K'.[21] These processes, however, appear to be definable in a useful way only on highly articulated structures like sentences or DRSs.

Let us assume that such a stage of belief integration is given. A more sophisticated definition of the information conveyed by ϕ to an agent A, whose background beliefs are characterized by a delineated DRS **K**, explicitly incorporates the idea that information is a function from one belief state to another: the information contained in a verbal message ϕ is given by the pair $<<\mathbf{K},\mathbf{K}'>, <\mathbf{W}_{\mathbf{K}}, \mathbf{W}_{\mathbf{K}'}>>$, where **K** is the antecedent delineated DRS, **K**′ is the result of updating **K** with K_ϕ (this updating may involve retraction or contraction as well as expansion) and where $\mathbf{W_K}$ ($\mathbf{W_{K'}}$) = $\cap$ K < $\mathbf{K}$($\mathbf{K}'$) Content(K). I'll say that ϕ has information for A just in case $\mathbf{K} \neq \mathbf{K}'$ and $\mathbf{W_K} \neq \mathbf{W_{K'}}$. This definition of the information conveyed by ϕ to A can be easily extended to yield a revised definition of information about a particular conceptual individual. Because of the complex nature of the operations of belief revision, however, I do not see (at present at any rate) any way of capturing the net semantic contribution of ϕ to A's beliefs, and thus the information conveyed by ϕ to A, other than with the ordered pair notation.

Exploiting the structure of A's beliefs in this definition of the information conveyed by ϕ for A also captures something of our intuition of what "information" it is that the agent has ready to mind. It is the information given by the structure of the belief state that the agent can access in belief and which he uses to formulate plans and conscious and deliberative actions. This notion of information is not closed under logical consequence, as the possible worlds characterization of information is. Thus, this notion of information may be more useful to theories of deliberative action and planning than the pure, possible worlds characterization. An algebraic notion also allows us to investigate a "growth of information under inference"; as one infers new conditions from the given information (together with other background beliefs as premises), the information one has at one's disposal grows; there are in a brute sense simply more conditions available for the agent to use consciously after such inferencing than before. This notion of using information together with background beliefs also provides us with a new sort of integration of belief objects, integration through inference.

IV

The *DR*-theoretic structures that I have sketched shed some new light, I believe, on various problems concerning belief. But these very same problems are also puzzles for a theory of information, and addressing them will further amplify some of the complexities of the notion of information. These puzzles typically arise when the process of interpretation and belief formation go awry.

The puzzles all feature tensions between internal and external connections. External and internal connections are independent, and the default rules for integrating new material within antecedent beliefs may yield internal connections incompatible with the newly acquired belief's external connections. One sort of incompatibility occurs when two discourse referents in a DRS characterizing a newly formed belief have the same external anchor but the agent has distinguished two conceptual individuals because he attributes very different, even incompatible properties to them. An example is Kripke's famous puzzle concerning Pierre, which has been discussed at some length within this framework.[22] A second type of incompatibility arises in cases of mistaken identity. These, too, can be very complicated.[23]

Yet a third kind of incompatibility is one where the agent supposes one conceptual individual to be the argument of a variety of conditions when there are in fact two real world individuals that each satisfy a portion of those conditions. Here is an example concerning two Smiths, one of whom was a famous politician, the other a famous pianist. Suppose that an agent A hears from person B that Smith is a very adept politician. Suppose also that this is the first time A has heard the name 'Smith.' If A accepts this statement, he will presumably introduce, on the analysis I have sketched, into his total cognitive state a conceptual individual x and the conditions 'named(x, 'Smith'), politician(x). Now suppose that later from another, reliable informant C, A hears 'Smith was a famous pianist.' Our rule of definite reference integration says to identify discourse referents introduced by occurrences of the same name if one has good reason to believe that the two occurrences are coreferential. So let us suppose further that A has seen pictures of the two Smiths, and that the two Smiths look remarkably alike. Given these assumptions, A is perhaps quite justified in presuming that the two occurences of the name are coreferential. In adding this information to his beliefs in the normal way, A should come to the belief that Smith is both a famous pianist and a famous politician. Of course if he then checks his conclusion out with C by saying: 'Oh, so Smith was both a famous pianist and a famous politician,' he will get a surprise when C replies, 'No, no, Smith never had any interest in politics.'

Such faulty interpretations may come to light both in the actions and words of the interpreters. These latter messages, even when false, however, convey information. So far I have talked only about integrating the content of the verbal message within antecedent beliefs. But clearly, a message often conveys information about the speaker as well. This is most clearly visible with messages that the recipient believes to be false, for then they do not convey information to the recipient about the situation they purport to describe. Consider, for example, the situation of C in the example above when he receives A's message 'Oh, so Smith was both a famous pianist and famous politician.' C does not accept A's claim, but it does convey information to him; it is good evidence that A holds certain beliefs — namely, that A believes that Smith is both a pianist and a politician, a belief that C takes to be incorrect. Nevertheless, there is a process of integration of beliefs going on here too — only what C integrates is a new element into his representation of A's cognitive state! This process is somewhat more complicated than the sort of integration investigated so far. For besides linking the discourse referent introduced by the occurrence of the name 'Smith' with a discourse referent that already stands for that individual in C's representation of A's total cognitive state, C must also link that discourse referent with the discourse referent representing Smith in A's total cognitive state; this last linkage is crucial for C to evaluate A's message as false. This additional linkage is another step in the interpretation of a verbal message and integration of the information it conveys.

The character of the linkages within C's representation of A's cognitive state are crucial when C takes it upon himself to correct what he takes to be a mistaken belief. If C has enough information about the source of A's beliefs (which might be possible if he has heard B talking with A), C will be able to diagnose A's doxastic difficulties. C will know that A employed a default rule about identifying any two discourse referents introduced by occurrences of the same name. The ideal remedy for A's false belief would then be for C to remark something to the following effect: 'No, there were two Smiths, one a pianist the other a politician.' A is lacking this information. Such a reply would pinpoint to A the source of his difficulties — his employment of the rule of definite reference integration. On the other hand, the information C conveys to A in the original story is less helpful; A has to consider more possibilities as to what went wrong (or what C thinks when wrong) when C says to him 'No, Smith was never interested in politics.'

The sort of indirect information discussed above is, when fully articulated, highly complex. To get such indirect information from a message containing a definite expression α, the recipient must assume that the speaker intends to refer to the same object that he does when using α. But the indirect information conveyed just by the use of a definite NP in a conversation, such as the one that would be necessary to correct A's mistaken beliefs about Smith, does not stop there. The speaker S must choose a definite NP that he believes will get his audience A to focus on the appropriate conceptual individual — the one that stands proxy for the object he intends to refer to. Suppose S chooses an NP that contributes a condition ϕ to DRS construction. S must first of all believe that ϕ is a property that picks out the individual he wishes to talk about; ϕ will be predicated of some discourse referent **y** for him. Now if this communicative act is successful, A will link a discourse referent bearing the ascription ϕ to some antecedently given discourse referent, say **x**. But A will then also infer that S intends to talk about **x**, so he will identify a discourse referent **z** in his representation of S's cognitive state that has the property ϕ with **x**. Similarly S will believe, if A does not protest, that A has a belief in which some discourse referent **z** bears the ascription of ϕ. So S will link **z** with **y**. In fact, nothing stops this process of embedding one representation of a cognitive state within another; the information implicit in a successful act of communication with a definite NP yields a mutual or common belief among the participants of the act, in which their representations of each other have this complex sort of internal connection.[24] It is this complex set of internal connections that underlie mutual beliefs that have a common focus, even in those cases in which, as in Geach's celebrated Hob-Nob example, there is no external correlate of that focus.[25] Further, this complex set of internal connections underlies the full indirect information conveyed by an utterance in successful communication. Once again, information depends on the agent's beliefs, but in the case of indirect information, much more complex beliefs than the first order beliefs of a single agent are involved.

So far I have looked only at the sort of links formed between individual discourse referents and how they affect the concept of information. The distinctions between internal and external connections to which NPs give rise among beliefs, however, can be largely replicated at the level of concepts, when we take into account the structural contributions of conditions to the description of beliefs. Recall that I introduced internal and external connections for individual-denoting NPs by considering how NPs should be interpreted in the formation of belief. The analogous question for conditions investigates how the semantic contributions of 'general terms' — common nouns, verb stems, adjectives, adverbs — interact with antecedent beliefs. So far, general terms have merely introduced conditions on individual-type discourse referents. But to explore their informational properties properly, we have to complicate DR theory considerably; we must introduce discourse referents for concepts.[26] A general term will now introduce a concept discourse referent, a condition on a concept discourse referent, as well as apply the concept discourse referent to discourse referent arguments. One condition on a concept discourse referent introduced by a use of a general term α is that the concept discourse referent **c** has the property of being names [α]. So 'named(**x**,α)' is one sort of condition applying to concept discourse referents as well as individual discourse referents.

The parallel treatment of general terms and proper names reflects similarities in their linguistic behaviour. For instance, general terms always contribute the same content whatever the modal context in which they occur. Furthermore, a version of the familiarity theory of definiteness also appears to apply to general terms. Suppose the recipient of a discourse is unfamiliar with a particular concept. That might arise in two ways; he might not know that a particular word expressed a certain concept, or he might lack the concept altogether. Let's suppose that the recipient is someone who is largely ignorant of physics, as a small child would be; if the speaker's answer to the recipient's question of why the sun shines uses concepts appropriate to the mathematical analysis of the process of thermonuclear fusion, communication won't have taken place or will be faulty. A discourse won't be felicitously received by a recipient, unless the general terms used are "familiar" in the sense that the concept-type discourse referents they introduce can be correlated with antecedently existing concepts in the agent's total cognitive state. As with proper names, the metalinguistic condition 'named(c,α)' where α is a general term is information that the recipient uses to get the right connections for the new belief; it should not be part of the descriptive content of the belief proper.

If we follow the suggestions of Putnam and Kripke, some concepts, those for instance introduced by natural kind terms, have their content determined by relations external to the agent's cognitive states. For a concept of this sort, we may employ an intension as an external anchor for the discourse referent introduced by the natural kind term.[27] But presumably not all concepts are so externally anchored; other concepts may acquire their content only because of their connections to other concepts. These connections are in effect higher order concepts. I have already used one such higher order concept 'named(x,α),' but presumably there are many others relating the concepts that are named by general terms of the language with each other and with properties of the perceptual system.[28]

The potential for incompatibilities between the external and internal connections also occurs for concept discourse referents, just as it does for individual discourse referents. For example, a conceptual analogue to Pierre's predicament is the situation of an agent who has, but fails to identify, the concepts of gorse and furze, because he associates incompatible properties with instances of gorse and furze.[29] But concepts have many other sorts of internal connections besides identity that are worth exploring. One very important connection concerns a causal link between concepts — event-types in particular.[30] Agents often ask why-questions about particular events when they want a causal explanation for the event's occurrence, and providing such links is an important part of explanation. Nevertheless, such explanations only make sense if the questioner *already* has made a causal connection between the event-types that the cause and effect instantiate. So, for instance, to use Salmon's example, suppose an agent A asks 'Why did the plane crash?' and C correctly responds, 'because it had too much ice on its wings.' Nevertheless, this response may not constitute an acceptable answer for A or remedy A's difficulty. If A has not already represented and accepted a causal connection between the event types of planes crashing and ice accumulating on the leading edges of their control surfaces, A will be confused by the response and most likely ask, 'why did the ice on the wings lead to the crash?'

Typically causal connections between concepts involve other concepts. Thus, a good understanding of the causal connection between ice accumulation on control surfaces and a plane's crashing will involve some appropriate set of concepts; that is, not just any connection between ice accumulation and crashing will do. But then, an agent may *misrepresent* the causal connection between two event-types. For instance, consider the following misrepresentation by A of the causal connection between icing conditions and airplanes crashing. Suppose that A believes that icing causes crashes because the ice makes the airplane too heavy to maintain altitude when the airplane is at or near gross weight. This is not a correct connection; the weight of the ice is not really a critical factor in aircraft accidents having to do with icing. But believing in such a causal connection is very dangerous. With this connection, A might think that as long as he is flying in a lightly loaded airplane, and he does not acquire too much ice, there will be no trouble. But suppose A encounters icing in a lightly loaded aircraft and is unable to get rid of the ice during descent to his destination. If A does not compensate for the higher stall speed that ice accumulation typically causes, the aircraft may stall and crash on the approach. A proper understanding of the effects of ice build up on an airplane's control surfaces would be quite different and involve the notion that ice build-up might deform the control surfaces and disrupt the airflow over the wing that generates lift.

As with internal connections between individual discourse referents, the interpretation of a message generating internal connections between concept discourse referents may also yield indirect information available to the recipient even when he rejects the content of the message. And as in the case of individual discourse referents, if an agent correctly captures the internal relations that operate within another agent's cognitive state, he may be able to more easily correct a false belief of the latter than if he does not. So, too, once concept discourse referents are introduced, it may be that a person so uses certain general terms that the recipient perceives a confusion about concepts; by paying attention to the internal connections between concepts, the recipient can offer a better explanation and resolution of the speaker's doxastic difficulties than when he does not. Let us consider again the case of A and his faulty beliefs about icing and suppose that A exhibits his confusion about icing conditions to a knowledgeable pilot. Now the knowledgeable pilot B is the recipient of A's stories about icing. B uses A's story to update his representation of A's cognitive state, even though he rejects A's claims about the effects of airframe icing. But the internal connections now evident in B's representation of A's cognitive state should exhibit to B A's faulty representation of the causal connection between the event-types of icing and crashing planes. Having pinpointed the problem in A's understanding, B can more easily convey to A a correct understanding of the effects of airframe icing.

Once the Pandora's box of concepts is opened, yet more complexities concerning information must be addressed. For instance, interpretation of a message appears to involve assigning to concept discourse referents various relations that "make sense" of the message or give it a "cohesion" that an intelligible discourse has but an unintelligible one lacks. Presumably, these relations will also affect the notion of information, but I don't know the nature of these effects at present. This suggestion points back to the main theme of the paper: the information encoded in a verbal message for an agent depends on connections between the semantic content of the message and the agent's antecedent beliefs. There are also external connections between the agent's beliefs and his environment that also affect the information content of a message. I suspect, though I cannot argue it convincingly, that an account of perceptual information would reveal many of the same complications. That is, perceptually gathered information would involve internal as well as external connections. In such an account of information lies a moral, I think, for those, like Dretske and Fodor,[31] who would attempt to use the notion of information in a reductive explanation of the content of mental states. The reductionist would seek to reduce the content of mental states (and hence all the apparatus developed here) to certain causal relations between the agent and his environment. The account I have tried to develop suggests that causal relations between components of the agent's representational system and his environment are only one part of the picture in an account of information and the content of mental states. At least as important in determining content are those internal connections that obtain between elements of the objects of belief within the agent's total cognitive state. Moreover, the independence of internal and external connections and the complexity of the connections themselves refutes the plausibility of any simple, reductive account of content. Indeed, the account of information I have been sketching suggests that such a reductive account may not be forthcoming at all.

V

Appendix

I give here a technical description of the DRS fragment and its interpetation used in sections I-III. The fragment underlying the remarks in section IV is quite complex; much, but not all, of it can be found in Asher (1986, 1987). For the purposes of illustration, it will be sufficient to take DRSs and the set of conditions to be recursively charaterized as follows:

(1) If ψ in an n-ary DRS predicate and $\mathbf{x}_1, \ldots, \mathbf{x}_n$ are discourse referents, then $\psi(\mathbf{x}_1, \ldots, \mathbf{x}_n)$ is an atomic condition.
(2) If $\mathbf{x}_1$ and $\mathbf{x}_2$ are discourse referents (of any kind) and $\mathbf{e}_1$ and $\mathbf{e}_2$ are event discourse referents, then $\mathbf{x}_1 = \mathbf{x}_2$, $\mathbf{e}_1 < \mathbf{e}_2$ are atomic conditions.
(3) If K_1, K_2 are DRSs and $\mathbf{e}$ an event discourse referent, then $K_1 \Rightarrow K_2$ and $\mathbf{e} - K_1$ are conditions.
(4) A DRS is a pair $<U, Con>$, where U is a set of discourse referents, and Con a set of conditions.

Define a DRS K to be *subordinate* to another DRS K′, just in case K occurs as a component of a condition in K′ or K and K′ are components of a condition of the form $K' \Rightarrow K$. Say that $K \leq K'$ iff there are $K_1, \ldots, K_n$ such that K is subordinate to K_1, K_1 is subordinate to $K_2, \ldots, K_{n-1}$ is subordinate to K_n.

Define a DRS model M to be a quintuple $<W, T, D, <, [\![\,]\!]>$, where: (1) W and T are non-empty sets (the set of worlds and times respectively); (2) D is a function from $W \times T$ into a family of non-empty sets ($D_{<w,t>}$ is the "domain of w at t"); (3) $<$ is a strict partial ordering on T, (4) $[\![\,]\!]$ is an interpretation function that assigns to DRS predicates functions from $W \times T$ into $P(\bigcup_{n \in \omega}(D^n))$. The domain at $<w,t>$ will itself be a pair of two disjoint sets — the set of individuals and the set of events existing at $<w,t>$. I will also invoke τ, a function from the set of events existing at $<w,t>$ onto T. Now define an *embedding* f of K *in* M *at* $<w,t>$ as follows (correcting an error in Asher (1986, 1987)):

(1) f is a map from U_K into $D_{<w,t>}$, if there is no K′ such that $K \leq K'$.
(2) $Dom(f) \supseteq U_K$ & f is a partial function from $\bigcup_{K_i \geq K}(U_{K_i})$ into $D_{<w,t>}$, if there is a K′ such that $K \leq K'$.

Define g to *extend* an embedding function f *to an embedding of* K in M at $<w,t>$ (written $_{g \supseteq K} f$) just in case $Dom(g) = Dom(f) \cup U_K$ and g is an embedding of K in M at $_{<w,t>}$.

Define *an external anchor for* K *in* M to be a partial function from U_K into $(\cup_{<w,t> \in W \times T}(D_{<w,t>})) \cup T$, such that if 'now' occurs in U_K then A(now) = the utterance time of the discourse, and if 'i' occurs in U_K A(i) = the speaker of the discourse.

Now we define *a proper embedding* f *of* K *in* M *at* $<w,t>$ (written $[f,K]^M_{w,t}$) with respect to a possibly empty external anchor A for K in M and satisfaction of a condition in M relative to an embedding function f at $<w,t>$ (written $M \models_{w,t,f}$).

(1) If ψ is an atomic condition of the form $\phi(\mathbf{x}_1, \ldots, \mathbf{x}_n)$, then $M \models_{w,t,f} \psi$ iff $<f(\mathbf{x}_1), \ldots, f(\mathbf{x}_n)> \in [\![\phi]\!]_{w,t}$.
(2) If ψ is an atomic condition of the form $\mathbf{x}_1 = \mathbf{x}_2$, then $M \models_{x,t,f} \psi$ iff $f(\mathbf{x}_1) = f(\mathbf{x}_2)$.
(3) If ψ is an atomic condition of the form $\mathbf{e}_1 < \mathbf{e}_2$, then $M \models_{w,t,f} \psi$ iff $\tau(f(\mathbf{e}_1)) < \tau(f(\mathbf{e}_2))$.
(4) If ψ is an atomic condition of the form $\mathbf{x}_1 <$ now, then $M \models_{w,t,f} \psi$ iff $f(\mathbf{x}_1) < f(now)$.
(5) If ψ is a condition of the form $K_1 \Rightarrow K_2$, then $M \models_{w,t,f} \psi$ iff $\forall g \supseteq K_1 f([g, K_1]^M_{w,t,f} \rightarrow \exists h \supseteq K_2 g[h,K_2]^M_{w,t,g})$.
(6) If ψ is a condition of the form $\mathbf{e} - K_1$, then $M \models_{w.t,f} \psi$ iff $\exists g \supseteq K_1 f[g, K_1]^M_{w,t,f})$
(7) If A is an external anchor for K in M, them $[f,K]^M_{w,t,g}$ iff (i) $f \supseteq g$; (ii) $A \subseteq f$; (3) $\forall \Theta \in Con_K\ M \models_{w,t,f} \Theta$.
(8) $[f,K]^M_{w,t}$ iff $[f,K]^M_{w,t,\Lambda}$.

NOTES

* I would like to thank Sally Ferguson-Ramzy, Ross Mandel, and Randy Mayes for comments on a previous draft of this paper. I also would like to thank the Center for Cognitive Science at The University of Texas for its generous research support.

1 Asher, "Belief in Discourse Representation Theory," *Journal of Philosophical Logic*, 5 (1986): 127-89; Asher, "A Typology for Attitude Verbs and Their Anaphoric Properties," *Linguistics and Philosophy*, 10 (1987a): 125-97; and Asher, "Belief, Acceptance and Belief Reports," *Canadian Journal of Philosophy*, 19:3 (Sept. 1989).

2 For much more discussion of the general approach of DR theory, the algorithm for DRS construction, and the correctness of definition, see Kamp, "A Theory of Truth and Semantic Representation," in J. Groenendijk, Th. Janssen, and M. Stokhof (eds.), *Formal Methods in the Study of Language*, Mathematisch Centrum Tracts (Amsterdam 1981), 277-322, Asher, "Belief in Discourse Representation Theory" and "A Typology for Attitude Verbs and Their Anaphoric Properties"; and H. Kamp, "Situations in Discourse without Time or Questions," manuscript (1983).

3 For a formal specification of DRS and the construction algorithm for generating them, see Kamp, "Situations in Discourse," (1981, 1983), Asher, "Belief in Dis-

course Representation Theory." I will in general follow the DRS formulation in Kamp (1983) and Asher (1986). When appropriate, I will use boldfaced letters of the appropriate type as metalinguistic variables for discourse referents.

4 See Asher N. and Wada H., "A Computational Account of Semantic, Syntactic and Pragmatic Constraints on Anaphora," *Journal of Semantics*, forthcoming.

5 Ed Stabler's comments on "algorithms" led me to reformulate slightly this aspect of the DRS construction algorithm.

6 The problem concerning conditions introduced by proper names is familiar from discussions concerning the scope of descriptions and modal operators.

7 The familiarity theory of definites has been developed within the DR theoretic framework and vigorously defended by Irene Heim in "The Semantics of Definite and Indefinite Noun Phrases" (Ph.D. thesis, University of Massachusetts at Amherst, 1982). Roughly, the familiarity theory of definites claims that it is a presupposition of a definite NP α that the recipient of a discourse containing α be familiar with α's denotation. Note, however, that this familiarity may be of a quite minimal kind). For further details see the discussion in Asher, "Belief in Discourse Representation Theory."

8 Barwise and Perry in J. Barwise and J. Perry, "Semantic Innocence and Uncompromising Situations," in P.A. French, T.E. Uehling, and H.K. Wettstein (eds.), *Midwest Studies in Philosophy VI: The Foundations of Analytic Philosophy* (Minneapolis: University of Minnesota Press 1981), 387-404, distinguish between a "value-free" interpretation of a definite description and a "value-loaded" interpretation of a definite description. Something like this distinction (though there are important differences) was noted earlier by Donnellan in K. Donnellan, "Reference and Definite Descriptions," *Philosophical Review*, 75 (1966): 281-304. The redundancy in (4) is avoided by supposing that the agent processes the definite description according to a "referential" strategy that employs at the level of discourse referents something like the value-loading strategy of interpretation in Barwise and Perry. The principal contribution that the DR-theoretic approach makes to singular terms is to combine in a fruitful way the distinction between novel/familiar long employed by discourse theorists with more familiar distinctions of scope. The proper analysis of names within DR-theory offers an illustration of this.

9 While conceptual objects differ from real objects, they have a very close connection. From the agent's phenomenological perspective, the individuals that constitute the subject-matter of his beliefs are no different from the objects he supposes to exist in the world around him; the imaginary saboteur of the reactor at Chernobyl seems just as real to the agent who believes in him as any other object towards which he might entertain an attitude. From the perspective of an observer of the agent, these conceptual individuals may sometimes be accurately described as "standing proxy" in the mind of the agent for real individuals like Chernobyl, but in other cases (especially in the case of false or confused beliefs) conceptual individuals may have no external correlates. But there may, in fact, be no phenomenological difference between those beliefs containing conceptual individuals standing proxy for real individuals and those beliefs containing conceptual individuals with no real world referent. A unicorn believer thinks in the same way about unicorns as he does about cats, at least from an introspective point of view, an example due to W. Rapaport, "Logical Foundations for Belief Representation," manuscript (1986). With general beliefs, the partial correspondence between belief objects and real situations as humans perceive them also breaks down, albeit for different reasons.

10 Much of this section reworks material found in Asher, "Belief in Discourse Representation Theory" and "Semantic Competence, Linguistic Understanding, and a Theory of Concepts," *Philosophical Studies* (1988); 1–36 .

11 Since DRT distinguishes between the DRS construction process and its model theoretic interpretation, the claim that the construction procedure describes a psychological process does not fall victim to problems that many have raised to a psychological reading of mode/theoretic semantics. See, for example, B. Partee,

"Montague Grammar, Mental Representations, and Reality," *Midwest Studies in Philosophy*, 4 (1979a): 195-208; B. Partee "Semantics: Mathematics or Psychology?" in *Semantics from Different Points of View*, R. Bauerle, U. Egli, and A. von Stechow (eds.) (Berlin: Springer Verlag 1979b), 1-14.

12 Note that since this deletion of information is performed prior to a stage of evaluation, the problems of interpreting conditions introduced by proper names within the scope of a modal operator examined in the last section will not arise.

13 F. Dretske, *Knowledge and the Flow of Information* (Cambridge, MA: MIT Press 1980). F. Landman "Towards a Theory of Information: The Status of Partial Objects in Semantics," Ph.D. Thesis, University of Amsterdam, 1986) also published by Foris, in Dordrecht, under same title (1986).

14 R. Stalnaker, "Assertion," *Syntax and Semantics*, 9 (1978): 315-32, anticipates this "contextual" theory of belief. See also R. Stalnaker, *Inquiry* (Cambridge, MA: MIT Press 1984). Stalnaker has in a paper entitled "Belief and Attribution," soon to be published, also advocated a contextual theory of belief. I would claim, however, that DRT offers a more useful articulation of the background context needed to interpret beliefs than can be found in pure possible worlds semantics.

15 This notion is borrowed from Heim's formalization of DRT in The Semantics of Definite and Indefinite Noun Phrases.

16 See J. Perry, "The Essential Indexical," *Nous*, 13 (1979): 5-21. See also J. Perry "Frege on Demonstratives," *Philosophical Review*, 86 (1977): 474-97. See also David Lewis, "Attitudes de Dicto and de Re," *Philosophical Review*, 88 (1979): 513-43.

17 See Perry, "The Essential Indexical."

18 Lewis, "Attitudes de Ricto and de Re."

19 So Lewis's two tests for property ascription *de re* fail here.

20 Again I am indebted to my commentator Ed Stabler, whose comments suggested that I should perhaps address issues of belief revision as well as of belief accumulation.

21 For a discussion of these operations, see P. Gardenfors and D. Makinson, "Revisions of Knowledge Systems Using Epistemic Entrenchment," in *Proceedings of the Second Conference on Theoretical Aspects of Reasoning about Knowledge*, ed. M. Vardi (Los Angeles: Morgan Kaufmann 1988), 83-95.

22 Kripke, "A Puzzle about Belief," in A. Margalit, ed., *Meaning and Use* (Dordrecht: D. Reidel 1979), 239-83. Asher, "Belief in Discourse Representation Theory" and Kamp, "A Theory of Truth and Semantic Representation," offer discussions of this puzzle.

23 For a discussion of one such case, see my "Belief, Acceptance and Belief Reports."

24 One might add that there would be in virtue of the conversation between A and S, a sort of external anchor between **x** and **y** — a link that I have called a "quasi-external anchor" in Asher, "A Typology for Attitude Verbs and Their Anaphoric Properties." Clark and Marshall have argued that such mutual belief is required for successful communication in H. Clark and C. Marshall, "Definite Reference and Mutual Knowledge," in A. Joshi, B. Webber and I. Sag (eds.) *Elements of Discourse Understanding* (Cambridge, Eng: Cambridge University Press 1982) 110-64.

25 I give a treatment of Hob Nob sentences in Asher, "A Typology for Attitude Verbs and Their Anaphoric Properties," but it does not take into account mutual belief.

26 The use of concept discourse referents that I have in mind is explored to some extent in Asher, "Semantic Competence, Linguistic Understanding and a Theory of Concepts." The first use of concept discourse referents in DR theory occurs in E. Klein, "VP-Ellipsis in DR Theory," manuscript (1984).

27 See Asher, "Belief, Acceptance, and Belief Reports," and H. Kamp (1987), "Reading, Writing and Understanding," a commentary on Hans Sluga's paper delivered at the APA central division meetings. Here is one place where I appeal to something like Dretscke's information theory, or rather those causal relations that it emphasizes.

28 See Asher, "Semantic Competence, Linguistic Understanding and a Theory of Concepts" as well as Asher, *Truth Conditions and Semantic Competence: Toward a*

Theory of Linguistic Understanding (Ph.D. thesis, Yale University 1982, for a discussion of these sorts of second order concepts. One second order concept that I studied in detail in the paper as well as my dissertation was the use of degrees of certainty and conditional probability functions. Restricting oneself just to one such second order concept, however, is perhaps too limiting.

29 This example is pursued in Asher, "Belief, Acceptance and Belief Reports."

30 One could translate this causal link between 1-place concepts for instance into a more familiar second order concept — that of universal quantification: $\lambda P \lambda Q \forall x(P\{x\} \rightarrow Q\{x\})$. Of course the familiar types of quantification certainly also must be mentioned as relations between concepts.

31 Dretske, *Knowledge and the Flow of Information*. See Fodor's article in this volume.

Comment
Discourse, Discourse Processing, and Information

E.P. Stabler, Jr.

"Discourse representation theory" (DRT) came out of Texas with the help of Hans Kamp, Nicholas Asher, and others and is now spreading around the globe. Asher illustrates the theory with the following example. In DRT, a mapping associates a sentence like "The meltdown at Chernobyl has ended" with a "discourse representation structure" (DRS) like the following:

e, e_1, v
e = the meltdown e at v named('Chernobyl,' v) e_1 end(e) e now

DRS1

In a discourse in which this sentence is followed by "Every European will remember it," a second DRS is created. This one actually contains a condition in which there are two subDRSs:

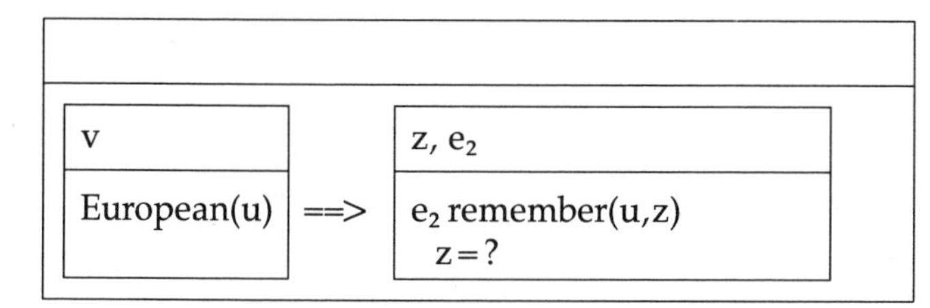

This second DRS is then "resolved with" the first, producing a third DRS as the result on page 58.

Each DRS K is a pair $< U_K \; Con_K >$ where U_K is a set of "discourse referents" (or "reference markers"), and Con_K is a set of "conditions." (U_{DRS2} is empty, and Con_{DRS2} contains only one condition.)

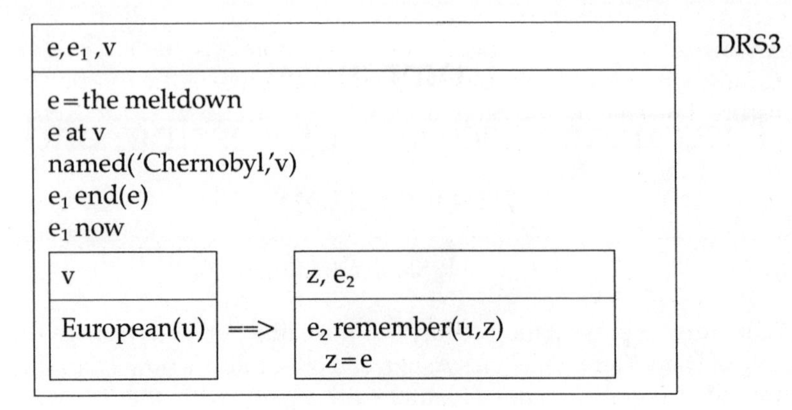

Asher makes a number of claims about this discourse representation theory. The following points are fundamental:

DISCOURSE

(1) Given a discourse, the construction of the associated DRS is carried out by a *DRS construction algorithm*. These DRSs provide a formal semantics for the discourse via a *correctness definition*.

DISCOURSE PROCESSING

(2) (UTI) DRSs describe not only the semantic content of discourse but also describe the content of beliefs.

(a) When an agent hears and accepts a discourse, the agent constructs the DRS for the discourse, and then integrates this object with antecedent beliefs.

(b) These contents are complex, and so there is actually a structural correspondence between DRSs and objects of belief.

INFORMATION

(3) When an agent accepts a discourse ϕ, the information conveyed by ϕ consists of the content added to the agent's cognitive state in integrating the DRS corresponding to ϕ. A purely syntactic account is also suggested for exploration.

So it is clear, even in advance of considering any of the detail behind these suggestions, that bold claims are being made. That is, regardless of how the details go, this picture is a bold and controversial one. It brings together a formal approach to linguistic semantics, a theory of discourse understanding, a representational theory of belief, and a theory of information and communication. I share Asher's enthusiasm for reaching a grand unification of these domains, but it is a risky business when even the rough shapes of each piece of the puzzle are obscure. I will quickly consider each of 1-3 in turn.

DISCOURSE

Asher says: "The construction algorithm maps a natural language discourse (or rather the sequence of its constituent sentences' syntactic structures) to a discourse representation structure. . . ." I can hardly believe he says this, but he repeatedly makes this claim, so let's assume for a moment that he means it. The obvious problem is simply that *there is no DRS construction algorithm*.

As Asher notes, the problem of integrating the content of a sentence with the content of a preceding discourse resembles the problem of integrating the content of a message with one's antecedent beliefs (1988:2). The latter problem has had centuries of attention, and still appears intractable. Asher proposes that a DRS construction algorithm is going to solve the former problem, in spite of the fact that he sees the analogy with the enigma.

Consider the following discourse:

> *Asher's favourite example concerns the meltdown at Chernobyl. Every European will remember it.*

If we imagine the construction of two DRSs, DRS1′, and DRS2′, I am worried about what the DRS construction algorithm decides about the reference of the "it" in the "resolution" of DRS2′ with DRS1′ to produce the complete DRS3′. In Asher's example, the "it" in "Every European will remember it" refers to the meltdown at Chernobyl, but in my example there is some ambiguity. Contrast

> *Asher's favourite example concerns the meltdown at Chernobyl. Every linguist at this conference will remember it.*

The proper mapping from this discourse to the appropriate DRS cannot be done algorithmically. In the first place, there is no single "correct" interpretation of my examples — they are genuinely ambiguous. My examples focused on referential ambiguity, but the properties that may remain undecided include lexical meaning and basic predicate-argument structures.[1] But in the second place, the context that usually disambiguates a discourse does so by providing a basis for a plausibility judgment, and this sort of judgment has not been formalized and may not be formalizable.

I should eliminate a potential ambiguity in the statement of what the DRS construction algorithm is supposed to do. Is it supposed to specify the "correct" meaning of a discourse, or is it supposed to specify the interpretation that a hearer would assign to the sentence? Notice that the normative and psychological projects are not united by the strategy of considering an ideal speaker-hearer, a hearer with no lapses in concentration or memory limitations, as might be the case in syntax, because the question of what a term really refers to is not generally a matter that can be settled by the state of the hearer, no matter how ideal he is. Two terms could be coreferential, as a matter of fact, even though the speaker or hearer had no idea that this was the case. It is clear, I think, that Asher does not intend his construction algorithm to resolve coreference questions in the "correct" way, but rather in the way a hearer would.

So now I can explain more precisely what I mean by the claim that there is no DRS construction algorithm (as I had better do since Asher has presented algorithms that he called DRS construction algorithms!). The claim is that no algorithm, or at least no algorithm we can imagine now, given an arbitrary discourse and an arbitrary representation of background knowledge, disambiguates lexical items, decides predicate-argument structure and coreference relations in the way an idealized, competent speaker-hearer would. Even for those who do not know the literature on induction, a little reflection on our plausibility judgments casts doubt on the prospects for making them algorithmic. The role of simplicity, parsimony, and charity in interpretation just does not look like the kind of thing that is going to be formalizable.[2]

There are a number of possible responses to the objection. Suppose that we agree that there is no DRS construction algorithm. One thing we might do — and I suspect that this is what Asher really was doing — is to treat the rules of DRS construction as if they were "algorithmic" by assuming that the rules have access to oracular decisions about coreference where they are needed. Kamp explicitly does this in some of his work, and for this reason he describes his approach as "fudgey" (1981:311). I think that as a general strategy for DRT, this is ill advised.[3] It obscures the very important distinction between what the discourse determines and what it doesn't. This strategy is also liable to mislead us into thinking that, once background knowledge has been used, there will be a particular DRS with determined coreference relations for any discourse, but it is clear that this is not the case. In fact, I think that a failure to attend to these points has led Asher to mistaken views about what the reasonable objectives of DRT should be. To support this claim I will devleop an alternative perspective on DRT from which it becomes clear that points (2) and (3) are untenable, at least in the form Asher has put them.

So suppose we accept my arguments and wisely agree to the following:

> (1′) *There is no DRS construction algorithm.*
> (1a′) *There is no unique appropriate DRS for a given discourse in a given context, unless DRSs can be noncommittal about coreference relations and predicate-argument structure.*

Discourse Processing

Adopting this perspective, we can see that some other dominoes have to fall as we move to a more defensible view of what DRT can do. Consider claim (2a):

> (2a) When an agent hears and accepts a discourse, the agent constructs the DRS for the discourse, and then integrates this object with antecedent beliefs.

What is the argument for this claim? I presume it is the one that is familiar from other parts of linguistic theory, and although Asher does not state it, I found it suggested by Kamp: "The role representations are made to play within the theory . . . places substantial constraints on their internal structure. . . . This is of particular interest if . . . discourse representations can be regarded as the mental representations which speakers form in response to the verbal inputs they receive" (1981:281-2). The argument is that since the structures of DRT actually correspond to mental structures involved in our interpretation of discourse, they should encode the distinctions that are essential in the account of our discourse competence.

Although the form of argument is familiar, so are a number of problems with (2a) as it has been presented. In the first place, the identification of the structures specified by DRT with the psychological structures formulated in discourse comprehension is too simplistic, for reasons that are familiar from discussions of the psychological reality of other formal theories of language and cognition. We could assume that DRT should allow that any DRS that a human formulates upon hearing a discourse is an appropriate DRS for that discourse; this makes Asher's claim (2a) true but trivial, and it is clearly not what he intends. The intended DRT must be one according to which there will be discourses for which no human will be able to construct an appropriate DRS. Suppose, for example, that the sentences in the discourse are grammatical and sensible but too complex to process in a reasonable amount of time. To allow for such cases, we need to abstract away from limitations on the memory and attention of the subject. This is not enough though, for certainly, even under this abstraction, the theory is going to have to allow for parameters of individual variation. It seems very unlikely that all linguistically competent speakers, English and Warlpiri, two years old or twenty, construct DRSs in accord with exactly the same principles.

I think Asher and Kamp are right to want a significant relation between DRSs and the theory of language comprehension and belief, but the relation must be a little more complex than they suggest. We have seen that DRT cannot algorithmically determine the appropriate DRS for any discourse, but it could aim to specify the relation between each discourse and the corresponding set of appropriate DRSs. Then the claim could be that, abstracting away from limitations on memory, attention, etc., the speaker selects among the possibilities specified by an appropriately parameterized DRT.

These familiar points apply whether we are considering psychological representations of discourse or objects of belief. The actual objects of belief and other propositional attitudes are subject to limitations that do not constrain discourses. These limitations must be kept in mind when formulating a claim about any formal model of human information processing. I cannot see any reason to think that DRT can avoid them.

> (2ab') *The simple identification of DRSs with psychological structures (in comprehension or in belief) cannot be maintained, but a suitably more complex relation should be sought.*

This point leaves most of Asher's picture standing. We can still regard DRT as a theory of structures that actually correspond to certain psychological states. Now let's consider whether the claim that DRSs correspond to our representations of discourse is actually supported. It is interesting to compare DRS structures with others for which a similar status has been claimed.

Because I spend so much time thinking about horribly complex syntactic structural representations, the thing that strikes me about the DRSs shown in Asher's paper is their wonderful simplicity: almost all of the syntactic structure of the discourse has been stripped out, as has even the order of the NPs and predications in the discourse, since a DRS K is just the pair consisting of the set of discourse referents U_K and the set of conditions Con_K (1988, 1986).[4] Only a little more structure is added by allowing conditions to contain DRSs in certain cases. These are called sub-DRSs. So let's begin by considering whether DRSs can be adequate with so little structure. First of all, notice that the structure that DRSs do have plays a significant role, not only in the definition of truth but also in the determination of coreference options. For example, in the sentence

Every farmer who owns a donkey beats it (a)

it is clearly possible for the 'it' be anaphorically related to 'a donkey', but the similar "donkey interpretation" is not available in

The farmer who owns every donkey beats it. (b)

Kamp explains the unavailability of farmers' donkeys to serve as the antecedent of "it" in terms of the structure of the associated DRSs. In particular, the discourse referent that satisfies the donkey predicate occurs in a DRS that is subordinate to the DRS in which "it" must be assigned a referent. This strategy is also used to define the available referents in Asher's extension of the theory to belief reports.[5] So we can raise the question of whether the DRSs have enough structure to define these options.

It seems to me that the evidence against the adequacy of such simplicity is really very good: subtle differences in details of syntactic structure influence coreference options. Consider the difference between Kamp's examples and the following[6]

A friend of every owner of a donkey beats it (c)

Here a donkey interpretaion of "it" is possible, but it is impossible or at least more difficult in

Your shouting at every owner of a donkey frightened it. (d)

Accounting for this range of cases has proven to be a very delicate matter, but one thing that all the promising approaches share is a recognition that coreference relations are influenced by the relative positions of NPs in a well-articulated syntactic structure. Kamp's treatment of the donkey sentences does not have any obvious generalization to this range of cases: certainly in example (c) the donkey discourse referent occurs in a structure subordinate to the pronoun, and yet coreference appears to be possible. In short, the absence of structure in DRSs appears to be problematic.

We did not really need to go to such complex examples as the donkey sentences. I did that just so I could refer to Kamp's interpretation of them in detail. But the case is stronger (as are my intuitions) when we consider simpler examples like the second sentence of Asher's first discourse:

Every European i *will remember it* j. i ≠ j

This has a required noncoreference relation. Here the fact that "it" is not personal blocks coreference with each European, but the structure itself would suffice even if by "European" we meant impersonal, asexual "European androids." The structural requirement of non-coreference *is not* represented in DRS2 and DRS3. Coreference is possible, since the DRS containing the European android referent is not subordinate to the DRS in which "it" must receive a value. A slightly different sentence makes the coreference optional,

Every European android's creator will remember it,

and another variation makes it mandatory,

Every European android$_i$ will remember itself$_i$.

No straightforward extension of the DRT approach handles all these cases.

Kamp's strategy seems to have similar difficulties in accounting for the subtleties of quantifier interaction that are illustrated by the following familiar contrasting pairs.[7]

Someone expects every NDP candidate to be elected.
Someone expects that every NDP candidate will be elected.

or

It is for this reason that I believe John was fired.
It is for this reason that I do not believe John was fired.

In each of these pairs, the first sentence is ambiguous in a way that the second is not, and the distinction comes from minor differences in syntactic structure. So I suggest:

(2a′) *Structurally determined coreference and noncoreference relations must be represented in some structure that guides discourse interpretation. DRSs do not have sufficient structure for this role.*

I do not see how a DRS can play the role Asher intends unless we either build enough syntactic structure into DRSs to reconstruct these influences on coreference and scoping relations. The prospects for accounting for these with the limited use of quantifier type and order seem dim. Another option is to turn the job of defining the available discourse referents back over to the syntax. The latter strategy is probably the more appealing, and approaches of this sort have been advocated by some linguists.[8].

Notice that this last strategy certainly does not imply that the restrictions on coreference and scoping relations must come exclusively from sentence syntax. In the first place, my claim has been only that sentence syntax is one of the important influences. Besides the apparent additional influence of background knowledge, there presumably are constraints that depend on *structural properties that span sentence boundaries*, and these would distinguish DRT from other theories of logical form. For example, notice that I had to modify Asher's example to illustrate ambiguity in discourse interpretation. I cannot think of a prior context that would make "it" refer, for example, to Chernobyl rather than to the meltdown in his discourse. The similar reference assignment seems to be determined in other discourses of similar structure [$_S$event_term Have V – ed] [$_S$ NP V it]:

[The dangerous game]$_i$ has ended.
No one liked it$_i$.

The best I can do by way of changing the reference assignment in such cases is to draw the emphasis from the subject to the event described by the sentence as a whole, as in

The meltdown at Chernobyl ended dramatically.
Every European will remember it.

Here the pronoun could refer to the ending of the meltdown, rather than to the whole meltdown event.[9]

Asher does not explore such matters in his paper and does not suggest even a single extra-sentential constraint on interpretation. The relevant point is that syntactic structure appears to determine some important discourse properties. Furthermore, keeping the determined discourse properties distinct from the properties that may be inferred from background knowledge clearly brings out the problem of getting all the determined properties properly accounted for, and it properly leaves open the possibility that even when background knowledge and context is used, important properties of the discourse may not be decided.

Let's turn to the most general statement of (UTI): that the structure of DRSs corresponds to that of beliefs. Notice that this correspondence could hold even if DRSs are not the objects that indicate syntactic restrictions on potential discourse referents. So here we can let our view of DRSs be guided by Asher's original remarks and examples. In fact, I think that this is the right view to take. There may be an algorithm that specifies the set of possible highly structured representations of the discourse, but then there is an inductive leap to the rather different representation of the intended interpretation of the discourse, and then there is another inductive leap to the representation of what the speaker-hearer comes to believe. It seems to me that DRSs cannot play all these roles at once. I will return to this point in a moment. For now, I will just adopt this picture. While the range of potential antecedents might be indicated at the highly structured level, as I have suggested, the representation of the intended meaning and the representation of the belief need only represent the sort of thing shown in Asher's examples, except that we still must add a representation of the nonidentities that the discourse implies. That is, even if DRSs are not required to capture all the influences on coreference relations that are determined by the syntax, still, when such relations are implied on any interpretation of the discourse, the representation of that interpretation better include them. So we have the following corollary of our previous point:

(2a") *Any representation of a particular, disambiguated, discourse must also represent required identities and nonidentities among elements of the discourse domain.*

For example, a representation of a discourse containing *John remembers it* had better represent the inequality of the referent of *John* and the referent of *it*. DRS2 and DRS3 do not suffice. Returning to the general statement of (UTI), the arguments that belief states are structured and representational are well known. Not everyone is convinced by them, but I am, and so I am not going to worry about this quite fundamental idea here. Looking then at the structure of DRSs, I do not find much to complain about, above the level of detail. Asher does not commit himself to, for example, a lot of structure hidden in each predicate, for example, as some decompositional views do. DRSs have the basic logical structure that is needed to account for the logical relations among beliefs, and I expect that Asher would agree that this structure may need to be enriched to allow for a proper representation of tense, modalities, or higher order propositions about relations. In fact, DRT does not depart very significantly from already familiar representational views.[10]

Asher mentions some different matters of formal detail, such as the special status of conditions introduced by definite NPs, but he does not formulate these proposals in any comprehensive way. For example, he does not explain exactly how he is going to account for the different contributions of the definite descriptions in the true reading of

The city named "Chernobyl" might not have been the city named "Chernobyl."

However, it is clear that DRT has enough machinery to underpin various approaches to this sort of problem.

Let's leave these details. I want to conclude with a couple of remarks about the information conveyed by a discourse.

INFORMATION

I have been referring, more or less favourably, to what Asher and Kamp call "UTI," "the principle of unity of thought and information." Asher says that this is the idea that "the very same structures that describe the semantic content of discourse ought to also serve in describing the content of our beliefs" (1988). This is clearly a version of a familiar idea, but I have some qualms about the name for it. The term "information" is used in many ways, but usually in connection with some materialistically acceptable notion of content (Dretske 1981). Of course, we need not use the term that way, so long as we are clear about what we mean. For example, in a recent paper Fodor has distinguished these "Standard" notions of information from what he calls "encoded" information (Fodor 1986). Let's consider what we've got for DRS contents.

Remember that Asher's first proposal was:

> (3) When an agent accepts a discourse ϕ, the information conveyed by ϕ consists in the content added to the agent's cognitive state in integrating the DRS corresponding to ϕ.

So let's consider these contents. To define the content of a DRS, Asher, in effect, provided the discourse with a partial model and defines the discourse to be true when the partial model is compatible with the world. Formally, a DRS is compatible with a world where it has a "proper embedding." Roughly, a function f is a "proper embedding" of K at world w just in case f maps the discourse referents of K onto objects in the world w in such a way as to preserve all the properties and relations specified by the DRS. (Cf. Asher 1986: sec. 4.) So the content of a DRS is defined to be the set of worlds at which the DRS has a "proper embedding." In these terms, Asher (1988) offers the following more formal definition of the information conveyed by a discourse:

> Suppose an agent with background beliefs K in accepting discourse ϕ goes into belief state K′. Then the information conveyed by ϕ is $\bigcap_{\kappa\in K}$content(κ) − $\bigcap_{\kappa\in K'}$ content (κ).

In other words, the change from K to K′ is presumed to involve the addition of information and hence a decrease in the set of indices where there are proper embeddings. We can then take the set difference between the set of indices where there are proper embeddings of K and the set of indices where there are proper embeddings of K′.

A number of technical difficulties spring to mind. For example, accepting a discourse does not generally involve this sort of addition of information. When I listen to the discourses of great philosophers, for example, I often find myself persuaded by impeccable arguments that many of my previously held beliefs ought to be rejected. This could result in my moving to a state K′ whose content is larger than or incomparable with the content of my original state K. How should we fix this? A technical trick might suffice to avoid the technical problem. For example, we could let the information conveyed by ϕ be a pair of sets $< \text{Del}, \text{Add} >$:

$$\text{Del} = \{i \mid i \notin \bigcap_{\kappa \in K} \text{content}(\kappa),\ i \in \bigcap_{\kappa \in K'} \text{content}(\kappa)\}$$
$$\text{Add} = \{i \mid i \in \bigcap_{\kappa \in K} \text{content}(\kappa),\ i \notin \bigcap_{\kappa \in K'} \text{content}(\kappa)\}$$

That is, in accepting ϕ we can reject some previously held ideas (increasing the set of indices of compatible worlds) and add some ideas that were not previously held (decreasing the set of indices of compatible worlds).

Is this good enough? That depends on what the notion of "content" or "information conveyed" is needed for. Notice that even this simple change is problematic for DRT, because although simple additions to K on the basis of ϕ might suggest that they were somehow a (formally representable) function of ϕ (or of ϕ together with background knowledge), we have seen that this is not generally the case. The adjustments in the total belief state that might be involved in "accepting" an arbitrary discourse ϕ are bound to be quite various and not formally specifiable, for reasons already noted. Even if we could specify the change though, some of the changes might be quite unrelated to what we would regard as the topic of ϕ and so it is not clear that we would want to call the whole change "information conveyed by ϕ."

This is another reason to favour sharp distinctions between syntactic discourse structures, representations of disambiguated discourse, and representations of belief. Not only are these separated from each other by the theoretical abyss of plausible inference, but relations spanning the abyss tend not to be neat, as shown for example by such things as Donnellan's "referential" uses of definite descriptions (1966).

The proposed notion of conveyed information has other, more familiar problems as well. For example, it seems to have a "logical omniscience" problem. While I might learn a lot from having a consequence of my prior beliefs pointed out, it appears that no information would be conveyed by a discourse that did only this, since the set of indices of compatible worlds presumably would not change.

Setting such problems aside, though, we can see that none of these proposals exhibits any obvious connection to any familiar notion of information that has been proposed by materialistically-minded philosophers. In this respect, the *DRT* proposals are in the same boat as earlier model-theoretic theories of meaning. These approaches all have the advantage that they are not plagued with the usual problems with information-based notions of content, such as a difficulty in providing a reasonable account of "misinformation." Asher's proposal can apparently account without difficulty for the acceptance of false discourses about fictional entities — in these cases we just add contents whose indices do not include the actual world. On the other hand, though, a materialistic reconstruction of such model-theoretic notions of content is problematic.

> (3′) *The model-theoretic interpretation of DRSs has not been provided with any reconstruction in terms of a materialistically founded notion of information, and it appears to be untenable in any case.*

Asher suggests that it might be worth exploring a second, very different idea about the information conveyed by a discourse. According to this view, we can identify the information conveyed by a discourse with the actual syntactic change from K to K′. It looks to me like this change will not in general be systematically related to the discourse, and so this will be problematic, but Asher points out that this approach avoids the omniscience problems, and that it can still serve computational purposes. But in this case we have no notion of content at all, no sense in which the *DRS*s are about anything in the world. I think we need a notion of content to make sense of cognition and communication, and the arguments for this are well-known (which is not to say that I think that they are obviously right — in fact, the issue is rather difficult).[11] However, Asher did not attempt to refute any of these arguments, and so I will settle for noting that they exist.

CONCLUSIONS

In spite of the various problems I have raised, I like the spirit of the work in discourse representation theory. I think that its greatest contribution comes not in making a new step towards the unification of information, interpretation and attitudes, but in a variety of technical contributions in the tradition that we usually call "logical form." It would be valuable to find constraints on logical form and principles of interpretation that span sentence boundaries.

NOTES

1 The familiar point that predicate-argument structure is not always determined by a

discourse is illustrated by examples like *John was murdered by Yosemite Falls*, in which Yosemite Falls could be either the agent or the location.

2 This point is not new! For example, Hans Kamp agrees with Barbara Partee on this point in the following passage (1981): "The strategies used in selecting the referents of anaphoric pronouns are notoriously complex .... Efforts to understand these strategies have claimed much thought and hard work, but, in its general form at least, the problem appears to be far too complex to permit solution with the limited analytic tools that are available at the present time .... Indeed I much incline to the opinion expressed, for example, in Partee (1975:80) that all we can reasonably expect to achieve in this area is to articulate orders of preference among the potential referents of an anaphoric pronoun, without implying that the item that receives the highest rating is in each and every case the referent of the anaphor." Hilary Putnam agrees with Jerry Fodor on this point in the following passage (1986): "In short, it looks like the problem of mechanizing — 'computationalizing' — the notion of sameness of reference judgements or sameness of possible reference judgements, and hence sameness of meaning judgements, cannot be carried out without carrying out what Jerry [Fodor], I think rightly, calls the too Utopian program of computationalizing scientific explanation, the principle of abduction, or inference to the best explanation. There is an intimate link between the question of what is and what is not the best explanation and the question 'Is he referring to the same thing?' or 'Has he stopped referring to the same thing?'" Cf. also Putnam 1988.

3 I do not mean to deny that there may be some special purposes for which the strategy would be appropriate. For example, Kamp uses this strategy because he wants to describe a model-theoretic approach to DRSs which is much simplified by the assumption that coreference relations have been decided (Kamp 1981).

4 Order is recoverable in Kamp's representations, since the elements of the unstructured sets are labelled with indexes that are related to surface order. Kamp does not specify the indexing function, though, and clearly does not intend this use of the indexes. In Asher's formalism, it is not clear whether order information can be reconstructed or not.

5 More precisely, in Kamp's 1981 account, a pronoun is related to a "suitable" discourse referent u that occurs either in the current DRS or in a "superordinate" DRS. DRS_i is "immediately superordinate" to DRS_j if either (1) DRS_j is the first member of a pair corresponding to a conditional or universal in DRS_i, or (2) $\langle DRS_i \Rightarrow DRS_j \rangle$ is such a pair. DRS_i is superordinate to DRS_j if there is a finite chain of immediate superordinates between them. Asher adopts essentially the same account (Asher 1986: 186n29).

6 Examples (b) and (c) are adapted from May (1985) and Haik (1984). Cf. also Aoun (1985), Aoun and Hornstein (1985), and Heim (1983).

7 These examples are discussed in, for example, May (1985) and Rizzi (1987). Asher apparently agrees that scope should be determined by the syntax in (1986:79), but then he proposes that the input to DRT might be the GB level of S-structure, at which scoping information is not explicit. Furthermore, the literature contains no defense of the view that all of the needed quantifier interactions can be elegantly expressed in DRSs. Leaving scoping to syntax removes one of the principal motivations for the particular structures of DRT.

8 Cf., e.g., Horvath and Rochemont (1986).

9 For discussion of such matters, see, for example, Sidner (1983), Webber (1983), and Dahlgren (1988). It is interesting to note that in these pioneering investigations, recency in discourse does seem to influence the availability of referents, but Asher's DRSs apparently lose recency information by treating DRSs as pairs of sets (and by treating "delineated DRSs" as triples of sets), as mentioned in note 5. It is also interesting to note that Kamp expresses serious doubts about the prospects for finding significant discourse principles of the kind I've tried to indicate — see the citation and passage quoted in note 8.

10 Of course, Asher presumes that the DRSs will all occur in an integrated system of

DRSs, a delineated DRS, and in that system certain coreference relations will be marked either with some sort of explicit equations or by the use of identitical discourse referents, but this is quite similar to old ideas, too.

11 In a very few words, most of the arguments seem to have the following structure. Although our ordinary beliefs do not seem quite appropriate for psychological explanation because of the dependence on context shown by the arguments of Putnam (1975) and others, it seems that they are really very close to what we need. Psychological states individuated "syntactically" or "structurally" seem to cut things up more finely than we may need for many psychological generalizations. I have some qualms about Dennett's analogy with the point that not all Turing machines that compute the same function share a syntactic description, but the basic idea is right. (I have qualms because the identity criteria for computations of the same function look difficult to pin down, and so I do not know whether the analogy can be made to stick under pressure. Cf. Stabler (1987).) For better presentations of such arguments see, for example, Fodor (1986), Pylyshyn (1984), and Dennett (1982). Field (1978) and others suggested that we can assume lots of syntactically distinct states so long as we can explain why a person in one state is nearly always in the other syntactically distinct states that represent the same proposition, but the provision of an adequate explanation of this kind is problematic.

REFERENCES

Aoun, J. (1985). *A Grammar of Anaphora*. Cambridge, MA: MIT Press — and Hornstein, N. (1985). Quantifier types. *Linguistic Inquiry* 16(4): 623-37

— (1986). Belief in discourse representation theory. *Journal of Philosophical Logic*, 15: 127-89

Barwise, J. and Perry, J. (1983). *Situations and Attitudes*. Cambridge, MA: Bradford/MIT Press

Dahlgren, K. (1988). *Naive Semantics for Natural Language Understanding*. Forthcoming

Dennett, D.C. (1982). Beyond belief. In A. Woodfield (ed.), *Thought and Object: Essays on Intentionality*. Oxford: Clarendon Press. Reprinted in D.C. Dennett, *The Intentional Stance*. Cambridge, MA: Bradford/MIT Press

Donnellan, K. (1966). Reference and definite descriptions. *Philosophical Review* 75: 281-304

Dretske, F. (1981). *Knowledge and the Flow of Information*. Cambridge, MA: Bradford/MIT Press

Field, H. (1978). Mental representation. *Erkenntnis* 13: 9-61

Fodor, J.A. (1986). Information and association. *Notre Dame Journal of Formal Logic* 27(3): 307-23

— (1986). *Psychosemantics*. Cambridge, MA: Bradford/MIT Press

Haik, I. (1984). Indirect binding. *Linguistic Inquiry* 15: 185-223

Heim, I. (1983). File change semantics and the familiarity theory of definiteness. In R. Bauerle, C. Schwarze, and A. von Stechow (eds.), *Meaning, Use and Interpretation of Language*. Berlin: de Gruyter

Horvath, J. and Rochemont, M. (1986). Pronouns in discourse and sentence grammar. *Linguistic Inquiry* 17(4): 759-66

Kamp, H. (1985) Context, thought and communication. *Proceedings of the Aristotelian Society*, 85: 239-61

— (1981). A theory of truth and semantic representation. In J.A.G. Groenendijk, T.M.V. Janssen, and M.B.J. Stokhof (eds.), *Formal Methods in the Study of Language*. Amsterdam: Mathematisch Centrum

May, R. (1985). *Logical Form: Its Structure and Interpretation*. Cambridge, MA: MIT Press

Partee, B. (1975). Bound variables and other anaphors. In D. Waltz (ed.), *Theoretical Issues in Natural Language Processing*, TINLAP-2. New York: ACM

Putnam, H. (1988). Much ado about not very much. *Daedalus*, 117(1): 269-81

— (1986). Putnam's response. In Z.W. Pylyshyn and W. Demopoulos (eds.), *Meaning and Cognitive Structure: Issues in the Computational Theory of Mind*. Norwood, NJ: Ablex

— (1975) The meaning of "meaning." In H. Putnam, *Philosophical Papers, Volume II: Mind, Language and Reality*. Cambridge: Cambridge University Press

Pylyshyn, Z.W. (1984). *Computation and Cognition*. Cambridge, MA: Bradford/MIT Press

Rizzi, L. (1987). Relativized minimality. Presented at the 1987 Linguistic Institute, Stanford University

Sidner, C. (1983). Focusing in the comprehension of definite anaphora. In M. Brady and R.C. Berwick (eds.), *Computational Models of Discourse*. Cambridge, MA: MIT Press

Stabler, E.P., Jr. (1987). Kripke on functionalism and automata. *Synthese* 70(1): 1-22

Webber, B. (1983). So what can we talk about now? In M. Brady and R.C. Berwick (eds.), *Computational Models of Discourse*. Cambridge, MA: MIT Press

CHAPTER THREE

Truth Conditions and Procedural Semantics

Robert F. Hadley

INTRODUCTION

A widespread assumption in both AI and Cognitive Science is that higher cognitive processes are not only modelable by computational ones, but fundamentally are such processes. The emergence of Connectionism in recent years has cast doubt upon the computational model in some quarters, but many see the connectionist model not so much as a threat, as an extension to, or as a program for the implementation of the computational model. Furthermore, it is widely conceded that the computational model of mind does not require us to say that mental processes are nothing but computer programs, for the execution of such programs may give rise to emergent properties. These emergent properties may account for such elusive phenomena as consciousness, and understanding (in the fullest sense), or so it is sometimes suggested. The question I wish to address here is whether the computational model of mind is consistent with a common view of semantics, namely, that the semantics of a set of declarative sentences is given by (or is identical with) the truth-conditions of those sentences. My concerns have much in common with those expressed by Putnam (1970, 1980), though my focus and proposals differ from his.

The thesis that semantics is somehow concerned with truth-conditions is common to a number of semantic theories, including those due to Tarski, Montague, Lewis, Cresswell, and Stalnaker. All those theories presuppose the existence of semantic functions, which are viewed *extensionally* as sets of ordered pairs. Typically, these functions map predicates onto denotation sets, or sentences onto sets of possible worlds. In all cases, the sets involved are either infinite, or so large that they might as well be infinite, given the resource limitations of the cognitive agents we now envision.

Now, it is no news that semantic theories based upon infinite or virtually infinite sets present difficulties for researchers in AI who wish to apply such theories in a computational setting. Frequently, these difficulties are sidestepped by bounding the cardinality of the extensional sets to computationally tractable numbers. However, this stopgap measure is obviously unacceptable to anyone seeking a plausible model of *human* cognition. An alternative solution, more plausible at first glance, is to suppose that the *extensional* functions in question are encodable as effective procedures. Procedures are clearly capable of representing many functions having infinite domains and ranges; yet they are finitely encodable and are clearly accessible to computational agents. Moreover, computational procedures have structure, and logically equivalent procedures typically possess different structures. I have argued elsewhere (Hadley 1988) that computational procedures possess just the right structural properties to serve as propositions and objects of belief. Unfortunately, as Fodor reminds us (1978), to assume that the mapping functions posited by classical semantics can be encoded as effective procedures is to endorse a wildly implausible form of the verification theory of meaning. After all, to assume that every sentence determines (through a compositional process, perhaps) an effective procedure capable of deciding the truth value of that sentence in a given world, or situation, is just to assume that absolute verification procedures exist for all sentences. It is difficult to find any verificationist who has endorsed such an extreme view, though as Fodor notes, the doctrine was accepted by prominent AI scientists during the 1970s.

But now, if we entirely abandon the notion that semantic mapping functions are computable, we are left with several difficulties, some of which are well known. To begin with, if sentential meanings are taken to be functions, and these functions are non-computable, it is difficult to see how a computational agent could ever *know* the meaning of a sentence. It is unclear what it would be for a purely computational agent to understand a non-computable function.

Now, it might be countered that cognitive agents never fully understand the meaning of a sentence. Rather, they grasp enough of the meaning to be semantically competent. Scott Soames, among others, has argued that it is a serious mistake to confuse semantic competence with semantics per se. However, it is not clear that accepting this distinction removes the immediate difficulty, since it seems true that finite agents know at least part of the meaning of many sentences, and it is problematic how a finite computational agent could know *part* of a non-computable function, or be semantically competent if semantics is *primarily* about non-computable functions.

In the next section I explore these issues in some detail. I argue that semantic competence presupposes the existence of certain kinds of *pragmatic*, computable procedures. These pragmatic procedures take as arguments particular situations or circumstances, and return fallible judgments concerning the truth of particular sentences. I refer to the results of these judgments as pragmatic truth values. In the following I review and extend arguments due to Wilks and Woods to the effect that, in the absence of effective verification procedures, there must exist pragmatic, computable meaning procedures if there is to be such a thing as semantic knowledge. I argue further that the existence of such procedures is a precondition of semantic functions and truth-conditions. On the theory presented here, denotations and truth conditions emerge as idealizations of pragmatically individuated sets.

In the section titled "Modified Procedural Semantics," I outline a semantic theory grounded upon a particular account of pragmatic meaning procedures. The theory includes an account of idealized truth conditions, for both observation and theoretical statements. The central thesis will be that the truth-conditions of a sentence are a function of how our pragmatic meaning procedures *constrain* the set of *circumstances* relevant to the truth of a sentence. As will be apparent, the theory attempts to integrate aspects of Quine's pragmatism with ideas developed by the later logical positivists. It also shares some of the spirit of Woods's theory of procedural semantics (1981, 1986). Indeed, the theory presented here may be viewed as a *kind* of procedural semantics (PS), subject to two caveats. The first is that "procedure" will be used in a somewhat extended sense. In computer science, *procedural* processes are commonly identified with formal symbol manipulations (symbolic computations). By contrast, I will take procedural processes to include not only symbolic computations, but also the kinds of analog processes exemplified in connectionist (PDP) nets and sense transducers. That is, if a mechanism consistently "computes a function" (or does so to a high degree of approximation) it will, for our purposes, be taken to "execute a procedure." No doubt this usage is somewhat non-standard, but I believe it captures *one* of the senses of "procedure" to be found in earlier treatments of PS. Moreover, the concept of a procedure seems to be our nearest analogue for describing the class of *processes* with which we shall be concerned.

My second caveat is that I am not defending the view that procedures tell the whole story about semantics; rather they are a necessary part of the story. Deductive and inductive inference are also part of the story, but they present no insuperable difficulties for the view that semantics is created by, and is largely accessible to, purely computational agents.

DIFFICULTIES FOR TRUTH-CONDITIONAL SEMANTICS

A principle concern of this paper is whether theories which presuppose the existence of objective truth-conditions are compatible with an essentially computational view of cognitive agents. In a later section I argue that these positions can be rendered compatible if we are willing to see truth-conditions as idealizations which arise as a complex function of both pragmatic meaning procedures and the way the world is. In this section I am concerned to show that the pragmatic meaning procedures, of *some kind,* must exist if we are to render coherent the view that denotations of common nouns exist, and that such denotations are knowable in a way that permits semantic competence. I focus upon the denotations of common nouns because their problems are inseparable from those which arise when individuating more complex denotations, and when individuating truth-conditions. Classical denotational semantics, and the possible-world semantics of Montague and Lewis quite explicitly assume the existence of denotations for nouns. Even the situation semantics of Barwise and Perry (1983), which approximately identifies truth-conditions with sets of situations, associates the *property* expressed by a common noun with the *collection* of objects to which the noun can truly be applied. There may exist technical differences between the use of 'property' in situation theory, and the use of 'denotation' in, say, Tarskian semantics. However, I do not believe these technical differences diminish the particular problems to be considered here. For this reason, Situation Semantics will not be explicitly addressed in what follows. Rather, I focus upon classical denotational semantics, and the possible worlds approach of Montague and Lewis.

Denotational semantics

Let us first consider classical denotational semantics. Typically, a denotational semantics will individuate the meanings of referring expressions solely in terms of the denotations or extensions of the expressions. This has the obvious drawback that it renders all coreferential expressions synonymous (e.g., 'unicorn' and 'pixie' would by synonymous). We will ignore this difficulty, however, since it can be substantially diminished by moving from *actual* to *possible* denotations. Rather, we focus upon certain foundational problems which arise for either formulation. The first of these problems is epistemological, the second ontological. Aspects of the epistemologial problem are discussed by Woods (1981), and Wilks (1982). In the next section I attempt to crystallize certain issues underlying the epistemological problem. Following that, the ontological problem is presented. To the author's knowledge, the ontological problem has not been explicitly addressed in the cognitive literature.

The epistemological problem

The epistemological problem arises if we assume that the meanings of such common predicates as 'cat,' 'chair,' 'ancestor,' and the like are knowable, or at least tacitly knowable. The problem is this: if denotational semantics is correct, *what could constitute* knowledge of the meaning of a denoting expression? In posing this question it is not being suggested that one who knows the meaning of say, 'cat,' could explicitly *define* that term. Rather, we are here concerned with the kind of knowledge which enables us to *use* words. This knowledge (or *know-how*) is largely tacit and falls within the domain of semantic competence. Now, I am not suggesting that a correct semantic theory should *explain* how semantic knowledge is possible, but it should not *preclude* the possibility of one's possessing at least substantial semantic knowledge of denoting expressions. Thus, it is important to enquire, what could constitute semantic knowledge if the meanings of referring expressions are denotations?

To begin, let us consider what it is to know the meaning of 'cat.' Like many denoting expressions, 'cat' possesses a denotation too large to be known by direct acquaintance. Since this is so, we may be certain that a finite agent's knowledge of the meaning of 'cat' does not consist of an acquaintance with all cats. Moreover, an agent who knows (even tacitly) what 'cat' means knows something more than that 'cat' applies to certain objects which the agent knows to be cats. That is, the agent must know something *more general* than that particular objects of his/her acquaintance are called 'cat'. If this generalized knowledge (which has often been regarded as *intensional* knowledge) is knowledge of *meaning*, it is difficult to see how this meaning could be equated with the *denotation* of cat.

Furthermore, we cannot *in general* explain the nature of semantic knowledge by suggesting (as early workers in computational linguistics have) that to know the meaning of a natural language expression is to *know how* to translate it into some other representation language. For, whether the target representation language is internal (e.g., Fodor's *language thought* [1975]) or external, and whether or not the target language is *innately* understood by an agent, the target language must have a semantics *known* to the agent, if the feat of translation is to be any indication that the agent has semantic knowledge. But, clearly, any attempt to define (or analyse) semantic knowledge by equating it with the ability to translate into a language whose semantics is *already known* is circular. (Lewis [1976] makes a similar point.)

Now, apart from PS, it is very difficult to discover *any* account of semantic knowledge which does not involve the kind of circularity described above. One can find such formulations as "To know the meaning of 'cat' is to know that 'cat' denotes the set of all cats." However, such formulations have no explanatory power, since one could hardly know that *'cat' denotes the set of cats* unless one already knew what cats are, and this is just the kind of knowledge we are seeking to elucidate. The underlying problem is that *any representational mode* of knowing what cats are presupposes knowledge of the semantics of some representation scheme. Since we cannot understand the structure of this kind of knowledge by reducing it to more of the same, it is reasonable to seek an explanation in terms of something we understand better. Procedural knowledge (or knowing how) *is* something we understand better (at least, since the advent of computational theory). Moreover, procedural knowledge, construed broadly enough to encompass analog forms of computation, appears to be the only form of knowledge capable of explaining how agents could succeed in *establishing* a mapping relationship between symbols and the external world. In saying this, I do not mean to imply that every mapping relation is *entirely decomposable* into effective procedures, but rather that mapping functions cannot be created by *finite* agents unless some subset of these functions is computable. Moreover, it seems undeniable that the mapping relationship which obtains between natural language and the world does depend, at some ultimate level, upon mapping conventions which *we* create.

It is illuminating to note that, whereas we have been unable to account for semantic knowledge of denotations purely in terms of knowledge by acquaintance or knowledge by representation, our prospects improve dramatically when we include knowledge of semantic procedures. Moreover, such procedures need not be decision procedures. If we merely assume that agents possess computational procedures which enable the agent to form *prima facie* judgments, concerning an object's membership in a word's denotation, then we can at least explain how an agent acquires a general sense of given denotation. If the agent possesses *pragmatic* procedures of this kind, we should also assume that the agent possesses procedures which enable him/her to adjudicate between conflicting *prima facie* judgments.[1] Otherwise, the agent may be unable, in principle, to determine the denotation in question, and as a consequence, would be unable to determine the meaning in question (assuming that denotational semantics is correct).

Now, it may be objected that the assumptions just described ultimately involve the kind of circularity previously encountered, since our assumptions require that the agent in question is endowed with procedures encoded in a language which the agent already understands. However, this objection overlooks the fact that we assume the agent compiles the given procedures into a set of low-level procedures which the agent understands only in an *operational* sense, i.e., the agent is constructed in such a way that, when invoked, the low-level procedures *cause* the agent to behave in appropriate ways. There is no need to suppose the agent *assigns* a semantic interpretation to these low-level procedures.

Perhaps a different objection will arise. Suppose it is conceded that cognitive agents must (at least tacitly) know semantic procedures *of some kind* if they are to be *semantically competent*. It may yet be objected that tacit knowledge of such procedures is not equivalent to knowledge of the semantics of a language. What agents know are not *meanings* but mere approximation methods for applying language to the world. It is simply a mistake to assume that denotations and truth-conditions are *really* knowable. Consequently, it has not been shown that denotational semantics is in any sense logically dependent upon the existence of the kinds of procedures described above.

To reply: I do *not* wish to defend the view that absolute truth conditions are *really knowable*, or that semantic procedures *constitute* the complete meanings of denoting expressions (although I eventually argue in "Modified Procedural Semantics" that such procedures *constrain* truth conditions by constraining the conditions which are relevant to the truth of a sentence). However, I do claim that denotational semantics *ontologically requires* the existence of semantic procedures, and that some of these procedures must be knowable (in the sense that we may *possess* these procedures, and that they give rise to our semantic "know-how"). The objector sketched above maintains that denotations, and consequently meanings, are not knowable. It is difficult to accept the claim that we do not, in some important sense, know at least part of the meaning of such common nouns as 'cat.' However, for argument's sake, let us adopt the objector's viewpoint. It can still be shown that denotational semantics presupposes the existence of semantic procedures. For, unless such procedures exist, the theory faces a very serious ontological problem.

The ontological problem

Perhaps the easiest way to see that denotational theory requires the existence of semantic procedures is to assume that no such procedures exist. We have already seen that one cannot individuate a virtually infinite set by exhibiting all its members. If we now assume that we lack all procedures for ascertaining whether a given candidate falls into some very large denotation, then what can possibly justify the premise, essential to denotational semantics, that a particular word denotes *some particular set* of objects, rather than some other set.

Workers in denotational semantics often proceed as though the *denotation relation* which holds between a primitive (or non-composite) denoting expression and its denotation were God-given, and could exist quite independently of the process by which the relation is created. However, since symbols do not *intrinsically* denote particular sets (not even symbols which occur in internal, *innate* representation languages), denotation relations must be established. But, in the case of primitive expressions such as 'cat,' such denotation relations cannot be established if the linguistic community entirely lacks procedures (including pragmatic procedures) for *identifying* instances of the intended denotation set.[2] Furthermore, if there is to be any distinction between correct and incorrect descriptions of experience, those identification procedures which enter into the creation of the denotation relation must bear a *logical* relationship to the meaning of the denoting symbol. Otherwise, we will have no reason for asserting that the symbol has a *correct* denotation.

Perhaps it will be objected that the required relation between a denoting symbol and its denotation could be explained in purely causal, rather than procedural terms. For example, we might attempt to analyse a denotation relation in terms of the causal relations which exist when objects of a certain kind elicit particular symbolic responses on the part of cognitive agents.

To reply: if causal explanations of denotation relations are to be possible, they must eventually appeal to a notion such as 'standard causal chains' or 'standard background conditions.' For a given object may provoke a given symbolic response, e.g., 'cat,' but do so via the wrong mechanism, e.g., becaue the subject's judgment has been affected by hallucinogenic drugs. In such a case there may well be a causal relationship between an external object and the subject, but the relationship is not of the correct kind. Any causal account of denotation must somehow distinguish the occasions when a symbol is linked by acceptable causal chains to its denotation, from those when it is linked to an appropriate object by an unacceptable causal chain, or to an entirely inappropriate object.

Now, the notion that certain causal sequences are acceptable, while others are not, is well captured by the concept of a computational procedure. Moreover, goal-directedness is an intrinsic aspect of procedures; procedures are executed *in order* to produce a result. This goal-directed quality is absent from purely causal accounts of meaning, but it is an essential aspect of symbolic representation. A procedure which possesses and labels sense experience has as goals the identification and representation of experience. Thus, the goal-directed aspect of procedures accounts, at least in part, for the intentionality of symbolic representation.

In light of the foregoing, I conclude that a causal theory of denotation will fail to account for the *correctness* and goal-directedness of the denotation relation *unless* causal explanations are enhanced to the point where they actually become procedural explanations. If this conclusion is accepted, we may formulate the ontological problem for denotational semantics as follows: unless the pairing of the *primitive denoting expressions* of a language with their respective denotations can be supported by a procedural foundation, we cannot render coherent the claim that denotation relations exist.

Now, it may appear that denotational semantics could find support in an entirely different direction, say, from a theory of natural kinds based upon *essentialism*. For example, one might argue that the denotations of natural kind terms (at least) are determined by the natural grouping of objects in the world, and that all objects within a natural grouping share a single essence which is discoverable by scientific means. Thus, the denotation of a natural kind term is simply a set of objects which share the same essence.

Now, we should note that natural kind theory does not pretend to explain how the denotations of non-natural class terms (e.g., "chair") are to be individuated, but even if we ignore that difficulty, it is clear that the proposal does not solve the problem. For, even if we suppose, as *natural kind* theorists do, that natural groupings of objects occur in nature, no natural class is automatically the denotation of *some particular symbol*. Unless denotation relations are established, denotations simply do not exist. This is not to deny that natural classes exist in nature, but to deny that such classes would constitute denotations in the absence of the creation of the relevant *external* relations. The creation of these relations presupposes the ability to form *prima facie* judgments to the effect that certain objects belong to the same class, and this in turn presupposes the existence of identification procedures of some kind.

At this point it may be helpful to formulate a concise statement of the foundational difficulties which arise for denotational semantics. A fundamental problem is whether denotational semantics is *consistent* with the fact that agents possess tacit knowledge of the meaning (on the theory — the denotation) of simple class terms (e.g., 'cat'). There appear to be only three hypotheses which could explain how an agent possesses knowledge of the denotation of a class term. They are: (1) We suppose that all members of the denotation-set are exhibited to the agent, say by pointing them out. This is clearly impossible for very large or infinite classes, such as the set of cats. (2) The agent represents the denotation-set to him/herself in some language (or fragment of language) whose semantics the agent already knows. This method obviously fails to address the problem of how an agent possesses semantic knowledge in the first place. It merely trades our initial problem for the problem of how the agent could know the semantics of the prior language. (3) We assume that the agent acquires (or is innately endowed with) a set of computational procedures which enable the agent to *determine* when an object falls into the denotation-set, or which at least enable the agents to make *prima facie* judgments that a given object falls into the denotation.

Now, assuming that the meanings of denoting expressions are denotations, only (3) explains how knowledge of such meanings is possible. Furthermore, even if precise knowledge of denotations is impossible (because the most we ever have is an approximate sense of denotations), only (3) appears *compatible* with the claim that (tacit) semantic knowledge exists. For hypothesis (1) is still hopeless when infinite (or virtually infinite) denotations are involved, and (2) still involves a vicious circularity. We conclude, therefore, that denotational semantics will be *consistent* with the fact that we possess semantic knowledge (and are semantically competent) only if we assume (3). Moreover, only (3) accounts for how we form a sense of the boundaries of a set of physical objects, and this sense of boundary is *logically* required to enable us to create the denotation relation which holds between a term and its denotation. *Thus the very existence of the denotation relation presupposes the truth of (3), which is an important component of procedural semantics.* This conclusion does not entail that PS is *part* of denotational semantics, but it does entail that denotational semanticists must concede the existence of particular kinds of semantic mapping procedures.

Possible-worlds semantics

I shall now argue that Montague-style (possible-worlds) semantics is similarly dependent upon the existence of semantic procedures. Consider first Montague semantics. In Montague's system (and in kindred systems such as [Lewis 1967]), the intension of a declarative sentence is formed through the composition of intensions which belong to the primitive parts of the sentence. Common nouns belong to this category of primitives, and the intension of a common noun N is taken to be a *function* F from each possible world W_1 to the extension of N in W_1. The question I wish to address is, what is the nature of the function F?

If F is taken to be a function in the purely set-theoretic sense, then F is simply a set of ordered pairs of the form (W,E), where W names a possible world, and E names an extensional-set in that world. The problems associated with individuating possible worlds are well known, and need not be recounted here. However, even if we ignore these difficulties, we are left with the problem of individuating the extension E, which occurs in each ordered pair (W,E) in F. Since any given extension E may be infinite in size, the difficulties involved in individuating these extensions are precisely those of individuating the denotations postulated by denotational semantics. Moreover, the question arises, how is the set of ordered pairs in F to be individuated? Presumably, this set, if not infinite, is too large to be explicitly displayed or stored. It appears, therefore, that the function F can only be individuated by means of some representation scheme, or via computational procedures whose meaning is understood *operationally*. In the former case, the *existence* of F will require the existence of a semantics for the representation language. Clearly, this approach leads to infinite regress unless the regression ultimately leads to a metalanguage whose primitives possess a semantics which can be individuated by a method *different in kind* from the one we have just considered.

On the other hand, suppose that the function F could be given as a computational procedure. Then, not only would we have solved the problem of individuating the ordered pairs in F, but the problem of individuating the various extensions (E) in each possible world would be solved. This is because a function which is encoded as a procedure *implicitly* defines the ordered pairs in the set-theoretic function. Thus, if intensional functions *can* be encoded as procedures, there is no need to suppose that we can produce an explicit enumeration either of possible worlds, or of extensions. Instead the intension is a *procedure* which can be applied to the possible circumstances which arise in the actual world we live in. In addition, we might suppose that F is encoded not as a procedure which explicitly returns an entire extension E as its value, but as a procedure which returns a new procedure, P, which implicitly determines the membership of E in the particular world W. In other words, the procedure F returns not an explicit extension, but a procedure capable of verifying membership in E, given world W.

Of course, it will be objected that to assume F and P may be encoded as computational procedures is to assume that strict verificationism is possible, that decision procedures exist for applying predicates to the world. This objection is valid as long as we retain Montague's premise that the intensions of complete sentences are functions which return absolute, unrevisable truth values for each possible world. However, if we retain this aspect of Montague's approach, it is clear that our semantics will be embroiled in precisely the difficulties which denotational semantics encounters in individuating extensions in the *actual* world. For absolute truth values exist only if absolutely bounded extensions exist. The fundamental weakness of both the Montague and denotational approaches is that both assume that it makes sense to talk of absolute extensions without providing a hint about how such extensions can be found in practice.

Perhaps a solution to the above dilemma can be found by relaxing the requirement that functions like P *infallibly* determine extensions for each possible world. If instead, we allow that P returns fallible decisions about membership in extensions, based upon *prima facie* evidence, then the intensions of entire sentences will, by the same token, return only fallible judgments about truth values.

Of course, this would require that we surrender the belief that the meaning of each individual sentence determines its absolute truth value in each possible world, but it does not require that we surrender a commitment to absolute truth as an ideal. We might suppose, for example, that absolute truth is determined in the limit by general procedures, which are not specific to each individual sentence, but which adjudicate between pragmatic, or *prima facie* truth values when conflict arises.

Modified Procedural Semantics

Thus far I have argued that, if denotation relations are to exist (either in classical denotational semantics, or in possible world semantics) then semantic mapping procedures must be associated with some common nouns. On the face of it, there appears to be no reason why a purely computational agent could not possess and use such mapping procedures. However, little has been said here about the nature of these procedures, and how they relate to the larger questions: what are truth conditions, how are they related to meanings, and what has meaning in general to do with computational procedures. In this section a theory of meaning is outlined which attempts to answer these questions, at least in part. It should become apparent that, although the complete truth conditions of most sentences are not knowable by finite agents, such truth conditions arise as a logical consequence of pragmatic meaning procedures (which *are* knowable by computational agents), together with standard forms of (deductive and default) *inference*. The theory to be described may be viewed as a modified form of procedural semantics (PS) which takes computational procedures as the logical foundation upon which the remainder of semantics must rest.

As previously mentioned, the modified PS presented here integrates aspects of Quine's pragmatism (Quine 1953) with elements of the verification theory of meaning. The verification theory is obviously related to PS, but the question naturally arises, what has PS to do with a Quinean view of meaning? Well, among other things, serious objections to standard PS have been made on essentially Quinean grounds (cf. Fodor 1978). Any defensible modification of PS should somehow accommodate those aspects of Quine's thesis which are now widely conceded to be correct. In particular, it should accommodate the thesis that the meanings of theoretical terms, at least, are inseparable from a larger network of beliefs (a background theory) in which those terms are embedded. Moreover, a defensible PS should come to terms with the fact that empirical beliefs are, to some degree, *revisable* in the face of recalcitrant experiences. Not all aspects of Quine's thesis are preserved in the theory which follows. I shall note the salient differences where appropriate.

Testable predicates

The modified PS which follows requires the notion of a "testable" predicate, and this notion, which is used here in a technical sense, requires some explanation. We might think of a testable predicate as one which comes *about* as close to expressing an objectively verifiable property as we ever get. For example, 'red,' 'cat,' '3mm long,' and 'house' are good candidates for testable predicates. Relational predicates may also be testable. 'Touches,' 'left of,' and 'colder than' are good candidates for testable relational predicates. The meanings of testable predicates are typically learned by having, among other things, appropriate kinds of sense experiences, and the application of such predicates to the world is typically *justified*, in part, by processing sensory information.

In calling such predicates "testable," it is not implied that we possess *infallible* procedures for applying these predicates to the world. As the critics of verificationism have repeatedly argued, no finite number of observations can conclusively establish that something is yellow. On the other hand, it is certain that we *do* manage to form *prima facie* judgments about instances of colours, shapes, sizes, and so on. I wish to suggest that our meaning rules for testable predicates may involve *some* calls to sub-procedures which *directly test* available sensory information, and calls to other sub-procedures which could involve us in an infinite search, *if they were all executed*. Normally, however, we do not attempt to execute the latter sub-procedures. Instead, we make the default assumption that they return success. For example, when we confront an object which *appears* to be producing sensory information appropriate to a yellow object (this is determined by a complex set of procedures which terminate) we normally make the default assumption that our sense organs are performing well, that the conditions of observation are standard, and so on. We may possess fallible procedures for investigating whether these background conditions obtain, but these fallible procedures presuppose the correctness of other sensory observations, which in turn raise new questions about *different* sets of background conditions, and so on *ad infinitum*. An initial default assumption, to the effect that background conditions are standard, bypasses this infinite regress and permits an empirical judgment (e.g., This is yellow) to be made.

Now, because empirical judgments, even judgments as intimately linked to experience as colour judgments, are founded upon default assumptions of the kind we have been considering, the *revisability* of experiential reports is assured. Furthermore, the existence of such default components seems to be a *necessary* precondition for the termination of empirical judgments. However, an equally essential precondition is the existence of terminating procedures which can process and *classify* sensory information as being "apple-like," "sweet-tasting," etc. There is no need to suppose that such procedures return clear answers for all possible inputs (i.e., the procedures can be partial functions), but the procedures which classify sensory-input must contain terminating, *non-default* components if the gap between symbols and sensations is ever to be bridged. For the gap (or the mapping process) cannot be bridged by a retreat into an endless succession of further procedure calls, nor can it be bridged by simply assuming that *all* sensory information has been detected by default processes. If *all* our procedures returned only default values, we would not be making genuine contact with experience.

In light of the above, I offer the hypothesis that the *use* of a testable predicate, T, may be determined by *open-ended* rules such as the following:

R1: *has(x,T) if appears-to-have(x,T) & p2 & p3 & p4.*

We may view R1 as a Prolog-like procedure which tells us that, to determine whether *x* has the property T, we must first determine that *x appears* to have property T. In R1, the predicate 'appears-to-have' is assumed to name a procedure, which processes sensory input and returns a value for the parameter T, say 'apple.'[3] Procedures such as *appears-to-have* might be *implemented* by means of one of the *wholistic* mechanisms postulated by connectionism (cf. Hinton and Anderson 1981). (I do not wish to suggest that connectionist models of data processing suffice to explain higher cognitive processes — far from it.) There is increasing evidence that connectionist mechanisms cope well with the problem of recognizing instances of family resemblance concepts. However, the theory I propose is not committed to connectionism. It may well be that more conventional approaches to object recognition, which employ weighted subsets of prototypical *features*, may ultimately prove more successful than the connectionist approach. In that case, the *appears-to-have* procedure might well be implemented by means of more conventional procedures.

Returning now to R1, the terms *p2, p3, p4* are intended to denote *background-checking* procedures. These background procedures are typically not-invoked, but are assumed by default to return 'true.'[4] When invoked, they check background conditions of observation, by invoking other rules which involve other testable predicates. However, it should not be supposed that *p1. . . . p4* are *effective* procedures. Rather, they may be regarded as collections of executable heuristics, something like the (previously mentioned) adjudication procedures. We might even suppose that not all of *p2. . . . p4* are *presently* defined. For example, *p4* may be left undefined until future scientific discoveries are made. Since *p4* is allowed to "succeed by default," this indeterminacy need not prevent the open-ended rule R1 from being used in our present state of incomplete knowledge.

A principle of empirical meaning

Now, given the concept of a testable predicate as described above, we may introduce the notion of a testable sentence, which, in turn, forms the keystone of the definition of an empirically meaningful sentence, which we propose below. We define 'testable sentence' as follows. A testable sentence, Y, must satisfy (a), (b), or (c) below, where T is a testable predicate, and $c_1 \ldots c_n$ are constants denoting identifiable individuals:

(a) Y has the form 'c_1 is T', as in 'c_1 is *red*,'
(b) Y has the form '$c_1 \ldots c_n$ satisfy the relation T,' as in 'c_1 and c_2 satisfy the relation *is left of*.'
(c) Y is the negation of a testable sentence.

Given the above recursive definition, I propose the following recursive definition of an empirically meaningful sentence S. This definition assumes that a sentence may derive its meaning, in part, from a background network of sentences (or a background theory) *TH*.

> (P) S is *empirically meaningful* **iff** one of the following conditions holds: (i) S is a testable sentence. (ii) S is truth-functionally molecular and all of its sentential components are empirically meaningful. (iii) S is not truth-functionally molecular, and there is a set of sentences *R*, such that $R \subseteq TH$, $S \in R$, some *testable* sentences are derivable from *R* (by deductive or default reasoning processes), and if *any* member of *R* were deleted, the testable ramifications of *R* would be diminished.

The principle just proposed bears some resemblance to the verification principle proposed by A.J. Ayer (1946). However, important differences exist. For example, condition (ii) in (P) ensures that difficulties raised by Hempel (1950), concerning molecular sentences, are avoided. Furthermore, Ayer's verification principle involves the concept of completely verifiable observation statements, whereas (P) appeals to the notion of testable sentences which are not susceptible to conclusive verification. Also, among other things, condition (iii) provides for the possibility that testable sentences may be inferred through default reasoning rather than deductive reasoning.

The motivation behind (P) is that we want to require that each empirically meaningful sentence has *some* testable consequences, at least when that sentence is combined with other empirical sentences, and default inferences are permitted. Given both (P) and the prevailing body of scientific belief concerning electrons, it is not difficult to show that 'Electrons exist' is empirically meaningful. For there is certainly a subset of existing sub-atomic theory which taken together with background beliefs physicists commonly hold, will enable us to derive testable consequences from 'Electrons exist.' Even the sentence 'God exists,' often cited as a counter-example to the verification theory of meaning, turns out to be empirically meaningful according to (P). At least this is so given the beliefs people typically hold about God. Of course, the meaning will vary according to the background beliefs people hold, but in most theological systems God is taken to be the ultimate cause of all physical reality. In such systems, God functions as a theoretical entity which plays a role in explaining the world we experience. Moreover, experiences which are described as 'mystical' do exist, whether or not we agree that they are *really* mystical. They hypothesis that God exists, if accepted, would help to explain the existence of some mystical reports, at least against a background of religious belief. Whether explanations of this type are correct is not the issue here.

Meaning and relevance conditions

Let us now consider how (P) bears upon the question, "What constitutes the meaning of a given sentence?" For (P) may tell us whether a sentence *has* empirical meaning, but not tell us what constitutes that meaning. Furthermore, how does (P) illuminate the meanings of non-empirical (a priori or analytic) sentences? Let us consider these questions in reverse order. Recall that, for methodological reasons, we are adopting an essentially Quinean approach to semantics. And, in Quine's theory, all meaningful sentences are empirical to *some* degree, because every such sentence could enter into one's network of beliefs in a way which *affects* the testable ramifications of that network. Thus, in Quine's thoery, non-empirical sentences do not present a problem for principle (P) for the simple reason that they do not exist. Of course, if one rejects Quine's theory, some other account of analytic sentences is required. A non-Quinean account is given in (Hadley 1973) which is compatible with the general thrust of verificationism.

We return now to consider how (P) bears upon the meanings of individual sentences. Let us suppose that a given sentence S is empirically meaningful according to (P). This still does not tell us what the meaning of S *is*. To deal with this problem, we propose to distinguish between S's *pragmatic* meaning, which governs the *use* of S in everyday situations, and S's *relevance conditions*, which constitutes the set of empirical conditions *relevant* to the truth of S.

A sentence's pragmatic meaning may be regarded as a procedure, Q (or a set of procedures), which determine the sentence's pragmatic truth value in any possible world. We may conjecture that Q is a procedure which is determined through the composition of more elementary *pragmatic* procedures, which are in turn attached to the primitive components of the sentence (cf. Hadley 1988). The compositional process which determines the pragmatic meaning Q may be understood by analogy with the compositional process which determines the *intension* of a sentence within the Montague-Lewis framework. Unlike Montague and Lewis, however, we regard sentential intensions not only as functions in the set-theoretic sense, but as *abstract* procedures which might be implemented in a variety of ways.[5] Furthermore, we suppose that the intensions of referring expressions (including common nouns) are not functions which return exact extensions for each possible world, but abstract procedures which identify *prima facie* members of an extension, given a circumstance in a possible world. An essential aspect of these abstract procedures, which form a subclass of the pragmatic meaning procedures, is that they *always* involve either the immediate evaluation of testable predicates, or the invocation of *other* pragmatic procedures which eventually "bottom out" in testable predicates. Any judgment returned by a pragmatic meaning procedure is revisable, in principle, since the judgment is justified, in part, by the evaluation of testable predicates, which in turn involves default reasoning. Furthermore, even when decisions returned by subprocedures are not called into question, pragmatic meaning procedures are assumed only to provide *prima facie* evidence for membership in natural classes.

We turn now from pragmatic meanings to *relevance conditions*, which ultimately constrain the truth conditions of a sentence. To understand relevance conditions, it will help to introduce a term, 'T_S.' Consider the *total* set of *testable consequences* which a sentence S entails (deductively or by default) when embedded in a larger network of beliefs. Let 'T_S' denote this total set. In general, T_S includes those testable sentences which are both derivable from the larger theory (network) in which S is embedded, and which also have S in their *set of support*. Among the elements of T_S will be *conditionals* which state that if certain testable conditions obtain, then certain other testable sentences are also true. The existence of these conditionals is important, because many ramifications of theoretical scientific laws become apparent only when testable antecedent conditions are assumed.

Given this definition of 'T_S', we may define the relevance conditions of an *atomic* sentence as the *set of conditions* which would have to obtain in order to satisfy each testable sentence in the set T_S which exists for that sentence. That is, the sentences in T_S may collectively be regarded as a *description* of the empirical conditions *relevant* to the truth of the given sentence. It is important to bear in mind that for a given sentence S, the membership of set T_S may change (and hence the relevance conditions for S may change) when the background theory in which S is embedded is modified. Thus, we obtain the Quinean result that the conditions relevant to the truth of any empirical sentence are inseparable from the larger network of assertions (or beliefs) in which it is embedded. We should also note that, since a given T_S will contain both deductively valid and *default* consequences of S, the sentences in T_S *should not* be regarded as collectively describing the absolute truth conditions of S. One or more default consequences in T_S may fail to obtain without entailing the falsity of S.

Truth conditions

We have just seen that the total set of relevance conditions for an atomic sentence do not constitute its truth conditions. What then are the truth conditions of an atomic sentence S? The short answer, I conjecture, is that the truth conditions of S are a state of affairs which causes the occurrence of S's complete set of relevance conditions, when that state of affairs obtains. However, this answer requires some unpacking. What, for example, are we to take states of affairs to be? I shall not attempt to provide a rigorous account of states of affairs, but rather to sketch what I think we ordinarily suppose them to be.

Our common sense notion of a state of affairs bears considerable resemblance to what Barwise and Perry call a situation. It consists of relations which hold between individuals or sets of individuals. Individuals are primitive objects which exist quite apart from our experience or conceptualization of those objects. Relations which exist between objects are likewise just primitive aspects of reality. Extensional accounts of relations (which abound in model-theory) often ignore the crucial fact that the n-tuples in an extensional relation all have something in common, in virtue of which they are classed together. *What* they have in common is *not* just that they are in some abstract set. This extra shared quality, whatever it is, is next to impossible to analyse, but it is an objective feature of reality which may or may not accord well with our conceptual scheme. This, at least, is what I believe we normally take relations to be.

Now, if relations are objective aspects of reality, and relations can hold between sets of individuals, then states of affairs can have objective structures, independent of our perception of those structures. Furthermore, it seems at least possible, in principle, that sets of individuals in a state of affairs can be in a 1-1 mapping with the referring expressions in a sentence, and the objective relations in a state of affairs can likewise be in a 1-1 correspondence with the relational terms in a sentence. Thus, it seems possible in principle for there to be an isomorphism between the elements of a state of affairs and the major components of a sentence.

Now, the isomorphism just described is reminiscent of one aspect of Wittgenstein's early picture theory of meaning, but I am not going to suggest that we are actually able to establish the mapping just depicted. Rather, I note that it seems to be part of our concept of truth that such an isomorphism can exist. Since both sentences and states of affairs are finite objects, the *idea* of such an isomorphism seems acceptable, provided we *idealize* and ignore certain messy issues. However, the ways in which an objective state of affairs must be *linked* to a sentence, in order for the sentence to be true, involve much more than isomorphism. In particular, that state of affairs must stand in a special relationship to what I have called the *relevance* conditions for the sentence. It is this special relationship we now consider.

Recall that the relevance conditions of sentence S are defined by reference to the set T_S, the testable consequences of S, given a set of background assumptions R. Now, each sentence *i* in R will have its own set of testable consequences T_i against the background of the remaining sentences in $\{S\} \cup R$. Furthermore, each T_i defines a set of relevance conditions for the corresponding sentence *i*. Now suppose that to each sentence *i* in $\{S\} \cup R$ there corresponds an objective state of affairs, A_i, such that the elements of A_i can be in an isomorphic mapping with the elements of sentence *i*. And further suppose that the joint realization (occurrence) of all the separate states of affairs, $\{\cup A_i s\}$ *causally ensures* the occurrence of the collective set of relevance conditions, which must hold if all the separate T_is are to be simultaneously satisfied. Then, under these **ideal** conditions we may say that each state of affairs A_i constitutes the truth conditions of the corresponding atomic sentence *i*. Truth conditions for molecular sentences may be defined recursively, in terms of the truth conditions of atomic sentences.

The foregoing account of truth conditions clearly involves a strong idealizing assumption. For we have assumed that the total set of testable consequences, consisting of the union of all the separate T_is, is a consistent set. However, there is no assurance that even a single T_i is consistent, since a given T_i contains both *default* and deductively valid consequences. Does this mean that, on the present account, sentences do not have truth conditions at all? Not necessarily; certainly sentences do not have the kind of ideal truth conditions just described, but it is possible for a given state of affairs to *approximate* an ideal truth condition to a high degree. For, it seems entirely possible for a set of states of affairs, $\cup A_i$s, to simultaneously cause the realization of *nearly all* the relevance conditions determined by each of the separate T_is . On this supposition, being a truth condition may be a matter of degree. The more nearly a given state of affairs causes all of a set of relevance conditions to occur, the more nearly that state of affairs is a truth condition. Or perhaps there are states of affairs which are maximal, in the sense that no alternate state of affairs could cause a larger percentage of relevance conditions to occur. However, we have no reason to suppose that, for each atomic sentence, there exists a unique maximal state of affairs. Moreover, there may be trade-offs in maximality between the separate states of affairs which correspond to the remaining sentences in the background set R.

If the foregoing account is correct, or even roughly correct, then being a truth condition is a messy business. The messiness does not arise because states of affairs are themselves fuzzy objects (though they may be), but because the manner in which a state of affairs becomes tied to sentences is very messy. My account of how this messiness arises does not require the acceptance of a relativistic metaphysics, as Quine seems to urge. On the contrary, I have assumed that objective, structured states of affairs exist entirely apart from how we conceptualize them. My further assumption that the structure of a state of affairs can stand in an isomorphic relationship to the structure of a sentence involves another substantial idealization, one which occurs in several semantic theories.

The realization that idealization, simplification, and even slippage are involved in the process by which sentences are mapped onto experience is not new. Indeed these facts have been publicly noted by several of the authors in this book. What I wish to stress is that these processes must be much better understood if we are to account for how semantically competent agents manage to associate truth conditions with sentences. My principle concern has been to clarify the process by which pragmatic procedures serve to constrain a sentence's truth conditions, and to show that truth conditions *would not be truth conditions* were it not for the procedures which semantically competent agents use.

Discussion

We are nearly in a position to address the question, is semantics possible for a purely computational agent? Before doing so, it will be helpful to consider a possible objection to the theory I have outlined. The objection arises from the domain of set theory, and may be formulated as follows: the form of PS proposed here *seems* to require that all mathematical functions be computable (recursively enumerable), because functions, after all, are sets of ordered pairs, and I have argued the memberships of sets must be at least *prima facie* identifiable. Thus, I appear to be committed to the claim that all sets are at least *approximately* constructive, and this sounds very much like Intuitionism — a controversial thesis among mathematicians.

To reply, let us first note that Intuitionists are not the only mathematicians to insist that all sets be constructive. Putnam (1980), for example, has argued for a form of "liberalized intuitionism," which confines itself to the thesis that no set is *intrinsically* non-denumerable. Putnum presents strong arguments to show that every set has a constructive membership when viewed from the perspective of *some interpretation* of the formal characterization of the set. Putnam's arguments are not universally accepted, but they do seriously challenge the view that intrinsically non-constructive sets exist.

Moreover, even if we discount Putnam's arguments, I wish to stress that modified PS *need not* be committed to the view that *all* extensions or denotations are even *prima facie* constructive. Rather, it is committed to saying that the extensions of *primitive* terms, such as, 'lemons,' 'chairs,' 'left-of,' be approximately constructive. Now, if we examine the formal definition of any non-constructive set, say 'the power set of the natural numbers' we find that the syntax of the definition is composite. It is, of course, a common view among philosophers and semanticists, dating back to Frege, that the meanings of syntactically composite expressions are formed through the composition of more elementary meanings. In (Hadley 1988) it is argued that a Montague-Lewis approach to compositionality is entirely compatible with, and is clarified by, viewing the compositional process as the composition of *procedures*. A principle for the individuation of composite procedures is also presented. Space does not permit a review of the arguments presented there, but the reader is no doubt familiar with instances of composite procedures arising through the composition of elementary procedures.

Now, modified PS is committed to the view that the meaning of a non-computable set description, e.g., 'power set of the natural numbers,' is formed through the composition of elementary procedures, but we need not suppose that such elementary procedures are computable for *all* possible arguments. For example, the power set function (call it F) is computable when restricted to finite arguments, and the set of natural numbers is constructable, in the limit, by a Turing computable procedure (G). When we compose these two functions we obtain the non-computable result, F(G). Non-intuitionistic mathematicians believe that applying F to G yields a non-denumerable set. But it is arguable that whatever confidence we have in the existence of this set, and whatever meaning we can assign to 'F(G),' derives from our understanding of a *restricted form* of F, in which F is a computable function for deriving the power set of a finite set, and from our understanding of G as a method for generating the natural numbers. In other words our understanding of 'F(G)' appears to be parasitic upon our understanding of both F and G as generation rules. Thus, if we wish to side with non-intuitionistic mathematicians, and maintain that 'F(G)' denotes a *non-constructive* set, PS provides an explanation of how we are able to form *some sense* of the membership of this set. It is also interesting to note that we *do* possess a semi-decidable procedure for determining whether a given set, S, is in F(G); if S is finite we merely determine that S is a set of natural numbers. Otherwise, if S is countable (which can be determined only in the infinitely long run), we check that each element of S is a natural number.

Now the question naturally arises, what has this digression to do with our central question, how is semantics possible for a purely computational agent? Well, the relevance is twofold. First, the fact that computable, pragmatic meaning procedures can be composed in special ways to produce non-computable, structured procedures, helps to explain how computational agents can be semantically competent in abstract domains. Second, our digression may help to explain how a computational agent could form the notion that an external state of affairs may possess structure, and that this structure might be isomorphic to the structure of a linguistic object.

On the present theory, agents do possess pragmatic meaning procedures which enable them to form *judgments* to the effect that particular observable states of affairs exist. In addition, they possess other pragmatic procedures which enable them to recognize and reason about particular structures. The process of compositionality alone, when applied to the notion of a state of affairs and the notion of a structure, would seem to explain how the concept of a structured state of affairs could arise. Now, add to this the fact that computational agents may possess procedures which enable them to establish isomorphic mappings between distinct finite structures that they *do* observe, and we should be able to explain how a pragmatic meaning procedure could be attached to the expression, 'an isomorphism exists between an observable state of affairs, and an observable sentence.' If we now assume, as seems likely, that the word 'not' has a pragmatic meaning procedure attached, then the compositional process can be used to explain how *procedural sense* can be attached to 'the structure of a non-observable state of affairs is isomorphic to that of an observable sentence.' As it happens, the compositional process in this instance produces a non-computable procedure, but that is not so unusual.

By now, I hope it is at least plausible that the existence of pragmatic meaning procedures can be used to account for much of what we call semantic competence. Pragmatic procedures, together with standard inference mechanisms, and informal adjudication heuristics, can explain how we manage to form judgments about the truth of empirical sentences, and to revise judgments in the face of recalcitrant experiences. Moreover, it is at least plausible that semantic competence in mathematical and logical domains arises through the interaction of pragmatic procedures with the compositional process and our inference capacities.

What is more difficult to explain, from a procedural standpoint, is the existence of objective truth conditions. But these are difficult to explain on *any* semantic theory. For it seems clear that *even* if the existence of objective truth conditions is granted, such truth conditions must somehow get linked to rather arbitrary bits of language, and the process by which this linkage is established is going to be complex and messy no matter what our theory.

I have tried to argue that we cannot even make sense of truth conditions unless we can make sense of denotations, and that can be done only if we tell a procedural story. However, if we do adopt a procedural account, then we can at least explain how sets of truth conditions might be *constrained* by the very procedures which make semantic competence possible. It is true that on the account presented here, we have no guarantee that sentences, especially theoretical ones, have *unique* truth conditions. But I know of no semantic theory which can guarantee that, although many theories assume it.

Thus far, nothing has been said here about the phenomenon of intentionality, and how meaning procedures might be used to explain its existence. The concept of a procedure, as used here, does have an intentional aspect (cf. Pylyshyn 1980), insofar as procedures are goal-oriented, but I do not believe the goal-directedness of procedures entirely accounts for the general sense of "aboutness" which our symbolic representations have. To a degree, our sense of aboutness may arise as a consequence of our *experiences* of situations which *satisfy* pragmatic meaning procedures, but even our ability to experience situations as integrated gestalts eludes a purely computational explanation. Of course, it is possible on the face of it, that intentionality may be an emergent phenomenon, caused by the structure of and/or the patterns of activity in our brains. In any case, it would appear that we *must* suppose this to be so if we are to cling to the computational model of mind.

Now, perhaps it will be suggested that the existence of emergent phenomena helps to explain how truth conditions arise, or at least how truth conditions might be knowable. Well, emergent phenomena may help a bit, but it is unlikely that they can take us very far. For, from a logical standpoint, emergent mental properties are arbitrarily related to their causes; they bear no special logical relationship either to mental representations or to external states of affairs. Truth conditions, by contrast, bear a logical (or *satisfiability*) relationship to representations, and it is very difficult to see how this relationship, or even knowledge of this relationship, could be explained in terms of the accidental by-products of brain activity.

SUMMARY

In the foregoing there have been two overriding concerns. First, I have examined some of the preconditions which must obtain if sentences are to have truth conditions. Second, I have been concerned to develop a form of PS which reconciles the view that semantics is in some sense the creation of computational agents with the view that truth conditions do exist. I have argued that such a reconciliation is possible, provided we view truth conditions as idealizations, constrained by considerations of maximal coherence. In this respect, the modified PS described here supports a coherence theory of truth.

Let us now briefly review the reasons for saying that the semantic theory presented here constitutes a kind of procedural semantics. First and foremost, I have argued that denotations and truth conditions of empirical sentences exist only if terminating procedures occur as *part* of the meaning of at least some words (i.e., testable predicates). The terminating procedures in question may not possess simple names, but we may refer to them via such artifices as: 'the procedure which determines whether something appears to be an apple, or whether something tastes the particular way spinach tastes.' Terminating procedures such as these directly provide the means by which testable predicates are anchored to experience, and indirectly provide the means by which all empirical sentences are tied to experience. Without procedures such as these there could be no semantics, because denotation relations could not be established.

Second, I have theorized that common nouns, empirical relations, and derivatively, many empirical sentences possess pragmatic meaning procedures capable of determining *prima facie* denotations and truth values. It is because such pragmatic procedures exist that we are able to use language freely, without considering the infinity of ramifications a sentence may have, and without engaging in exceedingly complex chains of inference. While knowledge of such procedures is a precondition of semantic competence, the pragmatic procedures themselves constrain relevance conditions which in turn serve to constrain the truth conditions of sentences.

Third, even a highly theoretical sentence (which may lack a pragmatic meaning procedure) may be regarded as an axiom embedded in a large network of axioms. The entire axiomatic network, together with background deductive and default inference mechanisms, may be viewed as an intricate high-level procedure which constrains the conditions under which declarative sentences may be applied to the world. Any attempt to confirm the truth of a highly theoretical sentence in this network must engender a chain of inference which terminates in the evaluation of testable sentences. Such sentences, in turn, presuppose the existence of terminating procedures. Positive confirmation of any proper *subset* of the testable consequences of a theoretical sentence cannot, of course, entirely confirm the sentence itself (for, among other things, this would involve the fallacy of affirming the consequent). But, insofar as it is possible to confirm a theoretical sentence to some degree, we must possess *procedures* for deriving testable consequences from the theory which contains that sentence. The fact alone may not provide a strong case for calling our semantic theory *procedural*. However, when combined with the reasons presented in the preceding two paragraphs, it seems not only plausible but natural to call our theory procedural semantics.

Now, I do not wish to suggest that the theory outlined here is complete, or that difficult questions do not remain. It is hoped, however, that the salient outlines of the theory are clear, and that the *need* for a procedural approach to semantics has been demonstrated. No doubt alternative approaches to procedural semantics, which come to terms with the *pragmatic issues* we have considered, are possible. My overriding concern has been to demonstrate that *some* procedural account of semantics is both possible and necessary. Procedures may not tell the *whole* story about semantics, but they are a necessary part of the story.

NOTES

* I wish to thank James Delgrande, William Demopoulos, Philip Hanson, Howard Hamilton, Zenon Pylyshyn, Len Schubert, Brian Smith, and Susan Wendell for their helpful comments on an earlier version of this paper. The author gratefully acknowledges the support of the Natural Sciences and Engineering Research Council of Canada (grant no. A0899) during the period of this research.

1 There is no need to suppose that such adjudication procedures constitute precise algorithms. Rather, in human cognition they appear to take the form of heuristics, which are applied in a somewhat haphazard fashion. One goal of these heuristics is to preserve the global consistency of a belief system, while maximizing the degree of agreement between beliefs in general and particular judgments based upon *prima facie* evidence. Another goal is to minimize the complexity of and the number of revisions required by our theory. To a substantial degree these two goals conflict, and no principled strategy appears to exist for resolving these conflicts. Nevertheless, in practice we do make pragmatic decisions required to resolve these conflicts, and it is reasonable to suppose that this process can be explained procedurally.

2 Note that even if something like Fodor's language of thought thesis is accepted, the denotation relation which holds between symbols in the innate language and their external denotations must be explicable in principle. If there is no process by which linguistic agents, considered merely as organisms, can identify objects as similar, what could possibly explain the origin of the denotation relationship? But if such processes do exist, they can be explained procedurally.

3 If a value for 'T' is supplied to the *appears-to-have* procedure, then that value is compared with the value *returned*. If a match is found, the next subgoal, *p2* will be tested. Otherwise, failure is returned for the top-level goal *has* (x,T)

4 It should be noted that 'appears-to-have (x,T)' is a terminating procedure which returns judgments which are *not* revisable in the way in which judgments such as 'has (x,T)' are revisable. Since the 'appears-to-have' procedure makes no default assumptions, and merely processes sensory input, my claim is that *if the procedure is correctly executed*, the judgment it returns is correct. This does not entail that I am committed to the positivistic doctrine that all sense-data reports are infallible, for it is always possible that we have incorrectly executed an effective procedure. The problem here is epistemological rather than ontological. We can never be totally certain that we have correctly executed a procedure, though we may suppose that there is an absolute truth of the matter. In this respect I depart from Quine's relativistic ontology.

5 The precise sense in which these procedures are *abstract*, and the relationship of these procedures to Montague's intensions, is described in detail in (Hadley 1988).

REFERENCES

Ayer, A.J. (1946). *Language, Truth, and Logic*. London: Dover, 2nd ed.

Barwise, J. and Perry, J. (1975). Semantic innocence and uncompromising situations. In P.A. French, T.E. Uehling Jr., and H.K. Wettstein (eds.), *Midwest Studies in Philosophy, Volume I: The Foundations of Analytic Philosophy*. Minneapolis: University of Minnesota Press

— (eds.) (1983). *Situations and Attitudes*. Cambridge, MA: Bradford/MIT Press

Dahl, V. (1981). "Translating Spanish into logic through logic." *American Journal of Computational Linguistics* 7:149–64

Fodor, J.A. (1975). *The Language of Thought*. New York: Crowell

— (1978). Tom Swift and his procedural grandmother. *Cognition* 6:229–47

Hadley, R.F. (1973). *Convention and the Intensional Concepts*. Ph.D. thesis, Dept. of Philosophy, University of British Columbia

— (1985). SHADOW: a natural language query analyser. *Computers & Mathematics* 11:481–504

— (1986). Fagin and Halpern on logical omniscience: a critique with an alternative, *Proceedings of the Sixth Canadian Conference on Artificial Intelligence*. Montreal, May 1986

— (1988). Logical omniscience, AI semantics, and models of belief. *Computational Intelligence* 4:17–30

Hempel, C.F. (1950). Empiricist Criteria of Cognitive Significance: Problems and Changes, *Revue Internationale de Philosophie* 11:41–3

Hinton, G. & Anderson, J. (eds.) (1981). Parallel Models of Associative Memory. Hillsdale, NJ: Erlbaum

Johnson-Laird, P.N. (1978). What's wrong with grandma's guide to procedural semantics: a reply to Jerry Fodor, *Cognition* 6:249–61

Lewis, D. (1976). General semantics. In B. Partee (ed.) *Montague Grammars* (pp. 51–76). New York: Academic Press

Michalski, R.S. (1983). A theory and methodology of inductive learning. *Artificial Intelligence* 20

Miller, G.A. & Johnson-Laird, P.N. (1976). *Language and Perception*. Cambridge, MA: Harvard University Press

Putnam, H. (1970). Is semantics possible? In H.E. Kiefer and M.K. Munitz (eds.), *Language, Belief, and Metaphysics* New York: State University of New York Press

— (1973). Meaning and reference, *Journal of Philosophy* 70:699–711

— (1980). Models and reality, *Journal of Symbolic Logic* 45:464–82

Pylyshyn, Z. (1980). Computation and cognition: issues in the foundations of cognitive science. *Behavioural and Brain Sciences* 3(1):111–32

Quine, W.V.O. (1953). Two dogmas of empiricism. In his *From a Logical Point of View*. Cambridge, MA: Harvard University Press

Searle, J. (1980). Minds, brains, and programs. *The Behavioural and Brain Sciences* 3:111–69

Wilks, Y. (1982). Some thoughts on procedural semantics. In W. Lehnert and M. Ringle (eds.), *Strategies for Natural Language Processing*. Hillsdale, NJ: Erlbaum

Winograd, T. (1971). *Procedures as a Representation for Data in a Computer Program for Understanding Natural Language*. Ph.D. thesis, MAC TR-84, MIT Artificial Intelligence Laboratory

— (1985). Moving the Semantic Fulcrum. *Linguistics and Philosophy* 8:91–104

Woods, W., Kaplan. R., and Nash-Webber, B. (1972). *The LUNAR Science Natural Language Information System*, BBN Report No. 2378. Cambridge, MA: Bolt, Baranek, and Newman Inc.

Woods, W. (1975). What's in a link: Foundations for semantic networks. In *Representations and Understanding: Studies in Cognitive Science*. New York: Academic Press

— (1981). Procedural semantics as a theory of meaning. In Joshi, Webber, and Sag, (eds.), *Elements of Discourse Understanding*. Cambridge, Eng.: Cambridge University Press

— (1986). Problems in procedural Semantics. In Z. Pylyshyn & W. Demopoulos, (eds.), *Meaning and Cognitive Structure*. Norwood, NJ:Ablex Publishing

Comment
Can you save Procedural Semantics by Turning It Into a Theory of Semantic Competence?

Zenon W. Pylyshyn

BACKGROUND

Whenever I have thought about semantics in the past (which is no more often than I have to) I have always found myself thinking about the shortcomings of other people's theories. My mother taught me that if I have nothing good to say about a topic I should say nothing at all. So I have said nothing. Well, almost nothing; I admit I found myself on at least one occasion saying something to the effect that, as with the topic of consciousness, there is not much that we can say about semantics without appearing foolish.

The position I have taken on many of the great weighty foundational issues of cognitive science is that the scientist should avoid taking a position unless forced to do so by the scientific enterprise itself. I have also, however, recognized that taking a position is something that tends to happen to scientists in spite of themselves; so scientists, particularly in the human sciences, are invariably implicitly committed to one or another philosophical school, and this commitment does shape their theories. Nonetheless, I am in general prepared to defend a "principle of least philosophical commitment." I am, in particular, prepared to advocate philosophical agnosticism (or perhaps even boorishness) when the discussion turns to such questions as "What is consciousness?" and "What is meaning?", along with "What is life?", "Why is there human suffering?", and "Why do I agree to give papers on topics that I feel are better left to professional philosophers?"

Although I have tried to refrain from discussing what meanings are, and in virtue of exactly what do people (and sentences) have them, I have not refrained from claiming that cognitive science needs semantical notions, such as sameness of meaning and semantic coherence. In fact I go on at some length in my book (Pylyshyn 1984) about why we cannot state many important predictive generalizations of human behaviour unless we can speak of the contents of such mental states as goals and beliefs. But how can I stand here bald-facedly asserting that we need notions of semantics while refusing to say what meaning is, how it arises, and in virtue of what sorts of properties of organisms and the world do certain systems *mean* things. Well, it's really quite easy; when the topic turns to meaning I simply leave the room.

The reason this works, however, is that it is entirely commonplace in science to build one's theories on constructs that are widely accepted yet remain undefined, and in many cases even unexamined. In fact, the most basic scientific constructs are almost never defined. For example, biology does not define *life*, it studies its mechanisms; physics does not define *physical object* or *physical property*, it merely stipulates them and studies their laws. Of course, some constructs (such as ether or phlogiston) turn out not to have any meaning, let alone a precise characterization. But that's not the fault of the foundational analysis: no amount of analysis would have revealed the bankruptcy of these concepts: that was an empirical discovery. Conversely, a great deal of foundational discourse went into the topic of how there could be action at a distance. Not only was the problem never satisfactorily resolved within this debate, but it eventually just went away when people realized that a philosophical answer of the sort they had been seeking was never going to appear. Similarly, meanings need not be explicated in the way some people would like in order for us to be able to scientifically study their interactions and the way they are processed and used in determining behaviour. Of course, the very concept of meaning could turn out to be empirically bankrupt, but that will only be revealed by the relative success of theories that use semantical notions compared with those that don't.

So that's why I have not thought much about what meanings are. But, like it or not, the present occasion requires that I say something on the topic of semantics. So at the risk of appearing uncharitable to those who have thought about such matters much more than I have, here I go.

Hadley has laid out very clearly some of the reasons why one might be attracted to what has been called *procedural semantics*. If one is dealing with a computational agent (and I don't see any alternative to the view that we are such agents) then all relations that enter into the causation of such an agent's behaviour must either be nomological or they must be computational (or some functional abstraction over nomological principles). Thus it might seem that if semantic relations are to form part of our account of an agent's behaviour (and again I see no way out of this), semantic relations will themselves have to be explicable either nomologically or computationally, or both. There is, moreover, some reason to think that computing can deal with at least certain aspects of semantics; it can, for example, express certain truth-preserving transformations. So at least on the surface, we are led to look at computing in order to explicate semantics.

Hadley not only lays out the general attractions of PS, but also recognizes its shortcomings as a reincarnation of a verification theory of semantics. He also recognizes the shortcomings of other (e.g., possible world) theories of meaning. To remedy these shortcomings he proceeds to pick the tenable parts of each of these approaches and to collect them into a refurbished PS. It reminds me of our students' cafeteria eating habits: pick all the things that you don't dislike and make a meal of that result. This strategy does produce a collection of items with the property that none of the items is objectionable by themselves (such as mounds of chips and gravy and marshmallows and ice cream), but it does not always produce a combination that is edible, let along nourishing or esthetic. Whether or not it works in Hadley's case remains to be seen. All I will do in this paper is raise a few questions about some particulars of Hadley's attempt as it relates to the scientific program of cognitive science.

SEMANTICS AND SEMANTIC COMPETENCE: PHILOSOPHY MEETS COGNITIVE SCIENCE (AND WALKS ON BY)

The first thing I want to do is try to separate the philosophical from the scientific questions. I begin in good philosophical tradition by making some distinctions. Consider the following questions that have something to do with meaning.

(1) What is the *meaning* of (the English sentence) S?
(2) What does a person P know when he (or she) *knows the meaning* of S?
(3) What does P know when he *understands* the sentence S?
(4) What do we mean when we say that a person "understands the sentence S"?
(5) What does P know when he knows whatever it is that S expresses (i.e., when P knows the proposition expressed by S)?

Despite the apparent close relation among these questions — all of which are concerned in one way or another with the meaning of a certain English sentence — we must be prepared for the possibility that the anwers to these questions may be only distantly related, at least from the perspective of an empirical science of cognition. Presumably an empirical science is concerned with questions that have the property that different answers to them have different empirical consequences. This is not crass operationalism, but merely a statement of the different concerns of different scholarly endeavours.

Consider question (1). The problems this question raises are deep and formidable. The problem of how sentences relate to the world, or more accurately to possible states of the world, is one that has baffled philosophers for about as long as the mind-body problem has, and with the same exasperating evasiveness. Although it is a fascinating question, it is not clear what, if anything, answers such as those that appeal to truth conditions or possible worlds can contribute to the empirical science of psychology. That's because what a sentence means, or what a thought is about, has empirical consequences only insofar as it is instantiated in some way in a human organism — presumably in the person's brain. And then, strangely enough, it does not matter whether the sentence is true or not, only that it is believed. Because of this, *real* semantics — in the sense of the relation between cognitive states and world-states — does not matter for a psychology of mental processes. Let us call this sense of semantics $\textit{Semantics}_{\text{phil}}$ in recognition to that fact that it is primarily being studied by philosophers.

Of course to suggest that $\textit{Semantics}_{\text{phil}}$ may not matter to anyone but philosophers would be very misleading. As I said at the beginning, I have gone on at length about why semantical *relations* matter very much in cognitive science. Without semantical notions like "sameness of meaning" one cannot state a great many cognitive generalizations, such as that if I believe this building is on fire I will have a tendency, to put it mildly, to go somewhere else. We need sameness of meaning because this generalization, and an indefinite number of others like it, hold independent of the form that the particular thought takes; independent of what language, if any, one thinks it in, and independent of the particular brain state one is in, providing only that the brain state is a member of a certain *semantically defined* equivalence class, viz., the class of brain states that have the same meaning, or the same semantic content, as the sentence "the building is on fire."

But what is not clear, and what I suspect is not true, is that the notion of sameness of meaning we need is the one we get out of the kind of study that has traditionally been directed at answering question (1), i.e. out of a study of $\text{Meaning}_{\text{phil}}$ in the theory of $\textit{Semantics}_{\text{phil}}$. We don't, for example, want a notion of sameness of meaning (equality of $\textit{Meaning}_{\text{phil}}$) that cuts things as finely as the criterion of "having the same truth conditions," for reasons that are well illustrated by the Putnum twin-earth story and other similar puzzles. (We want to count Oscar and Elmer as having the same thought just in case it leads to the same behaviour in all possible circumstances, independent of the objective scientific fact that, in Putnum's story, one thought *actually* refers to H_2O and the other to XYZ.)

I don't mean that we don't want sameness of truth conditions as a criterion for when two sentences have the same $\text{Meaning}_{\text{phil}}$ (in $\text{Semantics}_{\text{phil}}$). We do, inasmuch as the relation between a sentence and the world is relevant for many purposes. Without this notion we would not know under what conditions an utterance (or a belief) was true, and we would not know under what conditions a person's intended actions, based on his beliefs, would produce the desired effects on the world. In other words, we do need $\text{Semantics}_{\text{phil}}$ wherever the veridicality of perception or belief are relevant, or when the actual objective achievements of an action are in question. But we don't need it for purposes of developing a theory of mental *processing*. We don't want $\text{Meaning}_{\text{phil}}$ to be part of our analysis of what it is for a person to *understand* a sentence, or of what a person *knows* when he has understood a sentence or of the state he is in when he is said to "know the meaing of the sentence," i.e., of questions (2) – (5).

$\text{Meaning}_{\text{phil}}$ is not the sort of thing that is, or can be *known* or *represented* in the mind. Whatever a person knows when he "knows the meaning of a sentence," it is surely not its truth conditions. I haven't a clue as to the truth conditions of the sentence "snow is white," let alone the truth conditions of "black holes are singularities in the space-time manifold," both of which I believe I understand. And *nobody* knows the truth conditions of "Snow is white is true if and only if 'snow is white'" or even "John believes that snow is white," otherwise we would not be having this discussion.

Naive intuition notwithstanding, what we come to *know* when we have understood a sentence may not be the "meaning of the sentence" at all, if by "meaning" we mean the sort of relation between symbol structures and their denotations in some possible world that is studied in $\text{Semantics}_{\text{phil}}$. In fact, it is surely a bad idea to describe someone as "knowing the meaning of sentence S" because there may not *be* a particular thing that the person knows when in that state. What a person knows in that state may be something like aspects of the speaker's intentions in uttering the sentence.

Now you might think that I am leading up to an argument against literal meaning, in order to replace it with a view (such as that advocated by Searle 1979), according to which a sentence does not have a meaning outside of a context of use. But that's not my point at all. I believe a sentence does have a meaning, and some aspect of this meaning is relevant to what we know when we understand it. There is something fixed that a sentence contributes to the context of understanding, otherwise it would not matter which sentence you uttered in order to convey some idea: you might just as well grunt or whistle. My point is that the relation between sentence meaning and understanding is far from being one which we might call embedding: analysing "P understands (or knows) the meaning of S" as "P understands (or knows) M," where "M" is the sort of relation between S and a possible world that $\text{Semantics}_{\text{phil}}$ is concerned with. Understanding is both more and less than "grasping" or otherwise "knowing" this sort of relation.

Understanding is an aspect of a communicative skill. What you come to believe when you are said to "understand S" or when you are said to "know the meaning of S" is some function of the "meaning of S" in the sense of $\text{Semantics}_{\text{phil}}$ *as well as* common sense reasoning about why S is being asserted at that particular time, and a lot of other things you might believe about the situation, the speaker, and the world at large. Although there is such a thing as "the proposition P that S expresses," and there is such a thing as "believing P," it's far from being the case that to understand S is to be in the state of believing P, the proposition that S expresses. It's not even to be in the state of believing that someone has uttered a sentence S *which expresses P*. The relation between S, M(S), and P(S) is much more obscure than that.

Notice first that the real big questions on the table are these:

(1) *What is the meaning M of sentence S?*
(2) *In virtue of what does S have the meaning M?* and
(3) *What role does M play in* ***understanding*** *S?*

So far, it seems to me that what I have been saying is in perfect agreement with Hadley's thesis. Where I feel some disagreement lurking is that Hadley appears to want to give an answer to (1) and (2) which will be more or less directly applicable to answering (3) by what I have called an embedding account. I frankly see little prospect of this sort of project succeeding because for one thing, the grain of resolution of "meaning" needed to answer (1) in order for the notion to serve the needs of $\text{Semantics}_{\text{phil}}$ is different from what is appropriate for it to serve the needs of an empirical cognitive science. In other words, I don't believe the same "M" will work in two cases. Although I am reasonably convinced of this claim, what I don't have for you is a positive alternative answer to (3). In fact I don't even know whether there is an answer: I consider it an open question whether M (i.e., meaning$_{phil}$) has *any* role to play in the analysis of understanding, although as I have already said, it surely is relevant to explaining when a sentence or a belief is true, and therefore under what conditions actions based on such beliefs are likely to produce the intended results.

HADLEY'S ATTEMPT TO BUILD A THEORY OF MEANING ON A THEORY OF UNDERSTANDING

Hadley has presented a detailed account of what he has described as an alternative "procedural" theory of semantics. However, as we examine the account, it looks very much like a theory of aspects of the process of understanding, rather than the theory of Semantics$_{phil}$ that we were promised. The account provides a way to integrate a mechanism for evaluating a predicate such as *appears-to-have* (x,T) (which appears to be identical with what I have elsewhere called a *transducer;* cf Pylyshyn 1984) with essentially heuristic procedures for establishing what he calls a "Testable predicate" for the extension of property T relative to "background-checking procedures."

These "testable predicates" provide the empirical constraints for some referring terms, while additional procedures provide the basis for abstracting over these empirical constraints and dealing both with abstract and compositional concepts. All this seems perfectly sensible, at least in broad outline, when taken as part of a theory of comprehension. Such a theory might even say something about how people integrate what they have understood with possible actions that they might take, including actions to check whether they ought to believe the sentences in question. Sometimes it seems that this is precisely what Hadley has in mind. For example, he talks about how "the existence of pragmatic meaning procedures can be used to account for much of what we call semantic competence." But what about the theory of semantics we were supposed to be working towards? We still do not have either a theory of what the Meaning$_{phil}$ of a sentence *is,* nor of how it relates to the objective truth conditions of the sentence.

One possibility is to take this story as offering not only the outlines of a theory of "semantic competence" but also of a theory of Semantics$_{phil}$. In that case it would be a straightforward verificational semantics, whose limitations Hadley readily recognizes. But here is where things get a bit murky. In places Hadley talks as though the procedural story really is intended to provide a semantic theory. For example, he tries to counter a criticism of verificational semantics by relaxing the requirement that "semantic procedures" infallibly determine the extension of terms in each possible world: they may return only "fallible judgments of truth values." Once again there is no argument if this is an account of what goes on in comprehension, and when people judge the believability of a sentence. But surely Hadley is not suggesting that sentences *only* have plausible meanings or pragmatically determined truth conditions; that they only roughly refer to states of affairs in some possible world, depending on pragmatic considerations. However much "plausible meaning" may make sense in relation to the goal sentence-understanding, that can't be an adequate account of meaning in general. A sentence refers to *some* state of affairs (or set of states of affairs) and not others independently of the plausible procedures by which people understand them. Unless you are prepared to take a highly relativistic line on meaning, you still need to explain what a sentence S means and under what conditions it is true, independent of what any particular agent believes.

At other times Hadley speaks as though his goal is to say what people know when they know the meaning of a sentence. But this too, it seems to me, is not something that will be answered by reference to "meaning procedures," but to a person's entire repertoire of beliefs. Even speaking of procedures as "knowable" is misleading.[1] We may in principle have access to some procedures (e.g., those that consist of explicitly represented rules). But those are a very small subset of the "procedures" we use, which in turn are a very small subset of the ways that many-one mappings can be carried out by the nervous system. By this definition it is very rare indeed that we ever "know" what a term refers to, unless we study the relevant piece of psychophysics. But it seems odd to say that we only know the meaning of terms when we know a rule for picking out their denotata. While that may be one way of knowing what a term refers to, it doesn't even include my knowing what "red" refers to because my being able to detect red does not involve executing a "knowable procedure."

It seems to me that the basic problem with Hadley's attempt to resuscitate PS is that he is trying to have his cake and eat it too. He is trying to give an account of $\text{Meaning}_{\text{phil}}$ in a way that can be embedded in some plausible notion of understanding. This is motivated in part by the desire to cash out such statements as "P knows the meaning of S" by making whatever constitutes "meaning" something that is "known." In order to do that Hadley has to posit something that serves both the function of being the "meaning" (or at least part of the meaning) of S and also serves as something that is computed in the process of understanding S, or at least partakes in this process in some transparent and direct way. He wants the theory of semantic competence to subsume a theory of semantics. That's why he eschews a "translation" theory of comprehension, such as that advocated by

Fodor (1975), according to which understanding a sentence consists of translating it into an already interpreted mentalese. Naturally if one assumes that the process of understanding generates *meanings* (in the sense of some objective $\text{Semantics}_{\text{phil}}$), then translation won't do, since a sentence in mentalese does not constitute a "meaning" in the required sense (i.e., a $\text{Meaning}_{\text{phil}}$). On the other hand, if one is not stuck with the assumption that understanding consists of generating such meanings, or even that understanding must appeal to "knowledge of $\text{Semantics}_{\text{phil}}$" then a translational view cannot be summarily excluded. I personally doubt that Hadley can have it his way; that he can integrate a theory of $\text{Meaning}_{\text{phil}}$ with a theory of understanding without giving up either a theory of objective semantics or a theory of computable semantic competence.

As Hadley says, what is hard to explain is the existence of objective truth conditions and the knowability of these conditions. I recommend giving up on the knowability of truth conditions. Although the truth conditions of a sentence are objective, fixed, and principled, what is known may be quite ephemeral and ad hoc. What is known may be a bag of tricks about how to use the terms, the memory of when one last heard them, what experiences they are associated with, what other beliefs one has about the same things, and so on. These add up to "understanding the meaning of S" because of constraints imposed by our transducers, the history of our experiences, and the requirement of internal coherence.

It's true that what we come to know when we have understood, or partially understood, a sentence must have *something* to do with the truth conditions of the sentence, just as what we perceive must have something to do with the truth conditions of our perceptual beliefs (i.e., with what is really out there). But the relation need not be either a simple or even a highly principled one.

Consider, for example, the case of vision. There is now beginning to be some understanding of what makes vision roughly veridical, and of the conditions under which vision is and is not veridical. Recent research in visual perception, within the cognitive science paradigm, has preceded in the following way. Physics, geometry, and optics have been used to provide an account of the relation between properties of the proximal stimulus (the retinal image) and the world. That relation is now being studied more carefully than ever by people working in the "natural computation" framework of vision research (e.g., Marr 1982; Richards 1988).[2] The study of visual perception, on the other hand, provides an account of the conditions under which vision is in fact veridical, and the nature of the mechanisms that make it so.

Here is an example of how the physical/geometrical/optical studies and the psychological studies come together. Consider the following problem which is solved, inter alia, by the human visual system. The mapping from a three-dimensional environment to a 2-D retina is non-invertable: there are indefinitely many 3-D configurations that lead to the same 2-D retinal projection so the inverse mapping, or the interpretation problem, is underconstrained. Yet the visual system is apparently able to invert the mapping, inasmuch as we generally see a unique 3-D layout corresponding to any 2-D image. Of course we may use other information, such as shading or movement or stereo, but these by themselves do not render the projection invertable: the image remains in principle indeterminate as to its distal 3-D cause.

It turns out that the 3-D interpretation can be made univocal if certain conditions (or "constraints") are imposed on the interpretation process. These take the form of "assumptions" about our sort of world, for example that it is almost everywhere continuous, that most of the image arises from surfaces that vary gradually in their distance from the observer, that most visual features in an image arise from irregularities on the surface of coherent, opaque, and rigid bodies, that light usually comes from above, and so on. By selecting the interpretations that are consistent with these general "constraints" the visual system can very often recover the veridical 3-D layout of a scene. Moreover, appeal to these natural constraints also specifies the conditions under which perception will not be veridical — when it will fail and produce illusions.

Perhaps there is an analogue here for the relation between semantics and semantic competence. Semantics (particularly a worked-out naturalized semantical theory, which so far we do not have) may give us a theory of the sentence-world relation (or of where the meaning of the sentence comes from), and a theory of comprehension may give us an account of how we relate sentences to our beliefs and our past experiences in such a way that our new beliefs turn out to be veridical reasonably often "in our sort of world." That, I believe, is entirely consistent with some of the work that is reported here at this conference within the framwork of the "informational" view of meaning. But notice that even if such a scenario could be worked out, it would still be a far cry from providing a theory of semantics based on a theory of semantic competence, or a theory of understanding. One theory would in no way be embedded in the other. Indeed, the semantical theory or the meanings which it explicated would not be something in the head, unlike the representations or the computational operations upon them. Meanings would no more be something in the head than would the natural constraints to which theories of visual perception appeal. Both the constraints and the meanings are objective properties of the world. As theorists we need to know what they are in order to explain why what *is* in our head works most of the time (and perhaps why it also *fails* to work under specified conditions).

That may not be what some people hope to get out of a theory of semantics, but that may well be all that we will get; which may account for why we have managed to get along without it for so long.

NOTES

1 It goes without saying that I am not using "know" here to mean "is consciously aware of": tacit knowledge is ubiquitous in cognition. However, there is a difference between behaviour based on knowledge and behaviour based on habits or nonepistemic mechanisms. Although this distinction is beyond the scope of this commentary, it is absolutely central to cognitive science. In (Pylyshyn 1984) it is characterized as the difference between behaviour that is determined by semantically interpreted representations and behaviour that derives from properties of the cognitive "functional architecture" alone.
2 Burge (1987) provides a clear summary of some of the earlier work in this tradition, which highlights some of its philosophical implications.

REFERENCES

Burge, T. (1987). Marr's theory of vision. In J.L. Garfield (ed.), *Modularity in Knowledge Representation and Natural-Language Understanding*. Cambridge, MA: MIT Press

Fodor, J.A. (1975). *Language of Thought*. New York: Crowell

Marr. D. (1982). *Vison*. San Francisco: Freeman

Pylyshyn, Z.W. (1984). *Computation and Cognition*. Cambridge, MA: MIT Press

Richards W. (ed.) 1988. *Natural Computation*. Cambridge, MA: MIT Press

Searle, J.R. (1979). Literal meaning. In J.R. Searle (ed.), *Expression and Meaning: Studies in the Theory of Speech Acts*. Cambridge, Eng.: Cambridge University Press

CHAPTER FOUR

Putting Information to Work

Fred Dretske*

Information isn't much good if it doesn't do anything. If the fact that an event carries information doesn't help explain the event's impact on the rest of the world, then, as far as the rest of the world is concerned, the event may as well not carry information. To put it bluntly, in the way positivists liked to put it, a difference that doesn't make a difference isn't really a difference at all. If an event's carrying information doesn't make a difference – and by a difference here I mean a causal difference, a difference in the kind of effects it has – then for all philosophical (not to mention practical) purposes, the event doesn't carry information.

Surely, though, this is not a serious threat. We all know how useful a commodity information is, how even the smallest scrap can radically alter the course of human affairs. Think about its role in business, education, and war. Or consider the consequences of telling Michael about his wife's passionate affair with Charles. Kaboom! Their lives are never again the same. A small piece of information dramatically alters a part (and – who knows? – maybe eventually the entire course) of world history. In light of such obvious examples, how can anyone seriously doubt the causal efficacy of information and, hence, its relevance to understanding *why* some things turn out the way they do?

MEANING AND INFORMATION

There is a small subset of the world's objects on which information *appears* to have this kind of dramatic effect. These objects are people, objects like you and me, who *understand*, or *think* they understand, some of what is happening around them. Talking to Michael has profound effects while talking to rocks and goldfish has little or no effect because Michael, unlike rocks and goldfish, understands, or thinks he understands, what he is being told. As a consequence, he is — typically at least — brought to believe certain things by these acts of communication. The rock and the goldfish, on the other hand, are impervious to meaning. Instead of inducing belief, all we succeed in doing by remonstrating with such objects is jostling them a bit with the acoustic vibrations we produce.

So appearances may be deceptive. It may turn out that the difference between Michael and a goldfish isn't that Michael, unlike a goldfish, responds to information, but that Michael, unlike a goldfish, has beliefs, beliefs about what sounds mean (or about what the people producing these sounds mean), and which therefore (when he hears these sounds) induce beliefs on which he acts. These beliefs are, to be sure, sometimes aroused in him by sounds that actually carry information. Nevertheless, if these beliefs in no way depend on the information these sounds carry, then the information carried by the belief-eliciting stimulation is explanatorily irrelevant. After all, rocks and goldfish are also affected by information-carrying signals. When I talk to a rock, the rock is, as I said, jostled by acoustic vibrations. But the point is that althogh my utterances, the ones that succeed in jostling the rock, carry information, the information they carry is irrelevant to their effect on the rock. The information in this stimulation doesn't play any explanatory role in accounting for the rock's response to my communication. From the rock's point of view, my utterance may as well not carry information. Subtract the information (without changing the physical properties of the signal carrying this information) and the effect is exactly the same.

And so it may be with Michael. To find out, as we did with the rock, whether information is doing any real work, we merely apply Mill's Method of Difference. Take away the information, leaving everything else as much the same as possible, and see if the effect on Michael changes. As we all know, we needn't suppose his wife, Sandra, is actually having an affair with Charles to get a reaction — in fact the very same reaction — from Michael. He will react in exactly the same way if his alleged informant is lying or is simply mistaken about Sandra's affairs. As long as the act of communication is the same, as long as what the speaker says and does *means* the same thing (to Michael), it will elicit the same reaction from him. What is said doesn't have to be *true* to get this effect. Michael just has to *think* it true. Nothing, in fact, need even be said. As long as Michael *thinks* it was (truly) said, as long as he thinks something with this meaning occurred, the result will be the same.

In saying that it is Michael's beliefs, not the meaning (if any) or information (if any) in the stimuli giving rise to these beliefs, that causally explains Michael's behaviour, I am assuming that we can (and should) make appropriate distinctions between these ideas — between information, meaning, and belief. Some people, I know, use these notions interchangeably. That is too bad. It confuses things that should be kept distinct. According to this careless way of talking (especially prevalent, I think, among computer scientists) information *is* meaning or (at least) a species of meaning, and a belief (or a reasonable analogue of a belief) just *is* an internal state having the requisite kind of meaning. So, for instance, anything that means that Michael's wife is having an affair carries this piece of information. If I *say* his wife is having an affair, then my utterance, *whether or not it is true*, carries this information. And if I enter this "information" into a suitably programmed computer, then, whether or not Sandra is unfaithful, the "fact" that she is unfaithful becomes part of the machine's "data base," the "information" on which it relies to reason, make inferences, answer questions, and solve problems. The computer now "thinks" that Michael's wife is having an affair. Michael doesn't even have to be married for the machine to be given the "data," the "information," that, as it were, his wife is having an affair. On this usage, the facts, the information, the data, are what we say they are.

It is perfectly understandable why computer scientists (not to mention a good many other people) prefer to talk this way. After all, it is natural to suppose that a computer (or a human brain for that matter) is insensitive to the *truth* of the representations on which it operates. Put the sentence "P" in a machine's (or a human brain's) data file, and it will operate with that data in exactly the same way whether "P" is true or false, whether it is information or misinformation. From the machine's (brain's) point of view, anything in it that qualifies as a belief qualifies as knowledge, anything in it that *means* that P is *information* that P. The distinction between meaning and information, between belief and knowledge, is a distinction that only makes sense from the outside. But what makes sense *only from the outside* of the machine or person whose behavior is being explained cannot (according to this way of thinking) help to explain that machine's (or person's) behavior. It cannot because the machine (or person) can't get outside of itself to make the needed discriminations. So, for practical explanatory purposes, meaning *is* (or may as well be) information.

Whatever the practical exigencies may be, something that makes sense only from the outside is, nonetheless, something that makes perfectly good sense. It certainly should make perfectly good sense to those of us (on the outside) talking about such systems. Something (like a statement) that *means* that Michael's wife is having an affair need not carry this information. It needn't carry this information either because Michael's wife is *not* having an affair or because, though she is, the words or symbols used to make this statement are being used in a way that is quite unrelated to her activities. I can *say* anything I like, that Mao Tse Tung liked chocolate ice cream for instance, and the words I utter will mean something quite definite — in this case that Mao Tse Tung liked chocolate ice cream. But these words, even if they are (by some lucky accident) true, won't carry the information that Mao liked chocolate ice cream. They won't because the sentence, used merely as an example and in total ignorance of Mao's preferences in ice cream, in no way depends on Mao's likes and dislikes. These words mean something, but they do not, not at least when coming from me, inform the listener. I might succeed in getting you to believe that Mao liked chocolate ice cream. I might, by telling you this, *mis*inform you about his taste in ice cream. But misinformation is not a species of information any more than belief is a species of knowledge.

So, at least on an ordinary understanding of information and meaning, something can mean that P without thereby carrying the information that P. And someone can believe that P without ever having received the information that P. Often enough, what makes people believe that P is being told that P by someone they trust. Sometimes these communications carry information. Sometimes they do not. Their efficacy in producing belief resides, however, not in the fact that the utterance carries information, but in its meaning (or perceived meaning), who uttered it, and how. No successful liar can seriously doubt this.

So if we distinguish between an event's meaning (= what it says, *whether truly or falsely*, about another state of affairs) and the information it carries (what it, among other things, *truly* says about another state of affairs), a distinction that is roughly equivalent to Grice's (1957) distinction between non-natural and natural meaning, the causal role of information becomes more problematic. What explains Michael's reaction to the verbal communication is his believing that his wife was having an affair with Charles. What explains his believing this is his being told it by a trusted confidant — i.e., his hearing someone (he trusts) *say* this, utter words with this meaning. At no point in the explanatory proceedings do we have to mention the truth of what is said, the truth of what is believed, or the fact that information (as opposed to misinformation) was communicated. If Michael *acts* on his belief, he may, sooner or later, confront a situation that testifies to the truth of what he believes (and was told). He will then, presumably, acquire new beliefs about his wife and Charles, and these new beliefs will help determine what further reactions he has, what he goes on to do. But still, at no point do we have to speak of information or truth in our explanations of Michael's behaviour. All that is needed is what Michael *thinks* is true, what he *thinks* is information, what he *believes*. Knowing whether these beliefs are true or not may be helpful in predicting the *results* of his actions (whether, for instance, he will actually find Sandra at home when he goes to look), but it is not essential for explaining and predicting what he will actually do — whether, that is, he will go home to look.

Appearances, then, *do* seem to be misleading. Despite the way things first looked, despite a variety of familiar examples in which information seemed to make a causal difference, we still haven't found an honest job for it, something information (as opposed to meaning or belief) does that constitutes *its* special contribution to the causal story.

TRUTH AND SUPERVENIENCE

It isn't hard to see why there is trouble finding a decent job for information. The information a signal (structure, event, condition, state of affairs) carries is a function of the way a signal is *related* to other conditions in the world. I have my own views (Dretske 1981) about what these relations come down to, what relations constitute information. I happen to think information requires, among other things, some relation of dependency between the signal and the condition about which it carries information. Signals don't carry information about conditions, even conditions which happen to obtain, on which their occurrence does not depend in some appropriate way. But it isn't necessary to argue about these details. I'm not asking you to agree with me about exactly what information is to agree with me that, whatever it is, as long as it (unlike meaning) involves truth, there is a special problem about how it can be put to work in a scientifically respectable way — how *it*, or the fact that something carries it, can be made explanatorily relevant.

In order to appreciate the problem it is enough to realize that since no signal, S, can carry the information that P is the case when P is not the case, you can change the information a signal carries by tinkering, not with the signal itself, but with the condition, P, about which it carries information. If it is possible, as it almost always is (by tinkering with the causal mechanisms mediating P and S), to change P without changing the character of S itself (i.e., without changing any of S's non-relational properties), then the fact that S carries information is a fact that does not (to use a piece of current jargon) supervene on its non-relational (intrinsic) properties. Signals that are otherwise identical can be informationally different. They will be informationally different when one occurs when P is the case while the other, a physical clone of the first, occurs when P is not the case.[1] But if this is so, and we assume, as it seems we must assume (more of this in a moment), that the causal powers of a signal are embodied in, and exhausted by, its non-relational properties (so that two signals that are physically identical are, in the same circumstances, causally equivalent), then the information character of a signal, the fact that it carries information, is causally inert. However useful it might be (because of the correlations it exhibits) as a predictor of correlated (even *lawfully* correlated)[2] conditions, the fact that a signal carries information will *explain* absolutely nothing. The fact that a signal carries the information that P cannot possibly explain anything if a signal lacking this information has exactly the same effects. This is not to say that signals bearing information cannot cause things. Certainly they can. It just means that their carrying information does not help explain why they cause what they cause. Distilled water will extinguish a fire — thereby causing exactly what undistilled water causes — but the fact that the water is distilled does not figure in the explanation of *why* the fire is extinguished. It won't because undistilled water has exactly the same effects on flames. And if the information in a signal is like this, like the fact that water is distilled, then the fact that a signal carries information is as explanatorily relevant to its effects on a receiver as is the fact that the water is distilled to its effects on a flame.[3]

Coupled with the idea that a symbol's meaning is a function of its relations to other conditions (the conditions it, in some sense, signifies or means), such arguments have led to the view that it is the form, shape, or syntax — not the meaning, content, or semantics — of our internal states that ultimately pulls the levers, turns the gears, and applies the brakes in the behaviour of thoughtful and purposeful agents. Semantics or meaning, the what-it-is-we-believe (and want) is causally (and, therefore, explanatorily) irrelevant to the production of behavior (which is not to say that it cannot be used to *predict* behaviour). I have merely extended these arguments, applying them to information rather than meaning or content. Since information, unlike meaning, requires truth, I think these arguments are even more persuasive when applied to information. I have, in fact, assumed up to this point that there is no particular difficulty about how meaning, either as embodied in *what a person says* or as embodied in *what a person believes*, could figure in a causal explanation. But this, too, has its difficulties. Since I am convinced that belief and meaning are notions that ultimately derive from the information-carrying properties of living systems, I think these problems are, at a deeper level, connected. I think, in fact, that the central problem in this area is the causal efficacy of information. If we can find a respectable job for information, if *it* can be provided a causal job to do, if it can be put to work, then the causal role of meaning and belief (indeed, of all the other psychological attitudes), being derivative, will fall into place. But these are issues that go beyond the scope of this paper and I will not return to them here.[4]

To put information to work will require understanding how the causal efficacy of a signal is altered by the fact that it carries information. A part of the task, then, is to see how the causal efficacy of S (the signal carrying the information that P) is changed or modified by the fact that, in a certain range of circumstances, it occurs *only when* P is the case. I say that this is *part* (not *all*) of the task since S's occurrence *only when* P is the case does not mean that S carries the information that P. It depends on what *else*, besides co-occurrence (by this I mean the occurrence of S only when P is the case), is required for S to carry the information that P. So showing the causal relevance of co-occurrence will not directly demonstrate the causal relevance of information. Nonetheless, since we are assuming, as a minimal condition on information, that S cannot carry the information that P unless P is the case, unless, that is, S and P *do* co-occur in the relevent conditions, demonstrating the causal relevance of co-occurrence will be an important step in showing that information has a causal job to do.

This isn't as easy as it looks. The job, remember, is to show how the co-occurrence of S with P, the fact that S occurs only when P is the case, makes a difference in S's causal powers. The job is not to show that S *together with* P causes things that S *alone* does not. That latter task is simple enough, but quite irrelevant. It is quite obvious that Tommy and his big brother can do things that Tommy alone cannot. And if Tommy never goes anywhere without his big brother, then Tommy will be a "force" to contend with on the school playground. Strictly speaking, though, the presence of his big brother doesn't enhance *Tommy's* prowess. It merely makes him part of *a team* that is feared by the other boys, a team whose presence is signalled by the appearance of Tommy and, thus, makes (via the beliefs of the other boys) Tommy an intimidating figure on the playground.

Such situations are familiar enough, but they are not what we are after. What we are looking for, instead, is a case where *Tommy's* punches carry increased authority because he (always) has his big brother with him.[5] How can Tommy derive added strength, increased causal powers, from the mere fact (a fact that may even be unknown to Tommy) that his big brother is always nearby? How can the (mere) fact that P is the case when S occurs, whether or not anyone — including S — realizes this fact, change *S*'s causal powers? Until we know how, we won't know how information can make a difference in this world.

INDICATORS AND ARTIFACTS

I think we can make a beginning at understanding how this is possible by thinking about how some elements are *given* a causal job to do because of what they indicate about related conditions. Suppose we want some particular kind of movement or change (call it M) to occur when, and only when, condition P exists. I want an annoying buzzer to go on when, and only when, passengers fail to buckle their seat belts. I want the fan to go on when the engine overheats but not otherwise. I want the light to come on when anyone, or anything, crosses a threshold. The way to get what I want is to make an indicator of P — something that will activate when, and only when, P is the case — into a cause of M. To design such mechanisms is a job for engineers. Find or build something that is selectively sensitive to the occurrence of P and make it into a switch for M. Find (or build) something that is sensitive to passengers with unbuckled seat belts. Make it into a buzzer switch. This device doesn't have to be very fancy — just a little electrical-mechanical gadget that will be activated by weight (on a car seat) and electrical contact (in the seat belt buckle). Joint occurrence of the right set of conditions (P) — the condition in which we want M to occur — is then made into a switch for, a cause of, M, the buzzer. If things work right, we now get M when, and only when, P: the buzzer sounds when, and only when, there are passengers with unfastened seat belts.

Building a gadget like this (and such gadgets are all around us) is an exercise in making a more or less reliable indicator, a structure exhibiting a more or less reliable correlation with P, into a cause of M — the response to be co-ordinated with P. By appropriate design, manufacture, and installation, the causal powers of an indicator, an information-bearing structure, an S which occurs *only when* P is the case, is modified, and, what is most significant, it is modified *because* this indicator (or the appropriate activation of this indicator) occurs when, and only when, a certain other condition (P) occurs. The properties of this internal structure (S) which are relevant to its selection as a cause of M, and hence which explain why it (now) causes M, are not its intrinsic properties — its size, shape, weight, colour, charge, and so on. These might help to explain *how* S is made to cause M, how it can be converted into a cause of M, but not why it is converted into a cause of M. It is, rather, S's relational properties that explain why it was selected (by the engineer designing the device) for this causal job. Anything, no matter what its intrinsic properties (as long as they can be harnessed to do the job), would have done as well. As long as the behaviour of this element exhibits the appropriate degree of correlation with P, it is a candidate for being made into a switch for M, the behaviour we want co-ordinated with P. If, furthermore, an element is selected for its causal role (in the production of M) *because* of its correlation with P, because it does not (normally) occur without P, we have (almost) a case of an element's informational properties explaining its causal properties: it does (or is made to do) this because it carries information about (or co-occurs with) that. It isn't the element's shape, form, or syntax that explains its conversion into a cause of M; it is, instead, its information-carrying, its semantic, properties.

This is all a little too fast, of course. We smuggled into the proceedings an engineer, with purposes and intentions of his own, soldering things here, wiring things there, because of what he knows (or thinks) about the effects to be achieved thereby. I therefore expect to hear the objection that deliberately designed artifacts do not demonstrate the causal efficacy of information. All they illustrate, once again, is the causal efficacy of belief (and purpose). In the case of artifacts, what explains the conversion of an information-bearing element (an indicator of P) into a cause of output (M) is the designer's knowledge (or belief) that it is a reliable indicator and his or her desire to co-ordinate M with P. To make information do some real work, it would be necessary to make the causal powers of S depend on its carrying information, or on its co-occurrence with P, *without* the intercession of cognitive intermediaries with purposes and intentions of their own. The information (correlation) alone, not some agent's recognition of this fact, must carry the explanatory burden.

Indicators and Learning

To see how this might be accomplished, simply remove the engineer. Since artifacts do not spontaneously change the way they behave (at least not normally in a desired way) without some help from the outside, replace the seat belt mechanism with a system that *is* capable of such unassisted reconfiguration. That is, replace the artifact with an animal — a rat, say. Put the rat into conditions — a suitably arranged Skinner box will do — in which a certain response is rewarded (with food, say) when, and only when, it occurs in conditions P. The response is punished when it occurs without P. Let P be some condition which the rat can observe — a certain audible tone, say. Let the response be the pressing of a bar. What happens? Given a hungry rat, enough trials, and a tolerance for stray errors, the rat learns to press the bar when, and only when, it hears the tone. A correlation between M (bar pressing) and P (the tone) begins to emerge. The engineer got the seat belt mechanism to behave the way he wanted it to behave, to buzz when a passenger failed to buckle his seat belt, by connecting wires in the right places, by making an internal indicator of P (an unbuckled belt) into a cause of M (buzzer activation). The same thing happens to the rat without the assistance of an engineer or, indeed, *any* intentional agent. An internal indicator of P (in this case a certain tone) becomes a cause of M (in this case bar-pressing movements) not through the intercession of an outside agent, but merely by having the response, M, rewarded *when* it occurs in the right conditions (P).

This kind of learning — discrimination learning — can only occur if there is some internal condition of the learner, call it S, that exhibits some degree of correlation with P (the external condition being discriminated) — unless, that is, there is some internal condition of the learner that under normal conditions carries information about the condition to be discriminated. Unless there is some internal condition that occurs when, and only when, P is the case, it will be impossible to get M, the behaviour, to occur when, and only when P, the discriminated condition, exists. You can't make a system do M when (and only when) P exists if there is nothing in the system to *indicate* when P exists, and having something in the system to indicate when P exists is, among other things, a matter of having something in the system that occurs when, and only when, P exists. So when this type of learning *is* successful, as it often is, there must be, internal to the learner, a condition S that, with some degree of reliability, occurs when (and only when) P exists. There must actually be, inside the rat, something that (under appropriate stimulus conditions) occurs when, and only when, the right tone is sounded.

Furthermore, for this type of learning to occur, there must not only *be* an internal structure, S, carrying information about P, it must, during this type of learning, actually assume control functions it did not formerly have. It must be converted into a switch for M, the behavior that (through this type of learning) becomes co-ordinated with P. Unless S is, through this type of learning, recruited as a cause (or partial cause) of M, there is no way of co-ordinating behaviour with the conditions on which its success depends. For S is, by definition, the internal element that signals *when* the conditions are right (when P exists) for the behavior (M) to achieve success (to be rewarded). So S must be made into a cause of M (at least something on which M depends) if learning is to occur. Otherwise it is sheer magic. Whatever it is in the rat — call it the rat's *perception* of the tone — that signals the occurrence of the tone, it must actually be recruited as a cause of bar-pressing movements in this kind of learning if the rat is to learn to press the bar when the tone occurs.

Not only must an internal indicator be enlisted for control duties in this kind of learning, the properties of the indicator that explain its recruitment (as a cause of movement) are its *semantic* or *relational* properties, the properties that do *not* supervene on its intrinsic neurophysiological character. What explains why the rat's perception of a tone causes it to press the bar, what explains this internal element's altered causal powers, is not the fact that it has certain neurophysiological properties — a certain electrical-chemical-mechanical profile or form. For it presumably had that form *before* learning occurred (the rat could hear the tone before it learned to respond to it in a particular way). Before learning occurred, though, this internal state, this perception of the tone, did not cause the same movements (or any movements at all). Hence, what explains why the rat's perception of the tone causes movements it did not formerly cause is the fact that it, the rat's perception of the tone, is, specifically, a perception *of the tone*, the fact that it is an internal state exhibiting the requisite correlation with those external conditions on which the success of output depends. What explains the perceptual state's new found causal power is, in other words, its semantic, informational or intentional properties — not what it *is*, but what it is *about*.

If this theory, sketchy as it is, is even approximately right, then we have found a place where information does some honest work: it does real work when living systems are, during learning, reorganizing their control circuits to exploit the correlations (correlations between what is happening inside and what is happening outside) that perception puts at their disposal. Whenever there is something inside a system that can, by suitable redeployment, affect output, then there exists an opportunity to put to work whatever information that element carries. Such information can be exploited if there is a range of behaviours whose benefits (to the animal) depend on their emission in those (external) conditions about which information is carried. If there is such a dependence, then, assuming the animal capable of modifying its behaviour so as to better co-ordinate it with those external conditions on which its success depends (capable, that is, of learning), the animal can exploit perceptual information by recruiting, as a cause of behaviour, the internal vehicles of this information.

I say that this is an honest job for information because, unlike the artifacts discussed earlier, it is information itself, the fact that S does not normally occur unless P, and therefore (other possible conditions being satisfied) the fact that S carries the information that P, that explains the recruitment of internal elements as causes of movement (those movements whose success depends on their coordination with P). At the neuroanatomical level it may be a mystery how such recruitment takes place, how learning actually occurs, but that it *does* occur, in some animals under some circumstances, is perfectly obvious. And it is its occurrence, not details about how it occurs, that demonstrates — indeed, requires — the causal efficacy of information.

There are, it seems to me, profound implications of this fact, the fact that information only begins to find a real causal and hence explanatory use in systems capable of learning. Only here do we find the behaviour of a system explicable in terms of the relational properties of the internal states that produce it. Only here do we begin to see something like *content* or *meaning*, properties (like information) that do not supervene on the intrinsic properties of the internal states that possess it, assuming a significant place in our explanations of animal and human behaviour. It is here, I submit, that psychological explanations of behaviour first get a real, as opposed to merely a metaphorical, purchase. It is only when information begins to do some real explanatory — hence, scientific — work that minds rightfully enter the metaphysical picture.

NOTES

* I wish to thank the Center for Advanced Study in the Behavioral Sciences, Stanford University, the National Endowment for the Humanities FC-20060-85, and the Andrew Mellon Foundation for their support during 1987-8 when this paper was written.

1 It is better to say that under such conditions they *can be* informationally different. Whether they *are* different will depend, not only on whether P exists in one case but not the other, but on whether there is, in the case where P exists,the required information-carrying dependency between P and the signal.

2 In "common cause" situations, cases where A, though neither the cause nor the effect of B, is correlated with B because they have some common cause C, we may (depending on the details of the case) be able to say that B would not have happened unless A happened and, yet, deny that A in any way *explains* (causally or otherwise) B. An explanatory relation between A and B, a relation that lets us say that B happened *because* A happened, requires *more than* counterfactual-supporting generalizations between A and B.

I disagree, therefore, with Jerry Fodor (1987: 139-40) that (to put it crudely) an adequate story can be told about mental *causation* without making intentional properties (like meaning or information) determine causal roles. It isn't enough to have these intentional properties (like meaning or information) determine causal roles. It isn't enough to have these intentional properties figure in counterfactual-supporting generalizations. That (alone) won't show that people behave the way they do *because* of what they believe and desire.

3 Jerry Fodor, *Psychosemantics*, 33, nicely illustrates this with the property of being an H-particle. A particle has the property of being an H-particle (at time *t*) if and only if a dime Fodor flips (at time *t*) is heads. If the dime turns up tails, these particles are T-particles. H-particles are obviously causally efficacious, but no one supposes that their causal efficacy is to be explained by their being H-particles.

4 For a full discussion see *Explaining Behaviour: Reasons in a World of Causes* (Cambridge, MA: 1988 Bradford/MIT Press).

5 Not, mind you, because he *thinks, believes, or knows* that his big brother is backing him up. For this would, at best, only demonstrate, once again, the causal efficacy of belief.

REFERENCES

Dretske, Fred (1981). *Knowledge and the Flow of Information.* Cambridge, MA: Bradford/MIT Press

— (1988). *Explaining Behavior: Reasons in a World of Causes.* Cambridge, MA: Bradford/MIT Press

Fodor, Jerry (1987). *Psychosemantics.* Cambridge, MA: Bradford/MIT Press

Grice, Paul (1957). Meaning. *Philosophical Review* 66:377-88

Comment
Putting Dretske to Work

Brian Cantwell Smith*

Dretske, I take it, argues as follows. He starts by asking a question:

(1) What is the explanatory role (if any) of information?

The question is made urgent by a problem:

(2) To be explanatory, something must be *causal*;
(3) Information, however, is a *semantical* phenomenon, and since (as is widely alleged)
(4) Causal efficacy is *intrinsic* or *local*; and
(5) Semantics is *relational* or *distal*;
(6) It follows that semantics can't be causal (from 4 & 5).
(7) Therefore, information can't be explanatory (from 2, 3, & 6).

For example, suppose signal S carries the information that P, and also engenders behavioural consequence M. Then (according to Dretske) S's carrying the information that P cannot be causally responsible for M. Why not? Because they're insufficiently (i.e., not counterfactually) correlated. If you change the content (i.e., make it not true that P) without changing S, S will go on affecting the world in the same way it always did, but will no longer carry the same information (since information is veridical). The two properties — carrying information, and causing effect — are, so to speak, too *disconnected*.[1] And if, as Dretske suggests, this disconnection robs information-carrying of any causal potency, what on earth use can information be? No use, perhaps.

Given question and problem, Dretske then proposes the following solution:

(8) Information derives its explanatory force from situations of *learning*.

The reason? Because during learning, acording to Dretske, intentional agents can respond differentially to something's carrying information (or can end up in different states depending on whether something carries information — or something like that). They can end up one way if S carries the information that P, in other words, and another way if it doesn't. As a result, even if at some later time S's carrying the information that P can't be locally responsible for its production of M, something else may be true, almost as good: S's carrying the information that P can still, at that later date, be causally responsible for the *fact* that S causes the production of M.

Maybe I've got the types wrong. Maybe facts aren't quite the sort of thing that can be caused. But the intuition isn't difficult to see. If, during the learning situation, S hadn't carried the information that P, then the agent wouldn't have learned in the way that it did, and therefore (later) S wouldn't have caused M.

So in this case it looks, according to Dretske, as if semantical properties (like carrying information) can be explanatory after all, because they can explain how symbols get to have the causal powers that they do. Maybe, that is to say, the problem really is solved. And note, too, *how* it would be solved. Dretske proposes to defuse (7) by eliminating (6), rather than (2) or (3). He claims, that is, that learning is a situation where semantical properties can, after all, be causally effective.

In this review I'll accept Dretske's question. Before considering his solution, however, I want to spend some time examining his formulation of the problem.

THE PROBLEM

By my reconstruction, Dretske's analysis rests on four assumptions (2-5). Two are relatively unproblematic, but two are going to cause trouble.

Start with those that won't. Assumption (3), that carrying information is a semantical notion, is surely right. People may disagree with Dretske's claim that information is *the* original intentional property, but that it is *an* intentional notion seems hard to doubt.[2] Similarly, assumption (5), that semantics is relational or distal, can hardly be denied. The spatio-temporal reach of "referring to Sir John A. Macdonald" is beyond the imagination of any known physical force. It doesn't follow, of course, from the fact that *some* semantic relations are remote (all, I take it, are relational), that *all* are remote. The three word term "the name 'Ichabod'," for example, seems to contain its referent right inside it. Nonetheless, the general fact remains: there is more to a system of symbols, representations, or "information-carriers" than can be found internally or intrinsically within them. If you want to determine the truth of the statement in the newspaper claiming that gold's been found in the creek behind your summer home, you don't get out the microscope and peer at the typescript; you go up North and look.[3]

Locality of causation

Dretske's other two assumptions, however, are less obvious. Look first at (4) — that causal efficacy arises from intrinsic (non-relational) properties. This may seem tautological, since it's hard not to accept the enduring locality of physical or material interaction. But it isn't necessarily so. While I'll agree that the locality of immediate, proximal effect is unassailable,[4] the locality of *causation* isn't nearly as clear, in part because of that notion's recalcitrance. It's famously possible to do something now, like planting a bomb, whose consequences don't happen for a long while, and still to call that long distance relationship "causal" (if you don't think so, I've got a job for you with an asbestos company). So for discussion I'll use the term "potency" to get at such immediate, local, almost physical properties as can engender discriminable effect, and leave "causation" as potentially more long distance.

Potency is thus somewhat like "first" or "proximate" cause. But not exactly. It's different because I mean it to include not only legitimate causes, but also any other putatively non-causal but still immediate influences, such as background or enabling conditions (such as gravity). That is, the aim is to corral the entire set of impinging forces that come together and — mechanically, as it were — give rise to a situation or event. Potency is thus both narrower than causation (due to the locality restriction) and at the same time broader (because of the inclusion of enabling conditions). It's important, furthermore, because something like potency (if not potency itself) is a necessary ingredient in the search for intentional foundations. More sophisticated notions of causation — ones that pack in such notions as relevance, long distance effect, enabling conditions, triggers, and so on — are ruled out, not simply because of their complexity, but because they stand as much in need of theoretic reconstruction as the semantical notions being defined in terms of them.[5]

Focusing on potency is important because it leads to a reformulation of (4):

(4′) *Potency* is intrinsic or local.

Rewriting (4) requires rewriting (6):

(6′) Semantics can't be *potent*. (from 4′ & 5)

This much isn't problematic. In fact (6′) seems a better distillation than (6) of the inexorable relatedness of semantic or intentional properties. But now a problem arises. In order to generate the problem (7), we also have to rewrite (2):

(2′) To be explanatory, something must be *potent*.

Do we want to do that? Or, rather: is Dretske prepared to accept that revision?

He should, of course, to generate his problem, since it is only the local (potent) version of causation, and its inherent proximity and immediate efficacy, to which the relational reach of semantics stands so pointedly in contrast. Furthermore (and more importantly), if he accepts this reformulation of his problem, he should also honour it in proposing solutions. That is, it would be unfair for Dretske to propose a solution (i.e., to claim that information is explanatory after all) by relying on a *non-local* form of causation. Similarly, if he discounts as *non*-causal something that is nonetheless fully and locally potent, on the grounds of its being, say, merely an enabling condition, that will also count against him. Dretske's goal (and we're playing by his rules) is really quite narrow: to demonstrate the local, potent effect of carrying information.

Identification of explanation and cause

Dretske's other problematic assumption is (2): the identification of explanation and cause. At issue is the relation between:

(a) *Explanatory* relations: between an event or situation A and other events B, such that A figures in the proper, naturalistic, scientific, intellectually satisfying, explanation of why B is the case; and
(b) *Causal* relations: between event or situation A and other events B, such that A causes B.

At the most general level, the two notions are clearly distinct (the etymology of "because" notwithstanding). Furthermore, to take just one obvious issue, if there are any explanations in pure mathematics, such as why there are only five regular convex solids,[6] then (at least on a Platonist construal) they must differ extensionally as well, since purely abstract objects presumably don't enter into causal relationships at all. So why are they being equated here? Dretske would presumably defend the move as constitutive of naturalistic reconstruction: whatever the case in mathematics (he would say), the goal is to give a *scientific* account of information, which (he would go on) means a causal account. That identification, furthermore, is made clear in his very first paragraph:

> Information isn't much good if it doesn't do anything. If the fact that an event carries information doesn't help explain the event's impact on the rest of the world, then, as far as the rest of the world is concerned, the event may as well not carry information. To put it bluntly, in the way positivists liked to put it, a difference that doesn't make a difference isn't really a difference at all. If an event's carrying information doesn't make a difference — *and by a difference here I mean a causal difference, a difference in the kind of effects it has* — then for all philosophical (not to mention practical) purposes, the event doesn't carry information.[7]

Similar statements permeate the paper:

> To put information to work will require understanding how the *causal efficacy* of a signal is altered by the fact that it carries information.
>
> Semantics or meaning, the what-it-is-we-believe (and want) is causally (*and, therefore, explanatorily*) irrelevant to the production of behaviour.
>
> How can anyone seriously doubt the *causal efficacy* of information and, hence, its relevance to understanding *why* some things turn out the way they do?

This is obviously a very strong position. Furthermore (this is going to matter), since it is a methodological commitment, not an empirical claim, it permeates Dretske's entire analytic stance.

As for reformulation (2′) — that to be explanatorily relevant information must be potent — nothing in the text directly supports it. On the other hand, even these few quotes suggest what his proposed solution makes clear: that the local, immediate, engendering of discriminable effect is exactly what he has in mind as a paradigmatic cause. So I'll call this a vote of acceptance.

Summary

Taken together, the four (revised) assumptions reveal Dretske's picture of the intentional landscape. One might have thought that local potency, long distance causality, and full-scale explanation lie on something of a continuum — ranging from the immediate or proximal to a realm of wide theoretic compass. On such a view, the salient naturalistic puzzle would be to show how such a continuum of "reach" or "inclusion" could arise. But of course that's not how

Dretske sees things. Instead, he presupposes a single, unproblematic, local notion of cause, and then identifies explanation with it. Having thus assumed that all three notions (potency, causation, and explanation) line up together, he asks whether information cannot also be drawn into the same small corner.

Dretske's Solution

Given this image of his project, let's turn to Dretske's proposed solution (8): that information's explanatory value arises in situations of learning. Just how is this supposed to go?

Structurally, as I said earlier, Dretske's plan is to deny (7), the claim that information isn't explanatory, by eliminating (6), the claim that semantics is causally impotent. That is, his solution will *be* a solution just in case it demonstrates a case where semantical properties are causally efficacious, after all. That much is clear. But there's still room for some tactical maneouvering. Conclusion (6) can be reversed by denying either of the two assumptions on which it rests: (4), that causation is local, or (5), that semantics is distal. Curiously (the plot thickens) Dretske's choice isn't clear. Sometimes he seems to opt for one, sometimes for the other.

The learning situation

The situation we're to imagine involves a rat, an audible tone, and a food dispenser. At an abstract level, we're looking for a case where a signal S's carrying the information that P *causes S to engender behaviour M*. So the example is parcelled up as follows: S is an internal state of the rat's brain, P is the sounding of the tone, and M is the pressing of a bar. Dretske must therefore show how the fact that the internal rat state (S) carries the information (P) that the tone is sounding explains why it (S, again) subsequently causes the rat to press the bar (M).

The intuition goes something like this. Even before the rat has "learned" in this way, it could (by presumption) still hear the tone. Dretske has chosen to register this by saying that the same state S occurred (at least in potential) before the learning situation, but did not then cause M. During the learning situation, however — through some combination of stumbling around and trial-and-error — things change appropriately, so that when it is all over, S *does* cause M.

There's nothing problematic with those facts, yet. But then Dretske makes his crucial move. He asserts that the change in S's causal powers comes about because of the semantic relation between S and P.

In order to understand this claim (I'll get to assessment in a moment), it's important to identify three relevant facts — part of what one might call the causal background (see Figure 1). The first has to do with S on its own: what causes it?[8] The answer is obvious: the tone's sounding — i.e., P. Turn on the sound, and S will result; turn it off, and S will go away. And they are causally coupled in the appropriate way. This is all true, furthermore, both before and after the learning situation. No information is needed.

Figure 1: The learning situation

The second question, symmetrically, is about what S causes — i.e., with its effects. Before learning, they don't amount to much of anything, we can suppose (maybe the rat simply turns the other way, if the tone is loud or unpleasant). After learning, on the other hand, S causes M — by hypothesis. Again, there's no need for information; the wiring (control circuits) will by now have been modified appropriately so that M simply happens.

The third question is about the effects of P. Here the answer is a bit trickier. Since (as I said above) P causes S, and (as I said just now) S causes M, by transitivity of "cause" it would be fair to say that, after learning, *P causes M*. It isn't an odd conclusion, either, if you think about it: the tone's sounding causes the bar to be pressed, in virtue of an easily imaginable sequence of appropriately potent states in the rat's head. The causal chain enters and leaves the rat, of course — but there's no reason to feel queasy about that. The rat simply comprises some "middle links" in a connected causal chain. Structurally, the situation is no different from Dretske's seatbelt case, where we would (perfectly happily) say that a person's sitting in the seat caused the alarm to go off, explained in terms of a sequence of potent relationships involving switches, wires, and electrical impulses.

So far, things seem rather simple. On the other hand, information hasn't intruded yet. And that's as Dretske plans it. Information, he claims, isn't implicated in causing S, nor in causing what S causes (i.e., M), but in causing *the change in S's causal powers* that takes place during the learning situation, leading it to (thereafter) cause M. His words: "the fact that S carries the information that P . . . explains the *recruitment* of internal elements as causes of movement." And he repeatedly talks of the "redeployment" of S's causal powers. According to Dretske, in other words, information has exactly the following causal role to play: to bring into effect the "higher-order" *adjustment* of the causal powers of a signal that carries it.

So the real thing we are being asked to believe — the putative solution that I want to examine — is the following claim: that whereas

S, M, and P are all causally efficacious states, with perfectly ordinary causes and effects (i.e., can on their own be satisfactorily explained without recourse to semantical notions), the *adjustment* in S's causal powers during the learning situation is the *causal* result (assumption 2) of the fact that it carries the information that P.[9]

Causal facts and causal explanations

I see three main problems with Dretske's solution. The first has to do, rather directly, with the causal structure of the situation. Think in particular about the learning scenario, complete with its pattern of behavioural modification. And then think about what Dretske doesn't: the food dispenser. Surely what causes the alteration is neither S itself, nor any information that S carries, but the (causally coupled, not just counter-factually correlated) *presence or absence of reward*.

There are at least two reasons to believe this is a better "causal" explanation of the net change than Dretske's. The first has to do with timing. Note that S itself can hardly be the cause of the change in its own causal powers — in part because (as Dretske himself says) it has the "same form" before and after learning. By the same token, however, S's carrying of information (which Dretske does single out as the relevant cause) is also temporally stable — true before, during, and after learning. If something is unchanged throughout a period, then it alone cannot be called on to explain why something particular happened during the middle.

What is needed, instead, is something (potent) that explains what is *different* about the learning situation. What might that be? The answer is surely obvious: the pattern of activity that includes the offering and retention of food. The rat's control circuits, speaking very roughly, are "dented" by the (perfectly potent) interaction between its hunger and its pleasurable reaction to food. *Nothing more need be said.*[10]

The second reason (why the reward system, rather than the information carrying, is responsible for the change in S's causal powers) has to do with the potent and quite palpable presence of P. The whole cast of characters, after all — S, M, and P — are all right there, front and center, flexing their efficacious muscles. Dretske sometimes seems blinded by his focus on the intentional fact (that, on his theory of information, S carries the information that P), and thus unable to see what may be much more important: the fact that S and P are not only both causally proximate (to S), but even causally coupled. In fact, as I've already pointed out, P is S's immediate cause. In light of all this potent proximity one is naturally led to wonder whether it isn't the *causal* relationships among S, P, and M (and the state of the food dispenser) that are doing the work, not the *semantical* relationships among them.

In fact — and this is really the essence of this first problem — you simply don't need intentional notions at all to explain, locally, what is going on in the situation Dretske imagines. Someone (the experimenter, presumably) causes the rat to be in the situation in the first place, wires up the tone and the food dispenser, arranges for the tone to sound appropriately, and so forth. And then, with the help of a little random exploration on the rat's part, together with a causally coupled tone and food dispenser, the entire situation can be (causally) explained in terms of the ordinarily efficacious states of bells, ears, air waves, neural circuits, food dispensers, and so on.

So far, in other words, it's not just that there's nothing for information to explain. There isn't even anything intentional going on.

Property identity and property overlap

So turn to that: the question of how intentionality enters the picture. This leads to the second problem. It has to do with just what Dretske thinks is the relationship between *carrying information* and *being causally coupled*.

In particular, one must not be misled by the fact that, in this special (learning) situation, *carrying the information that P* and *being causally coupled to P* are simultaneously true. That may be an important fact, but Dretske's conclusion doesn't follow from it.

The problem stems from a dilemma. Dretske can either claim

(a) That to carry information *is* to (have been) causally coupled, or
(b) That there is *more* to information than causal coupling.

Unfortunately, (a) is essentially vacuous. Reducing information to causation would automatically satisfy his desideratum of showing how carrying information can do work, but it would do so by evacuating the notion of information of all theoretical interest. *Carries the information that P*, on such a line, would simply be long-hand for *was caused by P* (as would *learning that P*). Premise (5), that is to say, would have been vitiated by the following trick: whereas *causing* something is local, *having been caused by* is distal. End of argument. But also end of interest. If intentionality amounts to no more than causation, then the naturalist's task wouldn't be to provide an explanation of information; it would be to eliminate the notion of information from intellectual inquiry.

So what about (b)? It too doesn't work, but for a more interesting reason. Dretske's analysis of information as causally dependent counter-factual supporting correlation rests on a necessary co-occurrence of information carrying and causal coupling (a co-occurrence that presumably happens early on, if not at the very beginning, of the information-carrying period). Quite strikingly, his current account of the explanatory value of information again involves a necessary co-occurrence, this time between information carrying and the complete causal coupling of S, P, and M that obtains during learning. However, in both cases the co-occurrence is (necessarily) a *sometimes* affair — something I'll call *property overlap*. And property overlap — even necessary (but still occasional) overlap — isn't property identity. The only thing that would even plausibly give you property identity would be constant co-occurrence. And, to belabour the obvious, it would only be property identity that would justify transferring the higher-order predicate "causally efficacious" from S's *coupling to P* to its *carrying the information that P*. (By analogy, crossing the country on Interstate 80 necessarily overlaps with being in New York City. And being in New York causes a [slight] increase in east coast smog. But it doesn't follow that crossing the country increases east coast smog. Being in New York is what does that.)

The importance of property overlap (even necessary property overlap) is that it opens up a middle territory between two properties being *identical* and two properties being *independent*. Consider another example: the size of Maine and the average size of a New England state. These two properties aren't what would normally be called independent. Not only is the average size not independent of the particular, but it would also be perverse (if not outright false) to claim that Maine's area was independent of the average. On the other hand, it's equally obvious that the two properties aren't identical.

So that's the second problem. It may be true that carrying information and being causally coupled necessarily overlap. It may even be *important* that they overlap (which I believe). Overlap, in fact, may even be *partially constitutive* of learning (though I'm less sure of this). But from none of these facts does it follow, even in the overlapped situations, that *carrying information* does any causal work.

Long-distance causation

Dretske's third problem, separate from subtleties of the learning situation, has to do with exactly what is being claimed. Remarkably enough, he doesn't distinguish the following two readings of his "solution":

(8a) Information is indeed explanatory, but only during learning situations.

(8b) Information, since it is causal during learning, is explanatory at other times.

Unfortunately, although he needs (8b), at best he has argued for (8a).

According to (8a), information does have a use: it lets people learn. Sure enough (the story would go), in ordinary, non-instructive situations — reasoning, say, or action — S's carrying the information that P wouldn't be able to do any work, and therefore might as well not even be true, which is too bad. Still, at least in a particular kind of instructive setting, information would have a definite, if limited, role to play.[11] This kind of learning, in other words, wouldn't be *what* gives information explanatory force; rather, it would be the only situation *when* it had explanatory force.

I take it that that would be an odd conclusion. First, if it were only during learning that information plays an explanatorily relevant role, then Dretske's own examples — of how strange it is to think that information is useless — would continue to be strange, since those examples don't involve learning. Second, by an argument analogous to the one given above about property overlap, it would be hard (read: impossible) for Dretske to argue that it was *carrying information* that was explanatory, instead of something narrower, such as *carrying information when still causally coupled during learning* (or even just: *being causally coupled*). And third, (8a) claims a modest role for information, but it's not a modest claim; if it *were* what Dretske had in mind, he would have admitted it (which he hasn't).

In fact, however, the text makes it clear that Dretske intends to endorse alternative (8b): that information is explanatory all the time, sustained (in some way) by its causal influence during learning. The intuition is presumably as mentioned in the introduction: that the causal coupling during the learning phase somehow causes the fact that the signal will (later) lead the interpreting agent to respond to it appropriately. That is, the intuition requires a long-distance notion of causal effect.

But this won't do, for a spate of reasons already cited. In order to avoid evacuating information of substance, there must either be (a) more to the long-distance reach of information than the mere historical record of prior causal coupling, or (b) more to long-distance causation than the footprint of local potency. But Dretske provides arguments for neither. All his efforts, as I've tried to show, have been dedicated to denying (5), the claim that semantics is causally ineffective. But this direction requires the opposite: eliminating (4), the claim that causality is local. And, as I suggested earlier, this isn't so easy. One would need an account of how explanatory force could stick to information-bearing signals, like pine sap to a wool sweater, long after the learning situation is over — an account, furthermore, that didn't invoke any notions of relevance, enabling conditions, triggers, or any of the other quandaries that plague attempts to develop an adequate theory of genuine causation, and yet amounted to more than a claim that there was once local causal connection. In addition, if that *were* how the argument was supposed to go, the whole problem would need restatement, since it was only in contrast to a *proximal* notion of causation (potency) that the original intuitions about semantics' relational nature had any bite.

DIAGNOSIS

Enough problems. Here's what I think is really going on.

Dretske, I believe, has three things: an insight, a claim, and a method. At least in its intuitive form, I believe the insight is correct (I also believe it is the same one that underlies his original account of information): that there is something crucial about the causal interaction between an intentional structure and its content.

In the present paper he locates this insight in situations of learning. In spite of this, however, so far as I can see, his analysis exploits nothing unique to learning, but would be true of any form of substantial participation in the subject matter, of which learning is merely an obvious case. For example, suppose while driving I maintain a sense of North by checking the dash-mounted compass every so often — more frequently when the roads are particularly twisty. There will be in my head the (maintained) information that *that direction is North*, for some internally oriented sense of "that direction." It overuses the term to call every instance of what is essentially an occasional servo mechanism "learning," but the properties of interest to Dretske still obtain: my internal orientation is differentially dependent on a causally connected chain of events tied to something that (at least in the arena of interest) is correlated with being North. And the differential result lasts. Perception is involved, of course — but action could be, too. The point is only that some form of causal coupling (that includes causal flow from world to agent) connects the agent to the semantic domain.

Based on his insight, however, Dretske states what I take to be an untenable claim: that (in consequence of this causal coupling) *semantical* relationships are endowed with *causal* powers. For all the reasons cited, I don't think this follows. In fact there seems to be a danger that

Dretske has the project backwards. He tries to show that an intentional relationship must exert causal force, whereas all that his examples demonstrate is that causal couplings are necessary ingredients in establishing intentionality.

The mistake, furthermore, is based on a common confusion. Dretske starts with the pervasive intuition that semantical relations aren't causally efficacious. And he concludes — as many people conclude — that this implies that thought, reasoning, computation, and similar intentional phenomena must proceed *independently* of their semantic value or content. This conclusion is reflected in the universal tendency to say (as Dretske does) that "you can change the semantic value, without changing the symbol."

This doesn't follow. If, as I've suggested, intentionality is inherently participatory (learning is just one example), then you can't necessarily perform that change. If some causal properties are even partially constitutive of intentionality, that is to say, then to change the semantic value may require changing those properties. This was the lesson of overlap. For example, the symbol might have to be propped up appropriately. Imagine a meter wired to a radar telescope indicating the exact current distance to the moon. We all know that this meter can fluctuate and flap around without being connected to the moon — as it was while being designed and tested, for example. But to make it indicate other things (as its designers and testers did during construction), you have to *wire it up to something*, since that's how it's got to be driven.

That is, there are cases — more or less, I haven't yet said — where various kinds of causal connection play an essential (but not total) role in engendering intentionality. But can we conclude from that that semantical relations are causally potent? *No!* Only that causality is — sometimes, and maybe even then only partially, but still, to some extent — constitutive of intentionality. That is, you can challenge the independence of semantical and causal relations, without thereby committing yourself to thinking that intentional properties are themselves causal — just as I am not my arm, though I am not independent of my arm, either.

The result, or claim, that causal coupling is partially constitutive of intentionality, is important, I believe. Terrifically important. It is legitimate to ask, however, whether it is *new*. Causal theories of reference, for example, are presumably founded on something like the same intuition. And so is the near-universal assumption (even Fodor believes this) that semantics is grounded in perception, action, and so on. What would be new would be a satisfying theory showing just what it came to, how it all worked. But none of this requires that we do what Dretske did: *reduce* information to causation. Nothing about how causation is partially constitutive of intentionality, that is to say, requires that we give up on what I take to be the deepest intentional insight of all:

> *Genuine semantical relations — including information — outstrip the locality of (whatever it is that we think we get at with the word) "causation."*

But does this maxim imply that information isn't explanatory, after all? *No, not that either!* And this brings us straight back to what I think is the deepest problem with Dretske's whole argument. The problem is with assumption (2): that explanation can reduce to cause.

Here, finally, we get to Dretske's method — and to the *reason* why he is forced to a causal reduction of intentionality.

If you restrict explanatory relations to causal relations, it's not just that Dretske has failed to show that information can "do work" — no one else could, either. That is, in the sense in which Dretske sets things up, information *can't* have any explanatory value. *But it isn't right to restrict things in this way.* In fact, to do so, I would argue, is not just reductionist, but fatally so. Fatal in that it doesn't make room for the recognition that local, causal couplings can, sometimes, lead to a system's overall exhibition of properties that *can't be locally identified.*

Various writers, worried about the relation between one level of explanation and another (often, between intentional and physical levels of description), have distinguished between *reduction* and *supervenience,* where in the first case the predicates and terms of the "higher-level" theory are directly translatable into the predicates and terms of the lower-level one, whereas in the latter they are not, even though the phenomena described by the higher-level account in some sense still "rise up out of" those described at the lower level. What Dretske has implicitly but nonetheless strongly done, by setting up his project in the way that he has, is to bypass the freedom offered by the supervenient line, and *endorse a full theoretical reduction of the intentional in advance*. Dretske, that is to say, is methodologically committed to reductionism.

Which I don't believe is right. For example, suppose a student asked you why two metronomes, mounted in the same (relatively rigid) base, and both obviously vibrating, weren't beating against each other, but were instead producing a single pure tone. By way of answer, it would be perfectly legitimate to say that they did so because they were synchronised. All that naturalism requires is that *you be able to tell a causal story that shows how that synchrony arose*. It does *not* require something stronger (and something impossible): that synchronisation can count in an explanation of the single tone *only if it can be shown to be a causally efficacious property* (which I assume it isn't).

So what are semantical morals? These three:

(1) Learning is important, but participation, of which it is a species, is what really matters.
(2) Causal coupling (and Dretske) notwithstanding, semantical reach still outstrips causal reach.
(3) Naturalistic explanation may rest on causal potency, but it doesn't reduce to it in anything like as direct a way as Dretske imagines.

Dretske, in other words, first reduces explanation to causation, and then tries (unsuccessfully) to do the same to information. I think he can't do the latter, and shouldn't do the former. So what should he do, instead? Well, there's what I take to be our collective homework. First, we need a naturalistically satisfying explanation of how intentionality manages to outstrip causation. With that in hand, we'll be ready to show how explanation can do so too.

NOTES

* Thanks to Kathleen Akins and David Chapman for comments on an earlier version.

1 Thus, a highway flare's *carrying information about an accident ahead* can't be causally responsible for your putting on the brakes. It can't be responsible because you would have slowed down, upon seeing the flare, whether or not there really were an accident. It's as if the accident itself, though intuitively relevant to why you slow down, is still too far away (in some appropriate sense) to be causally efficacious in making you do so.

2 In fact the situation is more complicated than this suggests. Dretske himself admits that the real issue — the question we're all finally interested in — has to do with the explanatory force of intentional or semantic notions more generally (including language, representation, meaning, content). He focuses specifically on information not only because he believes it is a relatively clear instance, but more seriously because, by his lights, it is the foundational case (as argued for example in his *Knowledge and the Flow of Information* [Cambridge, MA: MIT Press 1981]).

This intellectual cartography affects the assessment of Dretske's solution. To start with, the relevance of his specific analysis to the larger intentional question depends not only on specific concerns about its intrinsic viability (of the sort raised in this review), but also on whether information really does play the dintinguished role (among intentional phenomena) that he imagines. Furthermore, even if the reader agrees with this much — accepts, that is, that other intentional notions are derivative from information — a separate account of the explanatory value of those other intentional phenomena will still be required in cases where they don't manifest those properties of information on which Dretske's proposed solution rests. Representation is a good example: since it doesn't arise out of the sort of causal dependence that Dretske claims for information (cf. misrepresentation and representation of non-existent objects), and since Dretske relies on that causal coupling in showing information's utility, it follows that he hasn't even proposed an explanation of how representation can be explanatory.

3 Without a proper theory of intentionality and content, which of course we don't have, it is hard to make these claims precise. Still, the basic idea is simple: "meaning something" or "carrying the information that something" involves not only the sign, symbol, or information carrier, but also the referent, content, or interpretation. This will be denied, of course, by solipsists, nominalists, idealists, "social solipsists" (like Winograd), and any others for whom truth and reference amount to a social form of intersubjective agreement. Since Dretske's metaphysics are at least moderately realist, however, I want to address him from a comparable standpoint.

4 Ultimately based, I presume, on the locality of physics. It's possible, of course, that the rise of intentionality depends on the sorts of phenomena on which the EPR paradoxes shed such meagre light, but I doubt it.

5 My real worry about these more sophisticated notions is that they will turn out to be what one might call "post-intentional": themselves characterisable only in intentional terms (such as the suggestion that "causal" is a predicate on explanations, rather than being a pure ontological or metaphysical category), and therefore barred from playing a role in naturalistic reconstruction.

6 Consider a vertex. It must be formed by the intersection of at least 3 planar surfaces (2 would form just a sheet), each a corner of a regular polygon, such that the sum of the angles is less than 360° (360° would make the corner flat; more than that would start to buckle). Start with triangles (60° corners): 3, 4, and 5 are OK (tetrahedron, octahedron, and icosahedron), but 6 makes 360°, and more than that are precluded. Next comes squares (90° corners): 3 are again OK (cube), but by 4 you are already at the 360° limit. Similarly, 3 pentagons is OK (dodecahedron), but 4 is too large ($4 \times 108° = 432°$). By the time you reach hexagons, 3 (the minimum) is already at the limit (360°), and obviously no higher order figures will fit at all. So that's it: a total of five.

Now was that an *explanation*? Without taking on whole philosophies of science and mathematics, I would say it was. But it certainly didn't traffic in *causation* — at least not any form standing in obvious contrast to semantics.

7 The italics, in this and subsequent quotes, are my own.

8 Or causes it to be *activated* — it all depends on how you individuate mental states. Wherever possible I've tried to stick with Dretske's scheme of individuation.

9 In ordinary usage, the word "because" is ambigious as between *explanation* and *cause*, although the explanatory reading is probably more common. For Dretske, however, as codified in assumption (2), there's no room for two readings: explanation and cause have been identified. So when he claims that S's change in causal powers arises because S carries the information that P, he is committed to a causal reading: S's change in causal powers must come about as a causal effect of S's carrying that information. That is, should you have any tendency to distinguish the questions of *how* the change came about and *why* it came about, you should focus solely on the former. Dretske's claim that S's change in effect is explained by its carrying of information must mean that it was caused by that information carrying.

10 By analogy, suppose that instead of "rewarding" the rat by giving it food, the experiment was conducted with an electric rat, and the voltage were simply turned up when the tone was sounded (in fact turned up "by" the tone's sounding), so that the particular control path whereby the rat pushed the bar was "burned in" more deeply than others, and as a result (imagine some kind of internal contention network) was more likely to be chosen as a result of a series of "training" situations. What we would undoubtedly say caused the alteration in S's causal consequences, in this situation, over the course of the sequence of trials, would be the effects of the correlated higher voltages.

11 The situations that Dretske imagines, of course, are far from being *all* learning situations. Learning a skill, learning something on the evening news, learning a person's name — none of these would count as examples of what Dretske has in mind.

CHAPTER FIVE

Concept Formation and Particularizing Learning

Lee R. Brooks*

Cognitive psychology has devoted a large amount of its work on concept learning to understanding taxonomic categories, such as the groupings of everyday objects designated by familiar nouns, and "natural kinds" groupings of animals and plants. These categories appear to be a good testing ground for how we learn and use elementary generalities about the world around us. From countless exposures to dogs and cats and linguistic patterns, we evidently learn what is generally true of dogs and cats and the process of talking about them. As an approximate first story, it is reasonable to presume that this knowledge of general patterns is directly represented, that we have abstracted sets of prototypes or schemata that capture the experienced regularities. By this story, every time that we meet a new instance of a category, we understand it and elaborate it by referring to a central stock of information about that category. In turn, each instance contributes some small amount to the abstracted knowledge. The initial presumption, then, is that experience gives us an encyclopedia in the head with a consolidated entry for each concept, an entry that is consulted for and updated by each new instance of the concept.

In this paper I would like to argue that our conceptual resources are considerably more decentralized than is portrayed in this *abstractive* view, consisting in routine analogy to prior interpreted episodes as well as appeal to more abstract models and "automatized" identification procedures. Evidence that we will review suggests that for many everyday categories there is not a central core that is so stable and so available that we can use it to reliably characterize the operative knowledge that people have about that category. This is not denying that we do effectively have generalizations about the world; the importance of knowing general patterns in one's world is undeniable. However represented, conceptual knowledge serves crucial functions at every level of our cognitive economy. We can see a dog as a dog despite the animal being partially occluded by a table; we fill in unspoken assumptions about Fido, and have some reasonable expectations when a strange dog bares its teeth. The jobs of anticipation, correction, and filling in must be accomplished by some means. However, there is no reason why a single, decentralized store of knowledge is the only way of getting them done.

For a moment, imagine that most of our knowledge of a category is in the form of information about particular members of that category. When a new member of the category is met, we remember similarly encoded previous members of that category and base our presumptions about the new instance based on this local pool. If every known beast that seems similar to a newly encountered collie is loyal, brave, and true (as well as indulging in barking and eating meat), then these are probably reasonable presumptions about the new collie. If every dog that reminds us of slavering Fang was bad news on four feet, then the attendant information about eating meat takes on a new light. At a radical extreme (which I am not proposing), we might literally never directly store a generalization about the category and yet act as appropriately as if we had. The distributional properties of a concept (e.g., frequency, characteristic features, alternative likely attributes) can remain distributed in the knowledge base and be exploited by this process of forming immediate local models as the need arises.

The advantage of thinking about conceptual knowledge as distributed across instances is *not* the behaviourist predilection for being stingy in attributing abstractive ability to people, an ability which we undeniably do have. Rather, the advantage is that distributed knowledge gives us a principled way of dealing with interactions and variability among the instances of a concept. Our best generalizations about the world are usually at best approximations to the world, and if our specific experience suggests that there are occasionally important special circumstances, we need some way of responding to them. A well-established principle in cognition is that knowledge of the particular takes precedence over general principles; our particular knowledge about ostriches overrides the generalization that birds fly. The current suggestion is that this principle has far more work to do than is immediately apparent. As evidence that will be presented later indicates, there are interesting circumstances in which our knowledge of the particular is not a last resort but rather overrides quite adequate generalizations.

To anticipate an argument to which I will return later, this distributed knowledge of a concept could be thought of as analogous to the notion of a species. Clearly a species has a central tendency of attributes, but few biologists in this century have maintained that this is a result of all members of a species having been produced from the same template. Rather, individual animals are produced by individual members of the species that on average have a certain central tendency. Similarly, our behaviour with respect to different instances of a concept may show a strong central tendency without our having to assume that it is because all of these behaviours were elaborated from the same abstracted knowledge about that concept. New items can be coded by analogy to prior exemplars of the category that on average have a certain central tendency. In addition, as evolutionary biologists have stressed, there is as interesting a story in the *variety* within a species as there is in its central tendency. Creationism concentrated on the central tendency, the normative form of a species, and did not develop a story for the function of deviations from that norm. But in the eyes of the evolutionists, genetic variety was seen as a defense against uncertain future demands of the world. A variety of genetic variations in a species gave insurance that at least some members of that species would survive in a change in environmental pressure. Similarly, one protection that we have against peculiar interactions in the world is our knowledge of very similar past instances, instances that themselves may be the product of the same interactions and may be poorly captured by whatever generalizations we currently have. Possibly an important part of our knowledge of a concept, like the genes in a gene pool, is fundamentally distributed across individual instances.

The argument for the potential explanatory power of relying on distributed knowledge, on a pool of prior interpreted instances, has been made in psychology several times before (e.g., Brooks 1978; Medin & Shaeffer 1978; Hintzman 1986; Kahneman & Miller 1986; Logan 1988; Ross 1984, 1987). A comparable shift from an exclusive reliance on abstracted knowledge has been evidenced in other areas of cognitive science in work on case based reasoning (e.g., Shank 1982; reviewed in Kolodner 1988). What I would like to add is (a) an emphasis on the experimental power resulting from making an analogy with the research on episodic memory (Brooks 1987) and (b) that some of the requirements for coherence of concepts (Murphy & Medin 1985) can be met with a distributed knowledge base. These problems and arguments will be developed later. But first, let us turn to the evidence that suggests the need for complicating a reasonable first story: the abstractive view of conceptual knowledge.

THERE IS A PROBLEM TO BE EXPLAINED

There are several lines of evidence that indicate that an abstractive view of categories has difficulty explaining people's use of concepts in areas where it should do a good job. Let me briefly describe three of them.

Natural categories

When psychology's interest returned to everyday categories after an extended sojourn with artificial categories, natural taxonomic categories were given considerable attention. As ably argued by Rosch and her colleagues (e.g., Rosch 1978; Rosch & Mervis 1975), many everyday categories are not all-or-none: instead, some exemplars are taken as much better, more typical members than are others. For investigating these concepts, many investigators concentrated on goodness-of-exemplar ratings and on the listings that people gave of the attributes that are characteristic of these categories. By these ratings, robins were generally rated as more typical birds than were ostriches, and robins had more of the attributes that were rated as characteristic of birds. These ratings were generally taken to be fairly direct readings of knowledge abstracted over the many instances that people had experienced in their lives.

In an interesting set of studies, Larry Barsalou and his colleagues (reviewed in Barsalou 1987) demonstrated that goodness-of-exemplar ratings for a set of objects showed considerable variance depending on the concurrent task demands. For example, if subjects simultaneously had to give reasons for their rankings, then the rankings were very different than if they did not. Explicit properties given for good and bad category members tended to be ideal properties rather than the highly typical properties that might be expected from an averaged prototype interpretation of goodness-of-exemplar gradients. Furthermore, considerable variance was found both within and between subjects, again a finding that would not be expected if the ratings were mainly dependent on stable abstractions of natural structure.

Collectively, Barsalou's results suggest that the concept of knowledge operative at any given moment is not exclusively a steady organization, but rather is strongly influenced by the particular context and the purpose for which it is being employed. If Barsalou had found dependence on context and purpose solely for ad hoc categories (such as the category of all things you would take from the house if it were on fire), then there would be little surprise; but when such effects are also found for some of the most traditional natural kinds categories (such as birds or clothing), question is raised both about the goodness-of-exemplar rating task as a measure of stable prototypical structures and about the sufficiency of such knowledge structures in the first place.

Ad hoc norms

Kahneman and Miller (1986) and Kahneman in a paper presented to a previous meeting of this conference have demonstrated that in many domains the norms that people utilize to judge the surprisingness, adequacy, or value of current events do not seem to be as stable or as flexible as an abstractive account would suppose. Their evidence included attributions of causality as well as affective states such as surprise, frustration, and regret; all cases in which it was extremely implausible that stable norms could be accounting for the effects (they covered a sufficiently wide variety that I am not going to review them specifically here). To explain this evidence, they proposed that perceived objects and events recruit representations of similar objects and events that include a considerable amount of remembered interpretation. Norms for judging the current situation are then constructed from the elements provided by these representations and from elements in the current situation. The norms that are constructed are influenced by the availability of prior instances as well as principles that dictate the relative mutability of aspects of the current situation. In their phrase, the current experience selectively recruits its own alternatives that in turn help to shape the context in which it is eventually interpreted and evaluated. This backward, postcomputational generation of norms is a clear contrast to the precomputation of expectations and concept norms that has been focused upon in recent cognitive psychology.

Development of medical expertise

The third line of work that I will describe is particularly useful for the discussion that will follow in this paper. Recently, Norman, Rosenthal, Brooks, Allen, and Muzzin (in press) investigated the development of expertise in medical expertise. Many medical diagnoses, particularly in the highly visual domains such as dermatology, are good examples of "natural kinds," taxonomic categories. Textbook rules generally have sets of characteristic features that give cumulative evidence for the disorder, and these features often are not critically embedded in causal models linking the signs to the underlying disorder. The textbook rules sound remarkably like the natural kinds characterizations such as "it's a dog if it is a medium sized animal, normally domesticated, that walks on four legs, barks, and wags its tail — although no one of these attributes is essential."

An intuitively plausible position about the acquisition of expertise with such medical categories can be called the "independent cues interpretation": learners gain expertise mainly by acquiring knowledge about the specific signs or symptoms that characterize a disease and are best able to differentiate it from other diseases. (The term "independent cues" follows Smith & Medin 1981 in their classification of theories of concept learning). One consequence of an independent cues model is that performance should improve more rapidly on typical than atypical cases. Since typical cases possess more of the features which are characteristic of a category, these should be mastered with relative ease. Conversely, atypical cases have few features in common with a category and hence would require a high level of expertise to differentiate from other conditions. Based on this model, one could also expect that the difficulty of a particular case is related to the degree to which it embodies features which are predictive of a single disorder. If this is true, then cases which are empirically relatively easy for neophytes should be dealt with perfectly by experts. This independent cues interpretation then, is a straightforward application of an abstractive interpretation of conceptual knowledge to the acquisition of expertise.

The domain we studied was the diagnosis of common skin disorders on the evidence provided by colour slides. The stimulus materials were 100 slides, consisting of five slides chosen from each of twenty common skin conditions, with two judged by a dermatologist to be typical presentations and three atypical presentations. Six subjects were chosen at each of three levels of expertise in dermatology: first-year residents in family medicine, general practitioners, and practising dermatologists. Subjects were asked to diagnose each case as rapidly as was consistent with accuracy. The effect of expertise was strongly evident in a 3:1 drop in errors across expertise: the average error rates were residents = 44%, general practitioners = 33%, dermatologists = 14.5%.

To assess the effect of typicality, the number of errors was formed into the ratio of errors on typical items divided by total errors. This ratio was adjusted for frequency, so that an equal tendency to make errors on typical and atypical items would result in a .5 value. Learning to deal effectively with typical items before atypical items would result in declining values of this ratio with expertise. But in fact, the proportion of errors made on items designated as typical by the dermatologist was approximately constant over the three levels of expertise, despite the threefold decrease in overall errors (R = .40, GP = .42, D = .40). Whatever was improving across expertise was independent of typicality, and given that typicality is supposed to be strongly correlated with the predictors of a category, this was very surprising.

To assure ourselves that these results were not just a result of some peculiarity of the typicality ratings, we explored the effect of item difficulty on expertise. To make this comparison, we characterized the difficulty of each slide on the basis of the errors committed by residents: An easy slide had 0/6 errors by residents, a difficult slide had 6/6 errors by residents, with 5 intermediate levels of difficulty. We then examined the proportion of errors made by the general practitioners and the dermatologists on these difficulty groups (as defined by the performance of the residents). An independent cues model implies that (a) the difficulty of a case is related to the degree to which the cues present in the item support a single diagnosis, and (b) expertise is related to the knowledge of the appropriate combinations and weightings of these cues. Thus, we would anticipate that as expertise is acquired, proportionately more errors will be committed on difficult slides, so that the distributions or errors by slide difficulty would shift with expertise. If errors were a result of random processes such as inattention, the likelihood of an error by a GP or dermatologist should be unrelated to the resident item difficulty, and the distributions of proportions of errors should be flat. In fact, the distributions for general practitioners and dermatologists were very similar to those of residents, that is, an approximately straight line with positive slope. The data provided no evidence of the anticipated shift toward a greater proportion of the experts' errors being on the items that were most difficult for the residents.

In these data the traditional indices of category structure — typicality and average item difficulty, are roughly constant across a large change in overall accuracy. Remember the reasons why we might have expected a drop in the proportion of errors committed on easy and typical slides as one acquires expertise. If diagnosticians learn early on to deal with items that are on average easier, that display more and clearer signs of the disorder, and later learned to deal with the more difficult and marginal cases, then we would have found a decreasing typicality ratio across expertise. Conversely, if some of the slides lacked sufficient clear signs to make a diagnosis reliably, then we would expect even experts to continue making mistakes on these empirically difficult and atypical slides. Errors on such slides consequently would constitute an increasing proportion of total errors as the diagnostician became more expert, yielding a change in the distribution of errors with item difficulty as expertise increased.

We conclude that the improvement over the range of expertise observed in this study is not a matter of learning the identity of or more appropriate weights for the essential symptoms and signs. In other words, these data are incompatible with an independent cues interpretation of acquisition of expertise. In fact, the observed constant proportionality rules out any model that determines typicality, average item difficulty, and improvable error by the same information. Obviously, we are not claiming that there is no such thing as "independent cues" knowledge of the features of particular diseases, or that such knowledge is irrelevent to expertise. The subjects in this study had already acquired the "basics," generally by means of rules that were phrased in terms of independent cues. What we are claiming is these data constitute a case in which there is massive improvement that is independent of such knowledge. It is entirely possible that the acquisition of the basic definitional, independent cues type of knowledge is restricted to the initial stages of learning.

As in the case of Barsalou with natural categories, and Kahneman and Miller with norms, we are looking for a source of knowledge that is independent of the automatic abstractive mechanism that has been so influential in cognitive science in the form of schemata, prototypes, and frames. As indicated in the previous section, an attractive possibility is the variable contributions of previously experienced instances. Considering the medical material again, the previous occurrence of particular cases in a series may result in the increased availability of that case as an analogy for similar subsequent cases. This influence could be manifested in either the initial consideration of that diagnosis as an hypothesis or by a decreased probability of catching the error if the suggested hypothesis is in fact wrong. Because these cases might be available due to recency or because of context-specific cuing, these biasing influences could be highly variable. Such variable contributions could affect overall performance without necessarily changing the *average* difficulty of typical and atypical, easy and hard items. The contribution of these influences could be expected to decline with increasing expertise, either because the coding of the current cases becomes less variable, or because the analogous prior cases are more likely to be in the correct category however they are coded. In either case, the proportionate contribution of typicality or average item difficulty could be constant because the change across expertise was in the variable contributions of previously experienced instances.

ROLE OF PRIOR EXEMPLARS IN RULE-GUIDED DOMAINS

In each of the three lines of evidence just described, expectations from stable, abstracted knowledge interpretations of concepts did not fare well. The next step in arguing for the importance of distributed knowledge resources is to provide evidence that makes the contribution of prior instances plausible, even in domains in which there are known useful rules for categorization.

Specialization of perfectly predictive rules

Imagine that we gave subjects a perfectly predictive rule, such as that illustrated in Figure 1 for classifying imaginary animals. As long as the rule worked unproblematically, subjects might rely solely on the rule, making previous experience with other instances irrelevant. If, on the other hand, the classification of some instances was not predictable by

FIGURE 1: An example of the experimental materials used in the Allen and Brooks experiments. According to the rule given, the training items shown in the left column are a "builder" on top (long legs, angular body, spots) and a "digger" on the bottom (only long legs). The "good analogy" appears only in test and according to the rule is in the same category as the most similar training item. The "bad analogy" is similar to one of the training items, but according to the rule is in the opposite category to that item (spots in addition to long legs makes it a builder).

the rule, we could easily imagine that they would stand as special cases, and the application of the rule would be overridden in new instances similar to them. The issue here is whether change or specialization of a rule is occasioned by difficulty and error (e.g., Anderson 1984; or Holland, Holyoak, Nisbett, & Thagard 1986) or simply by the availability of prior instances that are retrieved by the current cues. Given the prevalence of explicit rules in formal instruction, the conditions for specialization of rule application are of interest.

Experiments relevant to this issue were designed by Scott Allen (Allen & Brooks, submitted). As illustrated, the stimuli to be learned were fictional animals that varied on five binary dimensions (e.g., 2 legs or 6, long neck or short, etc.) and which could be classified into one of two categories ("builders" or "diggers") by a three-dimensional additive rule: an animal was in a category if it had at least two of the three features that were predictive of that category. The animals were presented on a set of distinctive backgrounds, such as a snowy forest, by a pond, or in the desert. Each background was consistently paired with one digger and one builder, so that the backgrounds were not in themselves predictive of the category, but did help to make individual animals distinctive.

On the first presentation of each animal, the subject was required to apply the rule as rapidly as possible to predict the category; since this task was done with virtually no errors during the first trial, the subjects evidently could apply the rule. Eight animals were presented in training, and each of the eight animals was categorized five times. The intent of the training sequence was to give the subject perceptual experience with a specific set of "old" targets as well as practice with the categorization rule. As intended, the training task was quite easy, with virtually no errors being made and classification times on the initial slides getting down to the range of 1/4 second.

The test sequence was designed to assess whether the subjects had specialized the rule in the sense that old stimuli and stimuli similar to them (New Similars) were treated as special cases. To do so, some of the test items were animals that differed in one *relevant* feature from old animals. In the first half of the test sequence, this change left the new test animals in the same category as the most similar old animal (good analogies); in the second half of the test sequence, it changed the category of the new test animal relative to the most similar old animal (bad analogies). In general, the results indicated a considerable amount of specialization of the rule. The good analogies were classified at approximately the same speed and accuracy as the old items (12% error), while the bad analogies were classified more slowly and with considerably more errors (45%). If classification were being done solely by a speeded application of the rule, then one would not expect more errors to the bad analogies than to the good; in both cases the classification could have been made perfectly by attending to the relevant features — features that had been present in each of the eight possible combinations during training.

However, we do not mean to imply that the rule was having no effect — that it was simply abandoned at an early stage. This was made clear by the results of a control group that was not given the rule at any point during the procedure, but which was treated identically in all other respects. Subjects in this No Rule group had to rely on feedback during the training to determine the animal's category, but they were virtually errorless by the end of the training phase. These subjects were slightly slower and less accurate than the Rule group on the good analogies, but made 85% errors on the bad analogies. This result would be expected if they were responding solely on the basis of analogy to old items, but not if they had been implicitly abstracting out a rule. The result also indicates that the subjects in the Rule group had in fact been partially relying on the rule for accuracy, since they had been correct on a majority of the bad analogies.

For the current argument, a useful finding was that when the backgrounds on which the test animals were presented were re-paired with the animals, classification times increased but the errors to the bad analogies went down. This suggests that the animals had been integrated with backgrounds during training, such that changing the categorically irrelevant backgrounds disrupted the analogical basis of performance and encouraged more reliance on the rule. Including a number of these unfamiliar items in the test sequence decreased the influence of prior instances on the bad analogies because the whole retrieval context had been disrupted. Thus, the effect of prior analogies can be variable, depending on the retrieval circumstances and on irrelevant background features.

Finally, the effect of the bad analogies was reduced only slightly by instructions stressing accuracy. Even when subjects were told to be wary of bad analogies, their response times increased dramatically, but still resulted in twice as many errors on bad analogies as on good. The effect of similar prior items is not limited to circumstances in which the participants are pushing for speed. In this case at least, knowledge of the particular is not a last resort.

In all, this experiment shows a case in which experience with prior instances influenced performance on similar new items despite knowledge of an adequate, explicit rule. Knowledge of the rule clearly helped to maintain accuracy, but it did not prevent the influence of instance and context-specific knowledge. One could object that the training conditions were designed to produce perceptual specialization, but that is exactly the point. Conditions in which there is a considerable amount of distinctive item information and in which the task encourages integration between the focal object and the context are sufficiently common in the world that they should be investigated. Certainly many medical specialties deal with individual cases that are individually quite distinctive, a point which is the subject of the next set of investigations.

Effect of prior exemplars in a medical domain.

Dermatology was again the domain of these studies. They are a collaborative effort, with an initial report in Allen, Brooks, Norman, and Rosenthal (1988). In the first of these experiments, a group of general practitioners saw photographs of dermatological lesions and were given the task of rating them for typicality. In this initial phase, one photo of each of three different cases of the same disorder were presented simultaneously, together with the correct label for that disorder. Each of the eight disorders represented in this phase was among the twenty common dermatological disorders presenting in North American practice. The primary purpose of the first phase was to give the GPs experience with particular, correctly labelled cases of a variety of disorders. In Phase 2 of the experiment, the GPs were given a series of photos, each of which were accompanied by a list of three possible diagnoses. One of these diagnoses was correct, one was a relatively plausible differential (as determined by errors made to the same items in the Norman et al study described above), and the third was a relatively implausible differential. The GPs were asked to rate all three diagnoses for plausibility. Some of the photos used in this phase had already been seen together with the correct label in Phase 1 (Old). Some of the items were photos of new cases drawn from the same disorders represented in Phase 1 (Same Name). Finally, some of the items were drawn from 8 diagnostic categories not represented in Phase 1 (New).

If the GPs were uninfluenced by their prior experience with the cases shown in Phase 1, then we would expect them to rate the correct diagnosis as most plausible with about the same frequency in each of the three picture type categories. For the Old items, the correct diagnosis was rated as the most plausible on 81% of the trials, which is no surprise, given that those items had been correctly identified (on an average of) less than twenty minutes earlier. However, for the Same Name items the correct diagnosis was selected 67% of the time, which is significantly different than the 56% correct trials for the New category items. Apparently having considered the disorder earlier in the session increased the probability of selecting the correct diagnosis for another instance of that disorder.

The advantage enjoyed by the Same Name cases corresponds to a rather common (in our experience) and cynical anecdote about a physician who says "The most amazing coincidence happened. Last night I read about a newly identified disorder in the *New England Journal*, and the first patient I saw today had it!" The point of the anecdote, of course, is that increased availability in memory of the new disorder can bias the diagnostic task in favor of deciding that the current case is an instance of the disorder (potentially independently of the relative difficulty or typicality of the case itself). The fact that such an anecdote is in circulation suggests that people are quite aware of the possibility of such bias. However, this in itself does not indicate that the effect is frequent enough to produce an effect of practical size. Furthermore, going beyond the anecdotal level, the effect in this experiment occurred despite the fact that the physician always was given the correct diagnosis as one of the three alternatives. Whatever is meant by saying that the correct alternative is "more available" is stronger than saying that the physician just failed to ever consider it; the correct alternative was always one of the three that had to be explicitly rated. This feature in itself suggests that the remedy for these contextual effects is not as simple as to say "consider other alternatives to the first diagnosis that comes to mind."

However, the study just described does not directly demonstrate that the effect is due to available instances rather than increased availability of the category as a whole. A direct demonstration of the effect of prior instances was shown in a set of three studies done with a slightly different design. The subjects in these experiments were first given a set of rules for diagnosing six common dermatological categories. In the first phase, the subjects were given several trials of working through a practice set of twenty-four colour slides, four from each condition, diagnosing each slide and receiving feedback. The experimental manipulation involved the choice of slides used in the practice set. Two sets of practice slides were used, with the intention of biasing the diagnosis of later ambiguous items in the test series. For example, if one of the ambiguous items in the test series was a case of lichen planus on the wrist which might be confused with contact dermatitis, then one learning set would contain a similar lichen planus on the wrist, and the other would contain a similar appearing contact dermatitis on the wrist. However, in both cases there were an equal number of prior items in each disease category. Thus, if we observed differences between the two groups in performance on the critical test items, this must be a result of differences in the similarity of practice items rather than availability of the category as a whole.

In all three studies, the critical items were diagnosed in line with the most similar training item. The difference in accuracy between the test items when they had been biased by a previous similar slide in the same category and when the similar slide was in another category ranged from 47% to 14% across the three studies (all statistically significant). Some quick points about the studies:

(1) In one study the effects were obtained on general practitioners (14% difference), all of whom had had substantial prior experience with the categories. The effect of prior instances is not limited to low levels of expertise.

(2) Two of the studies were done on medical students to whom we had just taught the rules. The effects were obtained despite the fact that the rules (taken from standard dermatological texts) were sitting beside the students throughout the experiment.

(3) Three of the critical slides (all individually showing the effect) had previously been rated as typical of their disorders by three dermatologists. These effects are not limited to marginal items.

(4) One of the studies was done with a week between the training and the test phases of the experiment. The effect is not particularly short-lived.

(5) In two of the studies, the verbal description of the critical test items was in line with the prior similar item. This means that (for example) the same feature was being described as a vesicle, a papule, or omitted altogether depending on which of the training slides had been used for that subject. Thus, the usual (confirmatory) process of "checking" was not sufficient to eliminate the observed bias.

The cumulative effect of all of these studies is to demonstrate that there can be an effect of prior instances on later categorization of similar items despite (a) extensive prior experience in the field, (b) the presence of rules that are completely adequate (for the animals) or authoritative (for the medical studies). The categorizations made in both the medical studies and the studies with the artificial animals cannot be fully predicted by either the explicit rules or by any implicitly averaged rules, given that the effects are produced by analogy to single specific items in the current or preceding test day. Finally, the sizes of the effects are not marginal. Particular prior instances have to be considered as part of the resources these subjects were using to solve categorization problems. This raises the issue of how we should think of the representation of these kinds of taxonomic concepts. But first, there is one more theme that has to be addressed.

PROCESSING SPECIFICITY AND CONTEXT-SPECIFIC KNOWLEDGE

The major potential strength of considering the influence of specific prior instances is in the explanation of variability of performance. This has only been partly exploited in the research just described. The size of effect and range of generalization around an exemplar depends both on the way that exemplar was processed and on how those conditions of processing fit with the conditions at the moment of categorization. We cannot assume just because the learner can recognize having previously seen an item, that it is now generally available to guide current categorization. The extent to which prior instances were processed as whole integrated units and the tasks in which they were embedded are important conditions for determining when they will be available for nonanalytic generalization — it is these variations that will eventually give us the best argument for considering the influence of specific exemplars.

The importance of these processing variations is well illustrated in a seminal experiment by Jacoby (1983a) on word identification. Word identification is a task, like classification, that is often thought to be accomplished by well-practised routines abstracted over many prior episodes of identification. The purpose of Jacoby's experiment was to show the effect of a single prior processing episode on the later perceptual identification of a word; that is, on the later identification of a word that was briefly presented and followed by a visual masking pattern. The conditions in the first phase were designed to give subjects differing types of prior experience with the words that were later to be identified. In this phase, subjects were asked to do all of the three tasks illustrated in Figure 2. (a) *No Context*: On one third of the sixty trials, a set of four X's appeared on a video monitor. After three seconds a word appeared that the subject was simply to read aloud. (b) *Context*: On another third of the trials a word appeared followed after three seconds by its antonym. Again the subject was simply to read the second word aloud. However, since the antonyms were selected to be easily generated by the subjects, the word was being read in a context in which it was easily anticipated. (c) *Generate*: On the other third of the trials a word appeared that was followed after three seconds by a set of four question marks. The subjects had been instructed to respond by generating the antonym of the first word. In this case the antonym never appeared on the screen.

In the second phase of the experiment subjects were given a perceptual identification test; that is, these sixty words, together with

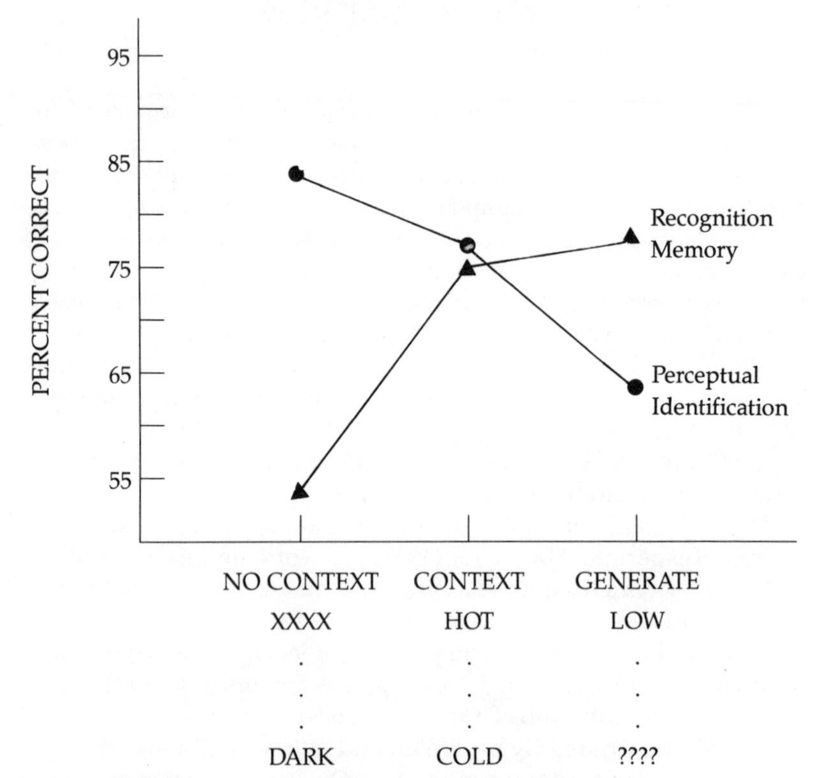

FIGURE 2: The training procedure and results of later test for Jacoby's antonym experiment (adapted from Experiment 2, Jacoby 1983a). The items used for both the later perceptual identification and the recognition memory tests were always the second member of one of the training sequences.

twenty new words were presented one at a time for about 35ms followed by a mask consisting of ampersands. As can be seen in Figure 2, variations in the way the words were processed in the first phase of the experiment made a considerable difference in later perceptual identification. The order of effectiveness of the conditions makes sense if we assume that the later perceptual identification task is sensitive to the amount and type of perceptual processing carried out in the first phase. In the Generate condition the target word was not presented visually at all, with the apparent result of little or no later facilitation of words from this condition over new words. In the Context condition, the target words were presented visually, but in the context of a cue that probably reduced the amount of perceptual processing necessary for identification. The No Context condition probably produced the maximum amount of practice in perceptually processing the target and clearly resulted in the maximum transfer to later perceptual identification.

The order of results is very different for a later recognition memory task. Subjects (the same subjects in one variant of the experiment, independent subjects in another) were given a list consisting of the same eighty words (sixty old and twenty new) and were asked to check off those words that had occurred in the first phase. In this case, as in many previous "levels-of-processing" experiments (e.g., Craik & Tulving 1975; Jacoby & Craik 1979), the best performance was obtained in the conditions that encouraged the most distinctive (usually meaning-based) elaboration. The Generate condition required the subject to work with the meaning of the word; the Context condition encouraged but did not require comparable meaningful elaboration; and the No Context condition did not even encourage meaningful or distinctive processing. As with the prior levels-of-processing results, the amazing aspect of these results is that the later retrieval of familiar words is so sensitive to these minor task variations. Subjectively one does not have the feeling of having processed the words in the "shallow" (No Context) condition any more or less distinctively; the fact that they are all meaningful words is quite apparent at the moment. Again, one might have expected for both the perceptual identification and the recognition memory measures that the processing of these familiar words might have been so automatized that few differences would have been found.

Let us now consider several implications of these results.

(1) *Rapid specialization of a well-practised skill.* These subjects obviously had general, definitional knowledge about the words. Since the words were reasonably common, the subjects would have had ample prior opportunity to have automatized a general, visual identification procedure. However, there evidently was still some benefit from having previously seen these particular words under these circumstances, embedded in this particular list (the amount of facilitation is decreased when most of the words in the test list are new; Jacoby 1983b — which was the precedent for our manipulation of a comparable variable in the artificial animal experiment shown in Figure 1). Furthermore, this specialization does not require a great many previous trials; a single prior trial was sufficient to produce a considerable benefit. This result is obviously very similar to the results on medical diagnosis reported previously in this paper.

(2) *Long-lasting facilitation (not temporary priming).* This perceptual facilitation has been demonstrated five days later (Jacoby 1983b)

and, in the case of pseudowords, a year later (Salasoo, Shiffren, & Feustal, 1985). It thus seems reasonable to describe the effect as due to episodic learning rather than to a relatively temporary priming of a stable node or identification routine. Again, this result foreshadows the effect of a prior dermatological case after a week's delay.

(3) *Specialization was not occasioned by extreme difficulty*. Obviously any model that stresses the importance of stable schemata provides for longterm changes of the schemata with changing circumstances. However, in keeping with a principle of economy, such changes are usually assumed to occur only when trouble is encountered. But, Jacoby's subjects could hardly have had much difficulty in identifying the words in the first phase of his experiment; the words were common English words, presented in the clear for about 1 second. As with the classification of artificial animals, specialization seems to be more at the behest of availability of local precedents than of experienced difficulty.

(4) *Perceptual facilitation is not reliably indexed by recognition memory.* There are at least two reasons why the dissociation of perceptual identification from recognition memory performance is of interest. One is that our intuitions are probably schooled mainly by the more "aware" types of memory tasks. A recognition memory task, for example, explicitly asks us to reflect on prior events, a process that, as we are all too aware, can be quite incorrect on the "unimportant surface details." As a result it seems implausable that we could be getting much help from memory of such details in specific prior episodes. However, we get a very different view of the potential contribution to be made by episodic memory for perceptual features from the performance measures in which the episodic contribution is incidental rather than explicitly noticed. As Jacoby's antonym experiment demonstrates, we can be quite unable to recognize such a contribution on a trial-by-trial basis (for further discussion on "remembering without awareness" see Jacoby & Witherspoon 1983). The second reason for interest in this dissociation is that it is tempting to use accuracy of recognition memory as an explicit index of the potential contribution of prior episodes to a categorization process (as was done, for example, by Reber & Allen 1978). Again, Jacoby's experiment is a clear demonstration of episodic effects on an identification task that are not well indexed by recognition memory. We need not be aware of the effects of specific prior episodes for them to have an effect.

An unaware influence of prior episodes is also illustrated by

Scott Allen's artificial animal experiment shown in Figure 1. Subjects from one of the experiments were questioned regarding which items gave them problems and why they thought they might be faster on some items than on others. They were shown pairs of items consisting of one good analogy item and one bad analogy item and told they answered faster on the good analogy item. They were asked why they thought this might be the case. Only ten of the twenty-four subjects ever mentioned similarity to an old item in the opposite category as a possible reason for the relatively poor performance on the BA items. Furthermore, those subjects who were counted as mentioning analogies were those who made any mention of analogies at any point in the discussion, often after considerable prompting from the experimenter. Thus, even using the most generous of criteria, over half the subjects failed to place the blame for their poor performance on the BA items on the fact that they were BA items. They generally picked features of the animal that seemed to be plausible causes. For example "this one had a long neck so I must have thought it had long legs and called it a builder" and "this one had long curvy legs and so I guess I got them mixed up with its body which is angular and called it a builder by mistake." Occasionally the subject couldn't find a plausible reason in the slide and would say "I guess I was trying to answer too fast," which of course is not a reason for *relative* decrements in performance. The lack of awareness of the error source as well as the subjects' inability to protect themselves from similarity to previous items when told about it suggests that analogy to old items may be an intrusive process that doesn't require a deliberate decision to occur.

(5) *Task specificity of transfer*. To return to Jacoby's experiment, the most important point for our current purposes is that the effect of instances is very dependent on the exact nature of the transfer task and the exact coding that was done when the instance was first encountered. Any account that relies on "the better it's learned, the better it's retrieved" cannot handle these results. For example, the notions of associative strength or of deeper or more intensive processing are by themselves insufficient. Rather, it is necessary to look more closely at how the specific requirements of the retrieval task fit the prior processing, again a theme common in the memory literature (e.g., the relations between recall and recognition). In the recognition memory task, for example, one of the learner's problems is to distinguish the occurance of the target word in the first part of the experi-

ment from its occurrence in (say) a previously read novel; after all, the subjects are not being asked if they have *ever* seen these common words before. For this purpose, elaborating one aspect of the meaning of the word in the experimental context might provide a more discriminable target in memory. No such discrimination between prior and experimental occurrences is necessary for performing the perceptual identification task. Instead, what seems more relevant is specific practice with the "bottom-up" processing of very similar cues in the prior episode.

It is this last point which provides the demonstration of the potential power of attending to specific prior episodes of processing when thinking about classification and identification processes. It is a common observation that there is variability in a person's efficiency and accuracy on classification tasks. But this variation is usually not addressed because it is not immediately interpretable in terms of stable conceptual knowledge. This processing specificity approach provides interesting possibilities for accounting for some of the variation, variation that (at least in the case of medicine) can be of real interest. Demonstrations of these same types of effects have been made in the case of picture identification by Jacoby, Baker, and Brooks (in press) and for a version of the word superiority effect in Whittlesea and Brooks (1988).

We have not yet fully applied the processing specificity paradigms to medical classification. We have had some indications of the importance of variations in retrieval context: if there were a lot of familiar old items in the test list in the experiment shown in Figure 1, then bad analogies had a stronger effect on the classification of new items than if most of the test items were unfamiliar. But the real test will be when we deliberately vary the mode of original processing in combination with the conditions of testing.

The Species Concept Of Concepts

Throughout this paper, I have been arguing for the substantial effects of specific prior instances on later classification and identification performance with similar items. What implications should these instance effects have on the way we characterize the knowledge underlying this performance? One possible reaction is that these effects of prior interpreted instances should be attributed to performance variables, leaving the core concept untouched. But what are we to say about the knowledge that was used by the dermatologists in making their diagnoses? If we insist on only allowing the stable knowledge about each disorder to be called the diagnostician's concept, we rule out whatever the source of knowledge is that allows a 3 to 1 reduction in error rate over that produced by residents. Recall that the average item difficulty as assessed on the performance of the residents did an excellent job of predicting the proportions of errors made on the same slides by the dermatologists, an effect which would not have occurred if one of the expertise groups were using systematically different average knowledge or if they had been mixing in different amounts of random processes. Further, the rated typicality of the items, which is often taken as a measure of the stable evidence provided by the item, was also independent of this 3 to 1 drop in errors. As argued earlier, these effects mean that improvable error, average item difficulty, and typicality cannot be produced by the same knowledge.

Suppose, as the instance effects and personal testimony of the physicians indicate, that some reasonable portion of diagnosis and case management are informed by analogy with prior cases, a knowledge base that *could* have effects independent of abstractive generalizations. If this were the case, we would rule too much out if we did not include these interpreted instances in the knowledge base that represents "the concept." It seems to me that if we try to maintain the unity of a conceptual core by only acknowledging stable, abstracted knowledge, we leave out too much of the knowledge that is used in the normal applications of that concept. The strength of the instance effects require us either to reduce the scope of influence of a personal concept, or to incorporate a large number of specific instances as part of the construct. Obviously, I am arguing for the latter course.

But, incorporating a large conglomeration of specific cases into the construct "concept" raises another problem. In what sense is a set of essentially separate prior cases sufficiently coherent to be called a single structure? How could such a collection give rise to the impression of coherence that we have about natural concepts? These questions were raised by Murphy and Medin (1985) and were one of the bases of their contention that natural language concepts should be thought of as having much of the integrated structure of theories. It is certain that an integrated theory could account for the impression of conceptual unity and the more efficient learning that usually accompanies concepts judged to be coherent. But, again there is no reason to presume that this is the only way of getting it done. I believe that conceptual coherence can also be handled by a distributed model and not suffer the difficulty in accounting for variability that an integrated theory must face.

Earlier in the paper, brief mention was made of an analogy between a species and a distributed conceptual representation. In both cases the output of the process could show clear central tendency without each member of the output being generated by the same proximal source. That is, new members of the species are generated by specific parents rather than by an overall template of the species, and by analogy, new concept members could be coded by analogy to specific prior instances rather than with reference to a single, integrated source of information about the concept. The idea of a gene pool is obviously useful for an observer in accounting for population statistics, but no one reifies this explanatory device into a literal mechanism for generating specific animals. By analogy, behavioural regularities for which the individual cannot give a rule, may also be explanatory fictions; the "unconscious" conceptual knowledge typified by linguists, anthropologists, and cognitive psychologists may be useful ways of typifying the person's knowledge rather than a description of a unified causative agent. When subjects say that they don't have rules, they may be right in a more fundamental way than just saying that they are not aware of them.

The point for the current discussion is that the distributed nature of the gene pool in a species is not generally taken as grounds for denying that a species is a coherent construct; there is considerable organization and coherence to a species that results in specific predictions on both a population and an individual level. On the same grounds there seems no necessary reason to say that a distributed concept is incoherent. To push the analogy, solution of common problems posed by the world and instance-guided coding could serve the roles of natural selection and interbreeding respectively. That is, not every individual nor every analogy is a success in solving problems posed by the world. Similarly, new individual codings are generated with reference to particular previous instantiations that are not identical with all other members of the group, but which in principle have the capacity of mixing properties with all other members of the group (with the limitations of the extreme members of a cline). We could think of various instances as acting in shifting groups in informing the coding of new instances as well as (breaking the analogy) the generation of new generalizations.

By a further reach we could think of context-specific retrieval as analogous to the isolation that produces genetic drift. If only a subpopulation of members of a concept chronically come to mind in a particular context, then the population of solutions may itself drift.

This process would most obviously produce a new "species" when the learner consciously distinguished the "two senses of the term" — that is, acknowledged that what originally seemed to be the same concept was now being used in two distinguishable senses. This analogy to genetic drift is strained, but something like "nonanalytic speciation" of concepts probably does happen. I, at least, sometimes find myself using observing differences in my own behaviour that seem to have been produced by using originally what was originally the same idea in different circumstances without a special effort at monitoring.

The obvious limit to the species metaphor is the existence of explicit, operative generalizations in our knowledge. The operation of analogies are often rejected by current analytic criteria (i.e., operative generalizations) rather than poor outcome ("natural" selection). The evolutionary analogy to such analytic editing, I guess, would be a species practicing eugenics on itself (not a totally impossible analogy, given social relations within species, but not one my conscience will currently allow). More important, clearly analytic intellectual functioning in general is poorly captured by the metaphor. Still, the benefit of invoking the metaphor is that by default too much credit is given to explicit analytic knowledge and its counterpart, apparent implicit generalizations. There is a tendency to believe the generality of any analytic statement, ignoring the frequency with which they can be overriden by specific prior examples and the tendency to invoke different analytic statements in different contexts.

The application of the genetic analogy to the particular issue of coherence is all that is new (to me at least). The general analogy has already been exploited in the work by Holland (1975) and by Holland, Holyoak, Nisbett, & Thagard (1986). The structure of genetic material and the resulting structure of a population were used as means of developing a progressively more fit population from which classification rules could be produced. Their emphasis seemed to be on the production of solutions to regression or classification problems in the form of successful weights or explicit inductive statements. My current interest is more in the adequacy, for some purposes, of the distributed knowledge as a model of a concept — a notion that is clearly inherent in their approach.

Regardless of the rhetorical value or limits of the "species concept of concepts" metaphor, any success of distributed knowledge models requires a modification of the concept of conceptual knowledge. The species metaphor is really an argument for redressing the current imbalance between discussions of analytic and of nonanalytic conceptual resources.

Co-ordination Of Analytic And Nonanalytic Thought

The relations between analytic knowledge and the nonanalytic, instance-based knowledge emphasized in this paper are dynamic. We are continually specializing general principles into context-specific understandings, which will become differentially available in similar situations. But we also occasionally crystallize this drift into new generalities. Conceptual consistency may be more a result of good policing than of different cognitive acts being generated from unitary cognitive models or context-free generalizations. Concentrating mainly on the accrual and use of our abstracted conceptual resources is too limited to account for the variation that has been observed and is of direct interest in some fields.

Ultimately we have to deal with the way in which these two modes of control and organization of knowledge are co-ordinated. This is obviously beyond any of the experiments that we have currently done. But in closing, let me speculate on two different forms of co-ordination that seem plausible:

(1) *Analytic control.* Sometimes we clearly limit the operation of analogies by current analytic criteria, by clear rules. In addition, deliberate analysis of currently available instances results in new rules which can also be used in analytic control. But even deliberate analysis is performed on particular materials, and similarity can influence the materials in the pool to be analyzed. Clearly the knowledge bases are unlikely to be exclusive in most applications, but at times the control can be dominantly analytic and deliberate.

(2) *Theories embedded in specific episodes.* The past instances described in this paper are clearly interpretations of events, not just the raw events themselves. This implies that most instances, as encoded, contain suggestions about modes of analysis, which is another way of saying that the similarity based retrieval of instances is likely to also be the similarity based retrieval of a theory. If similarity were solely based on theoretical criteria, this would be an uninformative circle. But since similarity judgments can also be influenced by surface variables that are irrelevant to the person's operative theories about the domain (e.g., the role of instances in dermatology), this process can produce real variance. Another way of saying this is that the operative form of knowledge may be a variety of successful instances interpreted in the light of different models — not just the bare and contending doctrines. Attending to what contexts are embedded in the coding of an instance could give

an important clue to which of the contending doctrines the individual will invoke in the current situation.

Seeing through the petty details to concentrate on the structure of a problem domain is one of the glories of human thought. But the other side of cognition demands respect for the details. Respect for details, "the small things that make the difference," is the guts of expertise. And, it is those guts of expertise that is poorly represented in the current preoccupation with economical and analytically powerful semantic nodes and mental models.

NOTES

* The preparation of this manuscript was carried out with the support of the Natural Sciences and Engineering Research Council of Canada.

The figures in this chapter were originally published in L.R. Brooks, "Decentralized Control of Categorization: The Role of Prior Processing Episodes," in U. Neisser (ed.), *Concepts and Conceptual Development: Ecological and Intellectual Factors in Categorization* (Cambridge, Eng.: Cambridge University Press 1987), 145, 153.

REFERENCES

Anderson, J.R. (1984). *The Architecture of Cognition*. Cambridge, MA: Harvard University Press

Barsalou, L.W. (1987). The instability of graded structure. In U. Neisser (ed.), *Concepts and Conceptual Development: Ecological and Intellectual Factors in Categorization*. Cambridge, Eng.: Cambridge University Press

Brooks, L.R. (1978). Non-analytic concept formation and memory for instances. In Rosch, E. and Lloyd, B. (eds.), *Cognition and Categorization*. Hillsdale, NJ: Erlbaum

— (1987). Decentralized control of categorization: The role of prior processing episodes. In U. Neisser (ed.), *Concepts and Conceptual Development: Ecological and Intellectual Factors in Categorization*. Cambridge, Eng.: Cambridge University Press

Craik, F.I.M. and Lockhart, R.S. (1972). Levels of processing: A framework for memory research. *Journal of Verbal Learning and Verbal Behaviour* 11: 671–84

— and Tulving, E. (1975). Depth of processing and the retention of words in episodic memory. *Journal of Experimental Psychology: General* 104: 268–94

Feustel, T.C., Shiffrin, R.M., and Salasoo, A. (1983). Episodic and lexical contributions to the repetition effect in word identification. *Journal of Experimental Psychology: General* 112: 309–46

Hintzman, D.L. (1986). "Schema abstraction" in a multiple-trace memory model. *Psychological Review* 93: 411–28

Holland, J., Holyoak, K., Nisbett, R., and Thagard, P. (1986). *Induction. Processes of Inference, Learning, and Discovery*. Cambridge, MA: Bradford/MIT Press

Holyoak, K. and Thagard, P. (in press). Analogical mapping by constraint satisfaction. *Cognitive Science*

Jacoby, L.L. (1983a). Remembering the data: Analyzing interactive processes in reading. *Journal of Verbal Learning and Verbal Behavior* 22: 485–508

— (1983b). Perceptual enhancement: Persistent effects of an experience. *Journal of Experimental Psychology: Learning, Memory and Cognition* 9: 21–38

— and Brooks, L.R. (1984). Nonanalytic cognition: Memory, perception and concept learning. In G. Bower (ed.), *The Psychology of Learning and Motivation: Advances in Research and Theory*, Volume 18. New York: Academic Press

— and Craik, F.I.M. (1979). Effects of elaboration of processing at encoding and retrieval: Trace distinctiveness and recovery of initial context. In L.S. Cermak & F.I.M. Craik (eds.), *Levels of processing and human memory*. Hillsdale, NJ: Erlbaum

— and Witherspoon, D. (1982). Remembering without awareness. *Canadian Journal of Psychology* 36: 300–24

Kahneman, D. & Miller, D.T. (1986). Norm theory: Comparing reality to its alternatives. *Psychological Review* 93: 136–53

Logan, G.D. (1988). Toward an instance theory of automatization. *Psychological Review*

Medin, D.L., and Schaffer, M.M. (1978). Context theory of classification learning. *Psychological Review* 85: 207–38

Murphy, G.L. and Medin, D.L. (1985). The role of theories in conceptual coherence. *Psychological Review*

Reber, A.S., and Allen, R. (1978). Analogic and abstraction strategies in synthetic grammar learning: A functionalist interpretation. *Cognition* 6: 193–221

Rosch, E. (1978). Principles of categorization. In E. Rosch & B. Lloyd, (eds.), *Cognition and Categorization*. Hillsdale, NJ: Erlbaum

Ross, B.H. (1984). Remindings and their effects in learning a cognitive skill. *Cognitive Psychology* 16: 371–416

— (1987). This is like that: The use of earlier problems and the separation of similarity effects. *Journal of Experimental Psychology: Learning, Memory, and Cognition* 13: 629–39

Salasoo, A., Shiffren, R.M., and Feustal, T.C. (1985). Building permanent memory codes: Codification and repetition effects in word identification. *Journal of Experimental Psychology: General* 114: 50–78

Smith, E.E., and Medin, D.L. (1981). *Categories and Concepts*. Cambridge, MA: Harvard University Press

Tulving, E., and Thompson, D.M. (1973). Encoding specificity and retrieval processes in episodic memory. *Psychological Review*, 80: 352–73

Whittlesea, B.W.A. (1987). Preservation of particular experiences in conceptual representation. *Journal of Experimental Psychology: Learning, Memory & Cognition* 13, 387–99

— and Brooks, L.R. (1988). The priority of particular experiences in the perception of letters, words, and phrases. *Memory & Cognition* 16, 387–99

Comment
Information and Concepts

Paul Thagard

This commentary concerns the relation between the nature of information and the nature of concepts. Most of the papers at the conference on which this volume is based discussed the topic of information, while Lee Brooks' presentation primarily addressed the question of the nature of concepts. After describing how these two topics are interrelated, I shall make some critical remarks about Brooks' proposal.

What is information? A scan through the literature suggests that there are at least three different notions of information to be found in the literatures of computer science, cognitive psychology, and philosophy. First, we have the communication-theoretic notion of information developed by Claude Shannon (1949). Information here is a property of a *signal*, as indicated by the equation:

$$I(s) = \log 1/p(s).$$

The information carried by a signal is the logarithm (to the base 2) of the reciprocal of the probability of s, so that the signal is deemed to carry more information the less probable it is. The use of base 2 logarithms means that information can be measured in bits. This notion of information has been put to epistemological use by Fred Dretske (1981).

In psychology, information theory using the communication notion of information was influential in the early 1950s, but the prospects of applying the mathematical notion of information waned with George Miller's (1956) highly influential discussion of "the magical number seven, plus or minus two." Miller drew together numerous studies that showed that the channel capacity of human thinking seemed to be very limited, with short-term memory, for example, restricted to around seven items. How, then, can people manage complex thinking with such restricted channels? Miller proposed that channel limitations are overcome because *chunks* of information, rather than bits, are what matter. By recoding stimuli into chunks, people overcome the informational bottleneck. Hence to understand thinking we need to look not just at bits of information but at the processes by which they are structured into chunks and at the nature of these encodings.

Around the same time, Newell, Shaw, and Simon (1958) were developing what they called an "information-processing" approach to thinking. Starting with the Logic Theorist, a program that proved theorems in formal logic, they developed ideas about how thinking could be viewed as directly analogous to the processes newly being developed on digital computers. It is clear from their writings that "information" has only the vaguest relation to Shannon's mathematical notion. Modern cognitive psychology is often referred to as "information-processing" psychology, again with little relation to the mathematical notion of information. In a recent cognitive psychology textbook, John Anderson (1980: 13) characterizes information as the "various mental objects operated on." Another textbook simply identifies information with "knowledge." Thus information-processing psychology, in keeping with Miller's notion of chunk but not with Shannon's mathematical notion, treats information as largely a matter of mental representation: computational structures in the minds of thinkers.

Although it has come to dominate cognitive psychology, the information-processing view has not been accepted in all quarters. Ulric Neisser (1976), following Gibson's "ecological" approach to perception, has been critical of computational models. He prefers to emphasize the presence of information in the *world*, advocating an *ecological* approach to the study of cognition. Similarly, the philosophers Jon Barwise and John Perry speak of information as a property of facts or situations (Barwise & Perry 1983).

It makes no sense to ask in the abstract which notion of information — the mathematical, the processing, or the ecological — is correct. Ideally, we would like to have a unified theory that could show how information can be found in (1) representations in minds and computers, (2) the world, and (3) channels of communication between minds and/or computers, and from the world to minds and/or computers. Such a unified theory would go a long way to providing answers to epistemological puzzles about how mental representations can possess content and be meaningful.

My purpose in pointing out these disparate views of information, however, is not to pursue this daunting project, but merely to situate Brooks' ideas about concepts in the context of concerns about information. His concerns about the nature of concepts have no relation to the communication-theoretic and ecological approaches, but have potentially important implications for concerns about mental representation. Brooks rejects influential information-processing accounts of mental representations based on centralized, abstracted models. He emphasizes instead the importance of specific, concrete episodes to thinking. Whereas for many cognitive scientists, the "various mental objects" postulated have primarily been general representations like rules and concepts, Brooks stresses the importance of particular instances.

For those readers not familiar with psychological controversies about the nature of concepts, let me briefly review. What is a concept, for example the concept *dog*? A survey by Edward Smith and Douglas Medin (1981) discusses three theories of the nature of concepts. On the classical view, concepts are characterized by necessary and sufficient conditions. The concept *dog* is defined by a set of conditions that apply to all and only dogs. Various experimental results, particularly the research on prototypes initiated by Rosch (1978), cast doubt on whether our mental representations are at all definition-like. Many researchers have argued that concepts should be understood as prototypes, specifying typical rather than definitional features. On this view, the concept *dog* is specified by sets of features belonging to a prototypical dog rather than a necessary and sufficient set of features.

Brooks' views seem closer to what Smith and Medin call the *exemplar* theory of concepts. Concepts are internally represented, not by definitions or by sets of typical features, but by instances of them. Your concept *dog*, for example, is established by the many past instances of dogs you have observed and stored away. I will discuss the adequacy of these views shortly.

In addition to the classical, prototype, and exemplar theories of concepts, we should also attend to two more recent computational accounts. The PDP (parallel distributed processing) view of concepts sees concepts as emergent entities in neural networks (Rumelhart & McClelland 1986). Concepts are patterns of activation over neuron-like units. The patterns arise as the result of training the network with examples: weights between units are adjusted to enable the network to categorize examples more accurately. The PDP view provides a computational account that cuts across the prototype and exemplar views of concepts. It is closer to the exemplar view by virtue of the role that training on instances plays in adapting weights on links between units to produce the patterns of activation, but since these patterns abstract considerably from the particular instances and can be expected to produce a kind of average or prototype, the PDP view approximates to the prototype view of concepts.

Also having mixed implications is another computational view of concepts developed by Keith Holyoak and myself (Holland, Holyoak,

Nisbett & Thagard, 1986; Thagard 1988). This view, implemented in the processing system PI, treats concepts as complex structures involving hierarchical relations such as kind-of, default rules stating typical features, and instances of the concepts. The concept *dog*, for example, is in PI a structure that has a kind relation to the concept *animal*, sub-kinds relations to *collie*, etc., default rules such as "If x is a dog then x barks," and instances such as Lassie. In accord with the view of concepts as prototypes, default rules are ones that state what holds typically of concepts. A dog that does not bark can nevertheless be a dog, although it is highly typical of dogs to bark. In addition, however, this view of concepts can potentially make use of examples, since instances of dog are stored with the concept as well.

In order to adjudicate between different theories of concepts, we need to have an idea of the full range of phenomena that a theory of concepts must account for. Let me attempt a list that will undoubtedly turn out to be incomplete. Psychological accounts of the nature of concept have unfortunately restricted most of their attention to problems of categorization, and it will be useful to look at a broader range of the functions of concepts.

What roles do concepts play in thinking? I see at least nine functions of concepts.

(1) *Categorization*. Our concept *dog* enables us to recognize things as dogs.

(2) *Memory*. Our concept *dog* should help us remember things about dogs, either in general or particular episodes that concern dogs.

(3) *Deductive inference*. Our concept *dog* should enable us to make deductive and inductive inferences about dogs, e.g., enabling us to infer that since Fido is a dog, he barks.

(4) *Generalization*. Our concept *dog* should enable us to learn new facts about dogs from additional examples, e.g., to form new general conclusions such as that small dogs bark with a high pitch.

(5) *Analogical inference*. Our concept *dog* should help us to reason on the basis of similar objects.

(6) *Explanation*. Our knowledge about dogs should enable us to generate explanations, e.g., saying that Fido barks *because* he's a dog.

(7) *Language comprehension*. Our understanding of sentences such as "Fido is a dog" depends on our knowing something about the concept *dog*.

(8) *Language production*. We need to be able to utter sentences like "Fido is a dog" or "Dogs are more friendly than cats."

(9) *Learning*. Our concept *dog* must be capable of being learned, perhaps from examples, or perhaps by combining other existing concepts.

We currently lack a theory of concepts that can handle all these roles. I want to show, however, how some of them are much less consistent with Brooks' view of concepts than others.

First, however, let me point to the most salutary elements of Brooks' discussion. He describes several experiments that show that concepts possess considerable flexibility and sensitivity to context. These results suggest problems for any view of concepts that characterizes them in terms of rigid defining or typicality conditions. Clearly, one cannot understand the full range of conceptual applications without taking into account how specific examples can be recruited to enhance processing in specific contexts.

There are additional lines of research not emphasized by Brooks that seem to me to support his general conclusion. Much research in cognitive science now concerns analogical reasoning, where past episodes, rather than general rules, are used to solve problems (see, for example, Holyoak and Thagard, 1989). Here, as in the research discussed by Brooks, specific instances are what guide problem solving. In philosophy and history of science, a related point was made by Thomas Kuhn (1970), who argued that scientific knowledge cannot be understood without appreciating the importance of *exemplars*, concrete examples that govern the thinking of a community of scientists. No theory of cognition could be complete without ascribing to concepts a role that includes the functioning of specific instances.

But it would be a grave mistake to argue on this basis that concepts should not be hypothesized to have general elements as well. The full-fledged exemplar and distributed representation views of concepts look very good for cases where specific instances are what matter, but seem much less plausible for some of the uses of concepts that appear to depend much more on generality: explanation, deduction, and perhaps language comprehension. Much recent research argues for the importance of the *theoretical* component of concepts, which becomes very difficult to capture if one is restricted to the instances view (Murphy & Medin, 1985). Brooks, of course, does allow for the presence of what he calls "analytic" as well as "non-analytic" thought (these terms have nothing to do with the analytic-synthetic distinction) so perhaps the difference between us here is merely one of emphasis: I agree with him that non-analytic thought has been relatively neglected in cognitive research, but still think he is exaggerating its generality.

Like most of the research on concepts done by cognitive psychologists, Brooks' discussion also seems unduly to concentrate on concepts represented by nouns. But many of our most important concepts are represented by verbs, like "walk," "eat," "select," "think," "give," and so on. Whereas it is not hard to imagine a major part of our concept dog being represented in memory as specific instances of dog, it is much harder to recall specific instances of walking, eating, selecting, and thinking. Many researchers in lexical semantics (for example, Miller & Johnson-Laird 1976) have tried to characterize some of the general features of verbs, for example their selectional restrictions. Such lexical knowledge seems very general and of a different kind altogether from knowledge of particular instances. It would be interesting to see whether concepts represented by verbs are also subject to as much context sensitivity as the concepts investigated by Brooks and others.

As a final criticism, let me say that Brooks' attempt to set up an analogy between concepts and species seems to me to be unhelpful. He wants to argue that just as a species is a collection of individuals, so is a concept. I have general reservations about such biological analogues for cognition (Thagard 1988: ch. 6), and this one strikes me as particularly inappropriate. Unlike concepts, which have to play roles in inference, explanation, and language comprehension, species do not have causal roles independent of their individuals. Unless Brooks can make a stronger case that the example-based reasoning that seems to affect categorization pervades all the other uses of reasoning, there is no reason to move over to a species conception of concepts.

In sum, Brooks has usefully pointed out some of the ways in which specific examples and episodes are important for the operation of concepts, but we must not neglect the general aspects of concepts that seem equally important in other kinds of information-processing.

References

Barwise, J., and Perry, J. (1983). *Situations and Attitudes*. Cambridge, MA: MIT Press

Dretske, F. (1981). *Knowledge and the Flow of Information*. Cambridge, MA: MIT Press

Holland, J., Holyoak, K., Nisbett, R., and Thagard, P. (1986). *Induction: Processes of Inference, Learning, and Discovery*. Cambridge, MA: Bradford/MIT Press

Holland, J.H. (1975). *Adaptation in Natural and Artificial Systems*. Ann Arbor: University of Michigan Press

Holyoak , K. and Thagard, P. (1989). Analogical mapping by constraint satisfaction. *Cognitive Science* 13:295–355

Kuhn, T. (1970). *Structure of Scientific Revolutions*, 2nd ed. Chicago: University of Chicago Press. First published 1962

Miller, G. (1956). The magical number seven, plus or minus two: Some limits on our capacity for processing information. *Psychological Review* 63: 81–97

– and Johnson-Laird, P. (1976). *Language and Perception*. Cambridge, MA; Harvard University Press

Murphy, G.L. and Medin, D.L. (1985). The role of theories in conceptual coherence. *Psychological Review* 92: 289–316

Neisser, U. (1976). *Cognition and Reality*. San Francisco: W.H. Freeman

Newell, A., Shaw, C., and Simon, H. (1958). Elements of a theory of human problem solving. *Psychological Review* 65: 151–66

Rosch, E. (1978). Principles of categorization. In E. Rosch and B.B. Lloyd (eds.), *Cognition and Categorization*. Hillsdale, NJ: Erlbaum

– and Mervis, C.B. (1975). Family resemblances: Studies in the internal structure of categories. *Cognitive Psychology* 7: 573–605

Rumelhart, D.E., McClelland, J.R., and the PDP Research Group (1986). *Parallel Distribution Processing: Explorations in the Microstructure of Cognition*, 2 vols. Cambridge, MA: MIT Press

Schank, R.C. (1982). *Dynamic Memory*. Cambridge, Eng.: Cambridge University Press

Shannon, C. and Weaver, W. (1949). *The Mathematical Theory of Communication*. Urbana: University of Illinois Press

Smith, E.E., and Medin, D. (1981). *Categories and Concepts*. Cambridge, MA: Harvard University Press

Thagard, P. (1988). *Computational Philosophy of Science*. Cambridge, MA: Bradford/MIT Press

CHAPTER SIX

Information And Representation

Jerry Fodor

This paper has two parts. The first part says: "Look, there is this terrible problem about information-based semantic theories" (often hereafter "IBSTs"). The second part sketches a solution to the problem that the first part raises. I'm sure the first part is right: there really *is* this terrible problem about IBSTs. I'm far from sure that the second part is right, but I guess I think it's worth considering.

A TERRIBLE PROBLEM ABOUT IBSTs

I want to take some things for granted.

First, I assume that information-based semantics is primarily an attempt to *naturalize* the fundamental semantic relations; that is, to say in *non*semantic and *non*intentional (with a "t") terms what makes a thing a symbol. IBSTs thus play the same role in our theories of language and mind that Hume's suggestion that Ideas are about what they resemble played in his. In both cases, the point is to break out of the "intentional circle"; to show how to replace semantical/intentional talk with talk couched in the familiar vocabulary of the natural sciences *salve* the explanatory power of semantic/intentional theories.

Second, I assume that mental representations are the appropriate domain for IBSTs. Hume was right: it's Ideas (in the sense of mental particulars) that have semantic properties in the first instance. Mental states (beliefs, desires, and so on) are typically relational, and they derive their semantical properties from the Ideas that are their immediate objects; forms of words in natural languages derive their semantical properties from the mental states that they are standardly used to express. So, for example, "It's raining" means what it does because it is the form of words that English speakers standardly use to express the belief that it's raining; and the belief that it's raining has the intentional content it does because its object is a mental representation that means that it's raining. The question thus arises what the fact that a certain mental representation means that it's raining consists in. An information-based semantics purports to provide naturalistic answers to this sort of question.

But though I assume that mental representations are par excellence the symbols of whose intensional properties information-based semantics provides a naturalistic theory, *none of what I'm about to say depends on that assumption*. Though I almost always have mental representations in mind when I speak of symbols,[1] all the following applies equally well to the view that it is forms of words that have semantical properties in the first instance and that therefore provide the primary candidates for naturalization. I am told that there actually are philosophers who hold this view. Since there is the duck-billed platypus, this may well be so.

Third, I take it that all IBSTs claim that the fundamental intensional property of a symbol is that of *carrying information*.[2] "Carrying information" is a relation that is best introduced by examples, so here are some popular ones. In typical cases: smoke carries the information that there is fire; a tree's rings carry information about its age; a falling thermometer carries the information that it is getting cold; utterances of the form of words "that is a platypus" carry the information that that is a platypus; and so forth.

Finally, I assume that it's common ground among IBSTs that "carrying information" is to be naturalized by reference to relations of *causal covariance* between symbols (viz. the information carriers) and things symbolized (viz. the things that the information carried is about). For our purposes, the Urversion of IBST is: As carry information about Bs iff the generalization "Bs cause As" is true and counterfactual-supporting. A couple of words about alternative formulations of IBST, and then we'll be ready to go.

(1) There's a (Stanford) version of IBST according to which *causing* is viewed as just a special case of constraining, the general principle being that things carry information about whatever they constrain or are constrained by. So, for example, smoke carries the information that there is fire; but also, fire carries the information that there will be smoke, and the proposition that P&Q carries the information that Q. I shall be concerned with this generalized notion of information-carrying only insofar as it overlaps the narrower one.

(2) Dretske has a very highly constrained notion of information-carrying according to which it's false that As carry information about

Bs unless the probability that an arbitrary A is B-caused is always one. The main argument for this constraint is this: Suppose we allow that As carry information about Bs even when the probability that As are B-caused is *less* than one; suppose, for example, we accept a probability of .9. Then there could be a situation where "P" carries the information that P, "Q" carries the information that Q, but "P&Q" does *not* carry the information that P&Q (because the probability that P&Q is less than .9).

But I think this argument is ill-advised. There's no reason why an IBST should assign informational content independently to each expression in a symbol system. It will do if contents are assigned only to the *atomic* expressions, the semantics for syntactically complex expressions being built up recursively, by techniques familiar from the construction of truth definitions. So, if we have a (naturalistic) way of saying that "P" carries the information that P and that "Q" carries the information that Q, we can leave it to a recursive schema to assign the information that P&Q to "P&Q." I will therefore stick with the version of IBST that makes As informative about Bs whenever the causal connection from Bs to As is backed by a counterfactually supporting generalization, albeit one that may be statistically imperfect.

So much for preliminaries: Now for the terrible problem. Here, by way of introducing it, is a very small argument against information-based semantics.

Information-based semantics is the idea that "means" is univocal in " 'smoke' means smoke" and "smoke means fire" (viz. "means" means "carries information about" in both cases). It is, in fact, because they hold that smoke really does *mean* fire, that informational semanticists are inclined to insist on how much meaning there is around. It's a favourite refrain of IBST theorists that there is nothing *all that* special about linguistic or mental representation; they are just species of the sort of causal constraint that is ubiquitous in the nonlinguistic, nonmental world.

But prima facie this can't be right. If "means" means "carries information about" in " 'smoke' means smoke" and "smoke means fire," then since *carries information about* is transitive, it would follow from these premises that "smoke" means fire. Which, of course, it doesn't. So something must be wrong.

I wouldn't be inclined to push this very small argument very hard except that it seems to me that its conclusion is perfectly clearly true. The point about "smoke" is that — most of the time, at least — it doesn't *carry information* about smoke; what it does is it *represents* smoke (stands for it; refers to it; means it in *that* sense). And similarly, *mutatis mutandis*, for the "mental representation" SMOKE. Notice that

— unlike *carries information about* — *represents, stands for, refers to,* and the like are *not* transitive. If the first word on this page is "Granny," then the expression "the first word on this page" represents (stands for, refers to) the expression "Granny." And, of course, the expression "Granny" stands for, represents, and refers to Granny. It does not follow that the expression "the first word on this page" itself stands for, represents, or refers to Granny; it doesn't follow and, indeed, it isn't true. The expression "the first word on this page" refers (etc.) to the first word on this page, and nobody's Granny is a word.

It is not, to put it mildly, clear how, or even whether, the relation between a symbol and what it represents is to be reduced to the relation between a symbol and what it carries information about. I want to spend some time rubbing this in.

Consider two ways in which a symbol — "platypus," as it might be — can be used to say something true. (I'm sticking to truths at this point because causal covariance theories have notorious problems with errors, falsehoods, mistakes, and the like. I'll have more to say about that in the next section.) Paradigmatic of one kind of case is the use of the symbol in platypus-spotting. A platypus gallumphs by and one says "Platypus!" (where, let's suppose, the utterance has the force: "There goes a Platypus!"). An old-fashioned sort of analysis of this situation would have it that the symbol-type "platypus" has an associated extension and one way to generate a true utterance (thought) is to apply a token of the symbol to something that is *in* its extension. You get a truth when you say *of* a platypus that it *is* one. I shall call the uses of symbols that generate such truths "labelling" uses — for want of something better to call them.

Now consider another way that "platypus" might be used to say something true. Somebody says: "The platypus has webbed feet." There is no question, in this sort of case, of applying a "platypus" — token to something that's in the extension of the "platypus" — type; perhaps it would be closer to say that such tokens serve to *represent* the extension associated with the corresponding type. The point is: when I use "platypus" to say that platypai are webfooted, I'm not applying the term to anything; rather, I'm using it to stand for the things that it applies to.

Universal generalizations aren't, of course, the only cases where the function of a symbol appears to be representation rather than labelling. Other examples are occurrences in existentials ("there are platypuses") occurrences in hypotheticals ("if there are platypuses, there may be anything"; "if there is anything, there may be platypuses") occurrences embedded to verbs of propositional attitudes ("No one can really believe that there are platypuses"), and so forth.

The point to keep your eye on is that in none of these cases is anything being *called* a platypus. Yet all these tokens of "platypus" are symbols; they are all paradigms of things that have bona fide intensional properties.

What, then, is the relation between cases where symbols are used to label and cases where they are used to represent? Notice that this is a question for which old-fashioned, unnaturalized semantics has a plausible answer: a symbol *has the same meaning* in both uses. Indeed, on Grandmother's view of semantics, symbols apply to things in virtue of the very meanings that they express when they are used to represent things. Or, to put it the other way around, the property that a symbol expresses when it represents is the very property that things have to have in order to be in its extension. So it's the very property that a thing has to have for an application of the symbol to it to be true.

But although I am an enthusiast for this grandmotherly analysis,[3] it isn't a story that a *naturalized* semantics is in a position to tell — reeking, as it does, of appeals to intensional notions like having a meaning and expressing a property.

OK, I can now say what the terrible problem is. It's that, although IBSTs provide at least a first fling at a naturalistic story about the cases where symbols are applied to things, they provide no story at all — or, worse, they provide a demonstrably wrong story — about representational uses of symbols. I now need to convince you first that this is so and second that it matters that it is so.

IBSTs say, in effect, that symbol tokenings carry information about what causes them; specifically, they carry the information that their causes have the property to which the tokenings are linked by a counterfactual supporting covering generalization. In consequence, there's a certain structural symmetry between the case where, as Grandmother Semantics would have it, *the symbol S applies to the object O in virtue of O's having the property P that S expresses*; and the case where, as IBSTs would have it, *the symbol S carries the information that O is P in virtue of a covering generalization that causally connects P-instances with S-tokenings*. In effect, Granny says that S applies to O because S expresses P and O is P. Correspondingly IBSTs say that S applies to O because O is P and Ss carry information about Ps. It is this structural symmetry that makes it not-utterly-unreasonable to identify the property that IBST says a symbol's tokenings *carry information about the instantiation of* with the property that Grandmother semantics says that the symbol *expresses*. So we get the much-wanted reduction: First of "the property expressed" to "the property about whose instantiation S-tokens carry information"; and then of "the property about whose instantiation S-tokens carry information" to the (naturalistic) notion of the "covering" property (viz. the property of Os in virtue of which "Os cause S-tokens" is counterfactual-supporting). Home free.

This is, no doubt, much too simple as it stands. For example, the relevant covering generalizations can't be of the form "Os cause S-tokens" simplicitar; some (e.g., unobserved) platypuses don't cause "platypus"s, and some (e.g., misguided) "platypus"s aren't platypus-occasioned. Almost equally depressing, *applying* "platypus" to a platypus doesn't uncomplicatedly reduce to having the platypus in question cause you to token "platypus". Think of the chap whose habit it is to use "platypus!" as an ejaculation expressing extreme surprise (where the rest of us would say "S'blood!" or "Blow me down!"). In this chap's case, a token of "platypus!" that happens to be platypus-occasioned would nevertheless not count as an application of the term to the animal, and would not count as the saying of something true. Other sorts of examples will, no doubt, suggest themselves; the moral is that, at best, only the right kind of platypus-caused tokens counts as applications of "platypus." Still, as previously remarked, I'm inclined to think that it's not utterly unreasonable to hope for a naturalistic account of labelling in terms of causal relations between symbols and the things they apply to.

There is, however, no plausibility at all to the proposal that *representing* uses of symbols be naturalized in this way. Clearly, there need be — to put it crudely — no platypus around, none in the local causal history of the tokens that do the representing — when, for example, I write "platypai astound me" or "would that there were not the platypus" or, more prosaically, "platypuses lay eggs." The proximal causes of such tokenings — the causes to which they are presumably connected by covering generalizations, *and which the tokenings therefore carry information about* — aren't platypuses; they're *thoughts*. It's, as it might be, recollections of Melbourne that cause me to token "would that I had a platypus!" or its Mentalese counterpart; and if, as we may suppose, the generalization that such recollections give rise to such tokens is counterfactual-supporting, then it's the thoughts, and not the platypuses, that these tokens carry information about.

I want to make this as clear as I can. It may be that there are causal connections to platypuses *somewhere* in the historical background of platypus-representing tokenings of "platypus". But they haven't the sort of causal roles that IBSTs construe information-carrying in terms of[4]. According to IBSTs, symbols carry information about the things that causally control their tokenings; and the current worry in a nutshell is that the representing tokens of a symbol — as contrasted with its labelling tokens — typically aren't caused by things that belong to its extension. If you are inclined to doubt this, notice the following: Whenever there is a true application of "platypus" there is (assuming IBST) an answer to the question "*which* platypus caused the tokening?" Viz., it's the platypus to which the token was truly applied. But which platypus caused the "platypus" in "Would that I had a platypus?" Answer: none.

As for why all this matters, the examples may already have made that clear: The typical use of symbols *in thinking* is representation, not labelling.

No doubt, I do sometimes think "Hello, platypus here"; thoughts of this kind are, I suppose, the usual product of perceptual processes. And, again no doubt, such thoughts must sometimes play a role in reasoning. But it is as silly to think of thinking as *primarily* consisting in labelling as it would be to think of talking as consisting *primarily* in crying "Gavagai!" Much — very much — of what goes on in thinking is a movement from representation to representation; we put to ourselves ways that the world might be, and then we figure out what follows. It's this sort of figuring out which, together with remembering, extends the reach of the mental life beyond what is locally given to perception; and, unlike labelling, we can do it with our eyes shut.

It wouldn't be very misleading to say that IBST gives us *at best* a naturalistic theory of representation in perception, but *no theory at all* of representation in thought. This helps explain the infestations of frogs by which IBSTs are repeatedly plagued. Say to an informational semanticist: "Please, how does representation work?" and you are likely to get a song and dance about what happens when frogs stick their tongues out at flies. "There is," so the song goes, "a state of the frog's central nervous system that is (1) reliably caused by flies (in normal circumstances); and which (2) is the (normal) cause of an ecologically appropriate, fly-directed response. This state resonates, as one might say, to flies; and it is, for this reason, a paradigm of natural intentionality." If you then ask how this story about frogs and flies is supposed to apply to the case where one thinks to oneself, as it might be, "alack for the ungainly platypus" you will be told, in effect, that sorry but the theory is out to lunch.

The point, of course, is that frogs don't *think* about flies except, perhaps, when in the course of arranging to ingest one. So the question of how the frog represents flies *in thought* doesn't much arise. Still, there had better be an answer to such questions in the case of mental lives that are richer than the frog's. And, as we've seen, the terrible problem is that IBSTs haven't the slightest idea what this answer might be.

ON THE NATURE OF REPRESENTATION

Recapitulation: We are assuming — in a spirit of be nice to IBSTs — that labelling occurrences of "platypus" are *adverbially* caused by local platypai. (What you put in the adverb position depends on which version of IBST you like best; candidates include "always," "normally", "ideally", "statistically significantly," "counterfactual-supportingly," etc; and you're allowed to choose more than one adverb.) But representational occurrences of "platypus" are typically not so caused since they are typically not caused by platypai at all. Tokens of "platypus" that occur in the course of thinking about Melbourne, for example, are typically caused by *thoughts of Melbourne*. Presumably such tokens carry the information that their causes are thoughts about Melbourne; anyhow, they don't carry the information that their causes are platypai.

Let's collect together the kinds of events that are *adverbal* causes of representing occurrences of "platypus"; call them T-events. Then it's a way of putting our problem that "platypus" tokens at large have *two* kinds of causes: the labelling occurrences are, by hypothesis, platypus-occasioned; and the representing occurrences are, by hypothesis, T-occasioned. Our problem is therefore: WHY DOES "PLATYPUS" MEAN *PLATYPUS* AND NOT *PLATYPUS-OR-T*? After all, what its tokens are *adverbially* caused by, and hence carry information about are events that are (not platypus instantiations tout court, but) *either* platypus instantiations or T-instantiations.

This way of formulating the representation problem exhibits it as rather like — but much worse than — a worry that has sometimes been raised about the treatment of *errors* in IBSTs, and that is known, in that context, as "the disjunction problem." Let's suppose that *true* applications of "platypus" are *adverbially* caused by platypai. Still, it must be possible — either in thought or in speech — occasionally to *mis*apply "platypus"; and, presumably, misapplications of "platypus" are ipso facto *not* caused by platypai.[5] Suppose that there are circumstances under which I would regularly mistake — as it might be — a cow for a platypus. Then, *prima facie*, the generalization "if platypus or cow-in-those-circumstances, then "platypus" token is true and counterfactual-supporting. So it appears that, on standard accounts of "carries the information" my "platypus" tokens carry the information that they are occasioned by *either* platypai or cows-in-certain-circumstances. From which it presumably follows that "platypus" doesn't mean *platypus*, and that cow-occasioned "platypus" labellings are *true*.

Now various people (including, for example, David Israel and the

Stanford crowd, Ruth Millikan, Georges Rey, Robert Stalnaker, and me)[6] have tried to get out of this by fooling with the adverb. The idea is that, though "platypus" tokenings may *sometimes* be cow-occasioned, still they aren't *adverbially* cow-occasioned. *Adverb* tokenings are, *ipso facto, always* occasioned by *platypai* (and are therefore always *true*). The problem, of course, is to find an adverb that will do this job *but that is not itself semantic/intentional*. "Normally" or "counterfactual-supportingly" presumably won't do since, at least in any unquestion begging sense of these terms, it looks as though error is a perfectly normal feature of the use of symbols, and there appears to be no reason why the statement that such-and-such circumstances regularly cause errors shouldn't be counterfactual-supporting. Other adverbs are, however, frequently proposed. Maybe you don't ever get errors in "ideal" or "ecologically valid" circumstances; or in circumstances where the mechanisms of belief fixation are behaving "in the way that God or The Forces of Selection designed them to..." etc. And maybe it's possible to say without circularity what "ideal" or "ecologically valid" or "designed to" means in these constructions. But I doubt it. Philosophers who pay for their semantics by drawing checks on Darwin are in debt way over their heads. Or so it seems to me. (For an extended discussion of these sorts of points, see Part I of my "A Theory of Content," forthcoming.)

And, anyhow, even if idealization, natural selection, and the like will break the disjunction problem for the case of *error*, they clearly won't help for the corresponding problem about *representation*. Idealizing away from sources of error won't work for representation because representational occurrences of "platypus" don't covary with platypai *even when they're true*. Nothing has gone *wrong* when thinking about Melbourne causes me to think about platypai; even Omniscience could entertain such chains of thought. Well, maybe you can idealize away from mislabelling; but surely you can't idealize away from *thinking*. (And unless you're prepared to suppose that evolution designed folks to label but *not* to think, God and Darwin won't help you either.)

The upshot is that — contrary to what a number of philosophers (including me) have supposed — the disjunction problem has nothing in particular to do with error. Or, rather, it has to do with error only insofar as misapplications of symbols share with lots of other kinds of symbol tokenings the property of not being caused by things in the symbol's extension. As long as there are such cases there will be a disjunction problem for IBSTs.

Now for some ancient history. Back in the days when I was still thinking of the disjunction problem as just the form that the problem of error takes in causal covariance theories of meaning, I published a book called *Psychosemantics* (one of the best-kept secrets of MIT Press). *Psychosemantics* contained a discussion of the disjunction problem, of which the following is the gist.

Platypuses and cows-on-occasions both cause "platypus" tokenings. But since "platypus" doesn't mean anything disjunctive, either meaning isn't reducible to causation or there must be something relevantly different between the ways the platypuses cause "platypus"s and the ways that cows-on-occasions do; some difference that would, inter alia, explain why tokens whose etiology runs along the second route are *false*. This difference is, of course, exactly what all the adverb-mongering theories are looking for when they suggest that cows, unlike platypuses, don't cause "platypus"s *normally*, or *under ideal circumstances*, or that *ecologically valid* cows don't. But, even when I was thinking of the disjunction problem just in the context of theories of error, these sorts of solutions struck me as pretty unconvincing. I tried, but I couldn't convince myself that, *ceteris paribus*, we're all infallible.

The idea that error is an accident won't wash; but there's an idea with which it's easily confused that may do the job: the Platonic doctrine that error is (ontologically) parasitic on truth. In particular, according to *Psychosemantics* it's true that some "platypus"s are causally dependent on cows, just as it's true that some "platypus"s are causally dependent on platypuses. But there is nevertheless a difference between the cases: The causal dependence of "platypus"s on cows is itself dependent on the causal dependence of "platypus"s on platypuses in a way that the causal dependence of "platypus"s on platypuses is *not* dependent on the causal dependence of "platypus"s on cows. Intuitively: But that one calls platypai "platypai", one's misidentifications of cows would not lead one to call *them* "platypai"; but that one uses "platypai" to label platypai, one would not use it to *mis*label cows. And this doesn't go the other way around. The use of "platypus" to label platypai does *not* depend upon its use to mislabel cows; you would call platypai "platypai" even if you never mistook a cow for one. It's because of this asymmetry that "platypus" means *platypus* and not *platypus or cow*. And it's because of this asymmetry that cow-caused "platypus"s are mislabellings.

So *Psychosemantics* said; but *Psychosemantics* suffered from the delusion that "platypus" tokens that don't carry the information that their causes are platypuses are ipso facto *false*. Whereas, we've now seen that there is a whole nother species of such tokens, viz. the representational uses. These are *not* ipso facto false, but they do raise a disjunction problem. So, clearly, there is something wrong with the *Psychosemantics* picture.

To summarize; Standard IBSTs offer us an exhaustive distinction between tokenings that are caused by what they apply to (platypus-occasioned "platypai") and false tokenings (cow-occasioned "platypai"). But, as it turns out, false tokenings are just a special case of tokenings of a symbol that are not caused by things in its extension. The taxonomy we are required to naturalize is therefore:

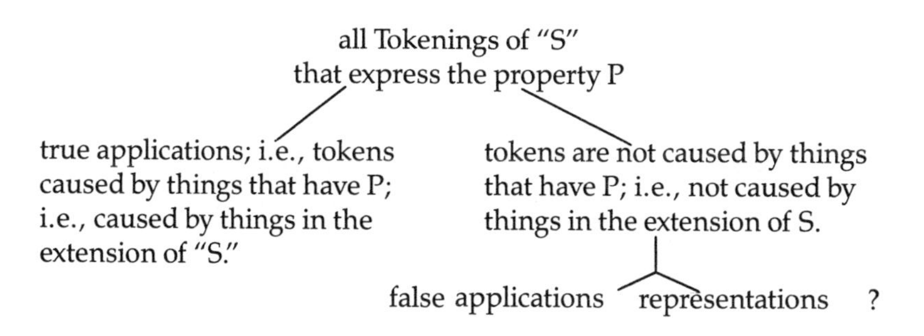

And there's a disjunction problem for every category on the right-hand side of the tree.

Here is what I now propose (but tune in for further revisions).

(1) I assume that IBST is right about true labellings: True labellings carry information about their causes. So, part of the story about applying "platypai" to platypai is that tokens of the symbol are under the causal control of instances of the property *platypus*.

(2) But the IBST story about true labelling doesn't, in and of itself, tell us what it is for "platypus" to mean *platypus*; it doesn't reconstruct the relation between a symbol and the property it expresses. Tokens of "platypus" are sometimes applied to cows (mislabelling) and they're sometimes deployed in thought (representation) and they do not thereby cease to express the property *platypus*. *The relation between a symbol and the property that it expresses* is not, therefore, directly reconstructed by a story about information carried, contrary to what IBST theorists have generally supposed. Outbreaks of the disjunction problem are indicative of an epidemic confusion between the analysis of the relation between a symbol and the property it expresses and the analysis of true labelling.

(3) Here's a first approximation to a story about the relation between symbols and the properties they express (for second and third approximations, see my *Psychosemantics* and Part II of "A Theory of Content"): Tokens of platypus mean *platypus* if (a) some tokens of

"platypus" carry the information that they are caused by platypai; and (b) tokenings of "platypus" that do *not* carry this information are asymmetrically dependent on tokenings that do.[7] I'm inclined to think that that's what there is to representation, and that, au fond, that's all that there is. No doubt it's wrong of me to be inclined to think this.

Note that:

(1) On this story, applications of "platypus" to platypuses express the property *platypus*; they carry the information that their causes are platypai, and all other tokenings of "platypus" are asymmetrically dependent upon them.

(2) Representational tokenings of "platypus" express the property *platypus*, and so do (mis)applications of "platypus" to cows; both are asymmetrically dependent upon there being (actual or possible) tokenings of 'platypus' that carry the information that their causes are platypai.

(3) Cow-occasioned tokenings of "platypus" do *not* express the property *platypus or cow* since it is platypus-occasioned "platypus" tokenings (and not platypus-or-cow-occasioned "platypus" tokenings) on which they asymmetrically depend. So the disjunction problem is solved. In this view, what was wrong with previous treatments of error in IBSTs is that they all tried to make erroneous tokenings of a symbol carry the same information that true labellings do. This doesn't work and it violates the basic tenent of IBST; viz. that symbol tokens carry information about their causes. According to the present version, by contrast, what platypus-occasioned "platypus"s share with cow-occasioned ones is (not the information they carry, but) *what they mean*: both express the property *platypus*. Granny wins again.

If any of this is right, then we now know what the relation is between "carrying the information" that so and so and meaning that so and so. A symbol that carries information about P-instantiation means P if its *other* information-carrying (the other causal covariances into which it enters) depends asymmetrically on its carrying information about P-instantiation. However, we still haven't got a theory of error. Or to put it another way, we know how false applications of "platypus" can mean *platypus* even though they aren't platypus-occasioned. What we don't know, however, is what makes them different from other tokens of "platypus" that also don't carry the information that platypai cause them; e.g., representing tokens.

Moreover (and here's where I depart from the intuitions that drove *Psychosemantics*), it seems reasonable to doubt that any variation on the theme of asymmetric dependency will throw light on this issue.[8]

The point about errors, after all, is that they are things that we want to *avoid*; and their asymmetrical dependency on true labellings doesn't, in and of itself, explain why this is so. (To put it another way, Plato's attempt to derive the *normative* objection to false beliefs from the *ontological* priority of true ones simply doesn't work.) This is immediately apparent when we notice that representational tokenings are also asymmetrically dependent on true labellings, though we feel no pressure to avoid *them*. The current position is that, having solved the disjunction problem, we need a story about error not because error raises the disjunction problem, but just because we need a story abut error. And it would be very desirable if the story we tell explained why calling something a mistake has the normative force that it does.

I don't have a worked-out naturalistic story to tell about error; but I propose to close this paper with some suggestions about how to get started on constructing one. For these purposes, I'll stick to erroneous applications — i.e., to mislabellings. I expect that there's an extension to nonlabelling tokens (for example, to a naturalistic story about what's wrong with "platypuses sing"), but I'm not going to worry about that here.

From the current perspective, the point to emphasize about error is that it involves *misrepresentation*. Here I'm just following Granny: What she says is wrong with applying "platypus" to a cow is that one thereby represents the cow as having the property which "platypus" expresses; viz. the property of being a platypus. Which, of course, the cow doesn't; only platypuses do.

Now, this formulation relys on two key notions: the notion of *applying* a symbol to a thing, and the notion of a symbol *expressing* a property. We have (see above) a story about what it is for a symbol to express a property; what needs working on is the notion of application. Here, in roughest outline, is how I think such a story ought to go.

Applying the symbol S to the object O isn't just a matter of having O cause a token of S (see the discussion on pp. 00–00). So, what more is required? An old-fashioned suggestion is that applying a symbol to a thing involves having a disposition to reason and to act in certain ways, both in respect of the symbol and in respect of the thing you apply it to. So, for example, if you apply "platypus" to Bossie and you accept (i.e., believe-true) "platypai lay eggs," then *ceteris paribus* you are prepared to accept "Bossie lays eggs" and *ceteris paribus* you are prepared to act towards Bossie in whatever ways you are disposed to act towards platypai and egg-layers.

That, surely, is what Grandmother would have said. The question is: are *we*, qua naturalists, allowed to say it? Answer: maybe. It depends on whether we can get a naturalistic construal of notions like believing-true, being prepared to reason/act in such and such a way, etc. Well *can* we? Answer: maybe again. There are, after all, ideas floating around for functionalist reconstructions of these sorts of notions, and I'm prepared to view these ideas as not altogether crazy.[9]

So, for example: there is a symbol (viz. "Bossie is a platypus") which expresses a property that is instantiated by and only by states of affairs that are constituted by Bossie's being a platypus. One way to apply "is a platypus" to Bossie is to have a token of this symbol in your "believe-true box."[10] The believe-true box is defined by reference to the causal roles of the symbols it contains; that is, a symbol is said to be in the believe-true box if certain of its tokens have certain causes and effects. Among the consequences of this functional characterization is that, *ceteris paribus*, if "Bossie is a platypus" and "platypai lay eggs" are both in the believe-true box, then so too is "Bossie lays eggs." A further consequence is that some symbols in the believe-true box have specified causal relations to tokens of symbols located in a "desire-true box" (as per your favourite decision theory.) Among the causal consequences of having "Bossie is a platypus" and "platypai lay eggs" in your believe-true box and having "I get to eat a platypus omelette" in your desire-true box is that *ceteris paribus* you beat the fields[11] around Bossie for platypus eggs.[12]

But, as a matter of fact, *ceteris paribus you don't find any* because, of course, Bossie isn't a platypus (applications of "platypus" to Bessie are ispso facto *mis*applications) and it is in the nature of platypus eggs to be laid by, and only by, platypuses. That, in microcosim, is what's wrong with false beliefs. Show me a man who applies "platypus" to a cow and (*ceteris paribus*) I will show you a man who doesn't get any platypus eggs.

According to this view, the Naturalistic answer to "what's wrong with false beliefs?" is primarily that false beliefs lead to abortive actions. And if somebody wants to know what's wrong with abortive actions, the answer is that they lead to not getting what you want. And if somebody wants to know what's wrong with not getting what you want, the answer is that explanation has to stop somewhere.

This explanation stops here.[13]

Notes

1 I shall often use quoted English expressions as names for their counterparts in the language of thought. Thus " 'rain' " will sometimes refer to the symbol that encodes the concept *rain* in Mentalese, and sometimes to the English word which translates that symbol. It should usually be clear from the context which is intended; and for most present purposes it won't matter. In fact, for most present purposes, one might as well assume that English *is* Mentalese.

For similar reasons, much of what I'm about to say about thinking holds just as well for talking and vice versa. In what follows, "say" will amost always mean *say or think*, and "think" will almost always mean *think or say*.

2 Carrying information is supposed to be an intensional relation in at least the technical sense that "a carries the information that S" and "S iff T" are supposed not to entail "a carries the information that T." Just *how* intensional "carries the information is" — whether, for example, it's as intensional as "believes that" — is moot.

3 It is sometimes said that Grandmother semantics will not do because "intensions don't determine extensions." But this is preposterous on the face of it. Nothing could be more clearly true — indeed, nothing could be more clearly a *semantic* truth — than that, on the one hand, "platypus" means platypus, and that, on the other, it is true for all values of *x* that "platypus" applies to *x* iff x is a platypus.

Philosophers get into no end of trouble by confusing the question whether intensions determine extensions (answer: of course they do) with the question whether words like "platypus" have *definitions* (answer: of course they don't).

4 If you take seriously the idea that carrying information about is a *statistical* relation (so that, for example, As carry information about Bs when there is a statistically reliable covariance between Bs and As), you could break the semantic relation between "platypus" and platypuses just by inflating the relative frequency of representating to labelling tokens; e.g., by saying (or thinking) "Up the platypus!" over and over again when there isn't a platypus around.

5 Or, if they happen to be caused by platypai, then they must be so caused in not "the right way"; i.e., not in the way that's required for them to count as labelling their causes platypai.

6 In an unpublished ms called "Psychosemantics." Not to be confused with the book of the same name, also by me, for a discussion of which see below.

7 One reason why this is *only* a first approximation is that you might have words which are *never* truly applied (i.e., never occasioned by things that they apply to). E.g., "unicorn." Presumably representational tokenings of "unicorn," and its misapplication to cows, are dependent upon counterfactual applications to unicorns. More precisely, they're dependent on the fact that "unicorns cause 'unicorns' " is counterfactual-supporting and (hence) can be true in the absence of unicorns.

Another reason is that, as stated, 3 is no good if "platypus" is ambiguous (as opposed to meaning something disjunctive); see *Psychosemantics*. Clearly there is a lot of tidying up to be done to make 3 work. That, however, does not mean that it's unworkable.

8 This rest of this paragraph is owing to a conversation with Paul Boghossian; I'm greatly indebted to him for making me think about the error problem in these terms.

9 I'm on record as holding that function plays no role in the determination of *content*. But, of course, that's quite compatible with holding that notions like *believing-true, accepting, inferring, acting on,* and other relations that agents bear *towards* contents should be functionally defined. For discussion, see *Psychosemantics*.

10 Other ways of doing so are distinguished by how Bossie is picked out (e.g., by a definite description, an anchored indexical), etc.

11 Here, "beat the fields for" means indulging in a form of behaviour which is caused in a certain way by interactions between what's in the accept box and what's in the desire box; which, *ceteris paribus*, terminates when a token of "here are some platypus eggs" appears in the accept box . . . and so forth. The general program for converting a belief/desire theory account of mental processes into a computational account is familiar from the literature in philosophy of mind. For early versions, see Fodor, TK, LOT; for a recent one, see Rey MWC.

12 This sketch of an account is "functionalist" about the notion of *applying* a symbol but, not, notice, about the notion of content per se. Applying a symbol involves a disposition to reason and act in certain ways; but representation itself requires only the asymmetric dependencies of some relations of information-carrying

upon others. This is important in relation to the — frequent — philosophical claims that content is per se holistic; for discussion, see *Psychosemantics*.

13 Unless perhaps it doesn't. It may well be that this story about what's wrong with making mistakes concedes too much to vulgar Pragmatism. For example, it suggests that a being that lacked practical concerns could have *no* reason for preferring true labellings to false ones; indeed, that such a being could find nothing to choose between mind/world *correspondence* and any other mind/world relation. There is, on this view, nothing special about truth except that, for some reason, true beliefs are the ones that it usually turns out that it pays to act on. For *what* reason, one wonders.

Deep down, I think I don't believe any of this. But the question what to put in its place is too hard for me.

Comment

Ali Akhtar Kazmi

Fodor views information-based semantic theories as part of a program which aims at naturalizing semantics. This is a threefold program: it attempts to show (1) that semantic facts about meaning, representing, and carrying information in a language are supervenient upon facts about linguistic usage and intentions of the speakers of the language; (2) that these semantic facts, as well as the corresponding intentional facts about speakers, are natural facts; and (3) that these facts are expressible in non-intentional and non-semantic vocabulary. Fodor characterizes a family of information-based semantic theories which identify the relation of meaning with the relation of carrying information about, and proposes to analyse the relation of carrying information about in terms of causal relations between tokens of symbols and objects. A simple version of one of these theories affirms (A).

(A) For any expression l which refers to a language, any expression e which refers to a non-indexical term of that language, and any expression m of English, ⌜e means m in l⌝ is true, if and only if, ⌜All tokens of e in l are caused by m[s]⌝ is true and counterfactual supporting.

(A) is true only if (1) is true.

(1) If 'English' refers to a language, and "Cow" refers to a non-indexical term of that language, and 'Cow' is an expression of English, then "Cow" means cow in English' is true, if and only if, 'All tokens of 'Cow' in English are caused by cows' is true and counterfactual supporting.

Clearly (1) is not true. Suppose, for example, that some tokens of 'Cow' in English are caused by horses and not by cows. Given this assumption and uncontroversial facts about English, (1) is true only if (2) is not true.

(2) 'Cow' means cow in English.

On the other hand, if (2) is true, then given uncontroversial facts about English, if (1) is true, then (3) is true and counterfactual supporting.

(3)All tokens of 'Cow' in English are caused by cows.

But (3) implies (4).

(4) All tokens of 'Cow' in English are caused by cows-or-horses.

And, if (4) is true and counterfactual supporting, then given uncontroversial facts about English, (1) is true only if (5) is true.

(5) 'Cow' means cow-or-horse in English.

Since (5) is false and (2) is true, (1) is not true, and hence, (A) is not true.

These objections to (A) illustrate two general problems faced by information-based semantic theories. Fodor describes these problems as 'the error problem' and 'the disjunction problem' respectively. The preceding version of the disjunction problem is avoided by the following revision of (A)

(B) For any expression l which refers to a language, any expression e which refers to a non-indexical term of that language, and any expression m of English, ⌜e means m in l⌝ is true, if and only if, ⌜All tokens of e in l are caused by m[s], and all m[s] cause tokens of e in l⌝ is true and counterfactual supporting.

However, like (A), (B) is in conflict with the apparently plausible claim that some tokens of 'Cow' in English are not caused by cows. Moreover, if it is conceded that there are possible circumstances in which (6) is true and counterfactual supporting, then another version of the disjunction problem arises once again.

(6) All tokens of 'Cow' in English are caused by cows-or-suitably disguised non-cows, and all cows-or-suitably disguised non-cows cause tokens of 'Cow' in English.

If there are possible circumstances in which (6) is true and counterfactual supporting, then given uncontroversial facts about English, in those circumstances (B) is true only if (7) is true.

(7) 'Cow' means cow-or-suitably disguised non-cow in English.

Advocates of information-based semantic theories have attempted to deal with these difficulties by exploiting the intuition that though, for example, suitably disguised horses may cause tokens of 'Cow,' there are circumstances, perhaps optimal, normal, or ecologically correct ones, such that it is true and counterfactual supporting that in these circumstances all tokens of 'Cow' in English are caused by cows and all cows cause tokens of 'Cow' in English, and that in these circumstances disguised non-cows do not cause tokens of 'Cow' in English. Thus, it is thought that some variant of (C) which is obtained by replacing 'in designated circumstances' in (C) with a suitable modifier such as 'in optimal circumstances' will not be exposed to the error problem and the disjunction problem.

> (C) For any expression l which refers to a language, any expression e which refers to a non-indexical term of that language, and any expression m of English, ⌜e means m in l⌝ is true, if and only if, ⌜In designated circumstances, all tokens of e in l are caused by m[s] and all m[s] cause tokens of e in l⌝ is true and counterfactual supporting.

Fodor's important observation is that objections to (A) and (B) arise not only in virtue of cases of misidentification, but also because of the existence of non-identifying tokens of terms. An ordinary utterance of 'Cows are mammals' in which the token 'Cows' is not caused by cows, but does not misidentify anything as cow either, is a counter-example to both (A) and (B) and gives rise to a version of the disjunction problem. It is, therefore, unlikely that any variant of (C) which is obtained by replacing 'in designated circumstances' in (C) with modifiers like 'in normal circumstances,' 'in optimal circumstances' or 'in ecologically correct circumstances' will avoid such counter-examples and the attendant disjunction problem.

Fodor has distinguished two sorts of non-identifying tokens of an expression, that is, those tokens of an expression which misidentify something, and the representational tokens which do not purport to identify anything to which the expression applies. Fodor argues that the semantic properties of both sorts of non-identifying tokens of an expression are inherited from the semantic properties of the identifying tokens of the expression. Fodor's view is characterized by (D).

> (D) For any expression l which refers to a language, any expression e which refers to a non-indexical term of 1, and any expression m of English, ⌜e means m in l⌝ is true, if and only if, ⌜ (i) Some tokens of e in l carry the information that they are caused by m[s], and (ii) tokens of e in l that do not carry this information are asymmetrically dependent upon tokens that do⌝ is true.

Fodor's proposal, like the previous proposals characterized by (A), (B), and (C), is part of a larger reductionist framework which aims at naturalizing semantic properties and propositional attitudes. Fodor assumes that asymmetric dependence is not a semantic notion, and that the relation of carrying information has a naturalistic reduction independent of the proposed reduction of meaning. On Fodor's view, a verb of propositional attitude, such as 'believe,' stands for a relation between agents and propositions, but an agent bears this relation to a proposition in virtue of bearing another relation to a mental representation which means or expresses that proposition.[1] On this view, because propositional attitudes are analysed in terms of the relation of meaning, naturalization of propositional attitudes requires reduction of the relation of meaning to non-intentional relations. The alternative strategy of reducing meaning to some intentional relations is not available in this framework.

There is some lack of clarity in the notion of asymmetrical dependence, and to that extent it is not clear what contribution clause (ii) of (D) makes to the content of (D); but clause (i) raises some difficulties for (D), as well. Given uncontroversial facts about English, (D) is true only if (8) is true.

(8) 'Prime' means prime in English is true, if and only if, (i) Some tokens of 'Prime' in English carry the information that they are caused by primes, and (ii) tokens of 'Prime' in English that do not carry this information are asymmetrically dependent upon tokens that do' is true.

Since primes presumably do not cause any tokens, they do not cause any tokens of 'Prime,' and therefore, no token of 'Prime' in English carries the information that it is caused by primes. Hence, if (8) is true, then (9) is not true.

(9) 'Prime' means prime in English.

This example illustrates a general difficulty that (D) shares with other information theoretic accounts of meaning. (D) and other informatic theoretic accounts of meaning are falsified by expressions which stand for abstract objects.

(D) is also falsified by vacuous terms. Given uncontroversial facts about English, (D) is true only if (10) is true.

(10) ''Unicorn' means unicorn in English is true, if and only if, '(i) Some tokens of 'Unicorn' in English carry the information that they are caused by unicorns, and (ii) tokens of 'Unicorn' in English that do not carry this information are asymmetrically dependent upon tokens that do' is true.

Since no tokens of 'Unicorn' in English are caused by unicorns, (10) is true only if (11) is not true.

(11) 'Unicorn' means unicorn in English.

Fodor suggests that the problem of vacuous terms may be resolved by appealing to the fact that the relevant generalizations involving vacuous terms are counterfactual supporting. He writes: 'Presumably representational tokenings of 'Unicorn,' and of its misapplications to cows, are dependent upon counterfactual applications to unicorns. More precisely, they're dependent on the fact that 'Unicorns cause 'Unicorns is counterfactual supporting and (hence) can be true in the absence of unicorns.'[2] It is not clear how (D) is to be modified in the light of this suggestion. But, even if this suggestion allows a resolution of the problem posed by contingently vacuous terms, appeal to counterfactual situations does not seem to help with the problem posed by necessarily vacuous terms such as 'Unicorn.' Since unicorns necessarily fail to exist, it appears that the relevant generalizations like 'Unicorns cause 'Unicorn', and counterfactual claims of the form 'Had there been unicorns, ϕ' are vacuously true.

As (D) is falsified by complex terms as well, it may be urged to restrict the range of the variable 'e' in (D) to expressions which refer to non-complex, non-vacuous, and non-abstract terms.[3] But even this restricted version of (D) appears to be problematic. Non-complex, non-vacuous, and non-abstract terms like 'Cow' may be introduced in the language by description, and their use entrenched, even though these terms are not synonymous with any description, and no tokens of these terms are caused by the objects to which they apply. The restricted version of (D) is incompatible with this possibility.

Second, (D) appears to have the consequence that necessarily coextensive terms are synonymous. If 'E' refers to an unambiguous term of some language L which is synonymous with an expression G of English, then (12) holds if (D) is true.

(12) ⌜(i) Some tokens of E in L carry the information that they are caused by G[s], and (ii) those tokens of E in L that do not carry this information are asymmetrically dependent upon tokens that do⌝ is true.

But, then, given Fodor's characterization of asymmetric dependence, it appears that for any English expression H which is necessarily coextensive with G, (13) holds as well.

(13) ⌜(i) Some tokens of E in L carry the information that they are caused by H[s], and (ii) those tokens of E in L that do not carry this information are asymmetrically dependent upon tokens that do⌝ is true.

And, if (13) holds, then given (D), ⌜ E means H in L⌝ is true, and hence G and H are synonymous in English. This result is unacceptable, if belief-ascriptions are interpreted as expressing relations between agents and the propositions expressed by their complement sentences. If G and H are synonymous non-indexical expressions, then, given this view of belief-ascriptions and compositionality, an individual satisfies ⌜x believes that α is G⌝ if and only if he satisfies ⌜x believes that α is H⌝. The restricted version of (D) is not committed to the view that any necessarily coextensive terms are synonymous, but it appears to have a weaker but still implausible consequence that non-complex, non-vacuous, and non-abstract terms are synonymous, if they are necessarily co-extensive.

NOTES

1 In *Psychosemantics* (MIT Press 1987) Fodor presents the following account of the nature of propositional attitudes: (i) For any organism O, and any attitude A towards the proposition P, there is a ("computational"/"functional") relation R and a mental representation MP such that MP means that P, and O has A if and only if O bears R to MP' (p. 17).

(i) is too strong, because it implies that every proposition has a mental representation; but it is also took weak, because it allows the relation R to vary across propositions and organisms. (i') avoids these problems. '(i') For any attitude A there is a ("computational"/"functional") relation R, and for any organism O and proposition P, there is a mental representation MP such that O has A towards P, if and only if, MP means that P and O bears R to MP.' But (i') is still too weak, because, like (i), it allows that an individual may not believe a proposition P even though he bears R to a mental representation of P.

In 'Propositional Attitudes' (*The Monist* 64(4) [1978]), Fodor suggests another conception of propositional attitudes which is articulated in (ii): (ii) For any attitude A there is a ("computational"/"functional") relation R, and for any proposition P there is a mental representation MP such that for any organism O, O has A towards P, if and only if, MP means that P and O bear R to MP.'

(ii) is unacceptable because it implies that if two or more individuals believe some proposition, then they bear R to the same mental representation of that proposition. This conception of propositional attitudes seems to presuppose that no proposition can be expressed by more than one mental representation. Otherwise, there is no apparent justification for the requirement that there is at least one mental representation of each proposition to which everyone who believes that proposition is related by R.

In *Psychosemantics* Fodor also remarks: "A cruder but more intelligible way of putting claim [(i)] would be this: To believe that such and such is to have a mental symbol that means that such and such tokened in your head in a certain way; it is to have such a token 'in your belief box' as 'I'll sometimes say' (p. 17). This remark suggests (iii). (iii) For any attitude A there is a ("computational"/"functional") relation R such that for any organism O and proposition P, O has A towards P, if and only if, there is a mental representation MP which means that P and O bears R to MP.'

Unlike (i), (i'), and (ii), (iii) is not exposed to the problems discussed above. For further discussion of these theses see Scott Soames, 'Belief and Mental Representation,' in this volume.

2 See 'Information and Representation,' in this volume.

3 A naturalistic reduction of the meaning of complex terms may then be thought to proceed in two stages. In the first stage, the restricted version of (D) would reduce

meaning of simple terms, and in the second stage meaning of complex terms would be shown to be dependent upon the meaning of simple terms in a naturalistically acceptable way. Success in the second stage of the project requires a naturalistic reduction of all term-forming operations in the language. As Scott Soames has argued in 'What Is a Theory of Truth?' (*The Journal of Philosophy*, 81 (8)), a Tarskian theory of truth and reference of complex terms does not accomplish this task.

CHAPTER SEVEN

Round Table Discussion

Nicholas Asher, Lee Brooks, Fred Dretske, Jerry Fodor, David Israel, John Perry, Zenon Pylyshyn, and Brian Smith

Hanson: Let me start things off by asking two related questions. This conference was called "Information, Language, and Cognition." One of the questions I'll ask is about cognition; the other is a fairly parallel question about language. So the cognition one first. Assuming that a cognitive state can have an intentional content that is relevant to the way in which we would want to appeal to that cognitive state in psychological explanations, can the cognitive state also have an informational content that is related in some principled way to its intentional content, and that illumines the explanatory efficacy of that intentional content? Perry, for instance, told us that intentional content is distinct from informational content; that while informational content is relative to constraints, intentional content is also relative to goals. But while he told us something about what constraints are, he didn't tell us what goals are. Presumably, goals aren't just further constraints, so I guess one way of getting at the relation between informational and intentional content in Perry's framework would be to ask what the relations are between goals and constraints. Perry also emphasized that it was one thing to carry information, another to have it. This raises a further subsidiary question of how, if at all, getting clear about the connection between information and intentional content is supposed to help us with this distinction and help us understand what it is for a cognitive agent to have the information that *p*. My second question parallels the first in the domain of language.

Assuming that a sentence token has a semantic content that helps to explain its effectiveness as an instrument of communication, does a sentence token also have an informational content that is related in some principled way to its semantic content, that helps to illumine the explanatory efficacy of the latter? Gentlemen, I turn it over to you.

Pylyshyn: Would you accept no for an answer?

Israel: Let me imitate John Perry. The questions were directed to him but I'll try to answer at least part of the first question anyway, though not with his wit or his Midwestern twang. Forget the word goals for the moment. Think of the ternary relation between a system of a certain kind, an environment of a certain kind, and a resulting state. Forget "goals," think instead just of the specification of that state. Now that looks to be a perfectly reasonable relation to be studied scientifically. I mean it doesn't offhand look like there's any kind of special mystery or problem about it.

Fodor: Except what it is.

Israel: Well, that's right. Let's study the relation, which I will dub the "works relation." It's the relation, remember, in which an entity of a certain kind, in an environment of a certain kind, operates in such a way as to bring about the specified end state. Forget the word "goal." That looks like, in principle, a relation that lots of sciences could study. It's very general, I agree, but there are lots of different instances, and lots of different sciences could contribute. Now I think that one claim that we have, and here is perhaps a way we differ from both Fodor and Dretske, is that there's actually no mystery at all, in general. There are lots of details that are complicated, but no mystery in general at all about intentional content. It is of course quite right that the story you tell in terms of mechanisms has got to be physicalist in some sense — you've got to be able to tell some story; but in particular, it's just a particular sub-relation of the works relation.

Fodor: But that doesn't make a lot of sense. We don't know how anything can mean anything.

Israel: What's the 'how-question'?

Fodor: We don't have a statement of anything like sufficient conditions in non-intentional and non-semantical language for something being meaningful. Let me show you why the problem of freedom of the will isn't a problem. Freedom of the will is just a certain triple, right? In a certain environment, a certain thing brings about a certain effect, that's all. I mean now of course the mechanism has to be physicalistic and so forth but freedom of the will, the willing, producing action by will, is just a case of that triple. So what's the problem?

Israel: No, the question is: What is the *general* problem about content?

Fodor: The *general* problem is that we don't have any idea of how anything in the world that's physical could have a truth value, for example.

Israel: Well...

Fodor: You mean I've been in the wrong business? Surely there is a real problem here, David. Quine and others thought it was such a bad problem that...

Israel: They're wrong.

Fodor: Well then, could you tell us what the solution is?

Israel: All I wanted to say is that we shouldn't get caught up with the word "goal." We think goals play a special role in a certain sub-relation of the work relation. The relation can be applied without any goal talk. We misled by using that word as if it were central to the story.

Fodor: What you want to do is to pick out the sub-instances of that triple which involve intentional relations. And I think that for *that* project, the goal talk is essential.

Israel: Ok.

Brooks: I really do feel like I've come in on a play that's been playing for many years. And furthermore it's been playing in some other theatre that I've been in. In psychology, the continental divide was a shift from an associationist program to a cognitive program. And the discussions that I'm hearing about information-based semantics keep reminding me of the associationist program. Associationism was straight positivism of some kind or another. And it seemed to me that the major complaint about this program, when the cognitive program replaced it, was that in fact any description of the world was badly underdetermined by the available data. Most of the action was in the descriptions themselves, which didn't have to be particularly closely related to constraints from the world. There are certainly things that could happen in the world that would violate any description, but you didn't get a lot of close positive constraint. Over the past two days this cognitive complaint has just sounded like a non-issue. Now maybe it was being said in other words and I didn't understand it, but it just seems that, all of a sudden, there was no under-determined process of description to worry about. I've not been hearing a lot of argument, the last couple of days, about those descriptions and the degree of freedom involved. Is that the point — that the intention is only loosely related to the information that's coming in? Maybe I just don't even recognize the issue. But it's as though all of a sudden it has become unfashionable again to talk about that sort of stuff. But I've spent too many years studying it!

Pylyshyn: Well, all this activity in the last couple of days has just confirmed my prejudice about the role that a theory of semantics would have in a cognitive science. The way one does semantics is, one starts off asking very hard questions about the relationship between states or sentences and the world, and then one goes about solving some other problem. In some cases, it's a problem about understanding, and sometimes it's a problem about information and how the information is carried by the signals, and sometimes it's the problem about how certain kinds of triples could be worked out. But the basic issue, I think, is never really addressed, the basic issue of naturalizing content. And it's fortunate that you can do cognitive science without solving that problem. Otherwise, we'd all be in a lot of trouble. I have the same feeling about consciousness because whenever consciousness comes up — and it usually comes up in meetings of philosophers and psychologists — it's the same kind of phenomenon, where people are prepared to give a theory of consciousness but what they give you is a theory of something else. And the relationship between consciousness and that other thing is never made explicit. And I've had that *déjà vu* through most of the talks in the last couple of days.

Dretske: I think that if people are just airing feelings of discomfort, I'll air mine. If we're going to do cognitive science, I think that what makes it something other than neurobiology is that some of the explanations are couched in terms not appropriate to neurobiology, in terms of the contents of the inner states, and that's what's going to do the explaining. So, if that's what cognitive science is, you've got to first figure out how these inner states get content and what that content is; how the inner states get assigned the content they get assigned, in terms of which they're going to do the explaining they're going to do. That's what the information base of semantics is about: that the way inner states get content has somehow to do with the way the inner states are connected with the outer states. But however that gets worked out, secondly, you've got to show how that content, however it got assigned, is somehow going to figure in, or be relevant to, the explanations you give of what the agent is doing. Those are two separate tasks that make cognitive science a distinct enterprise, it seems to me. (I was myself this morning, supposing that we were agreed about how you get the content assigned, and focusing on how to make it do some work.) Cognitive science thus postulates inner states, for which there are no clear behavioural criteria. So when I say someone believes something, there's something in there whose function it is to carry that information. Of course, since this is a philosopher sitting back making up the example, I can assign any states I want; but if I sat right down and looked at them or you, there would be nothing in your behaviour which would make it *certain* you were in that state. Your being at this cognitive state is as much a theoretical state as a quark having a charm.

Israel: That's right...

Fodor: But there is one further point. It would be reasonable at this point in the discussion to say, "Who cares whether philosophers are happy with this apparatus? The question is whether you can do science with it." But it's not quite like that. I mean maybe that *is* the right thing to say, but there are alternative enterprises in the air. For example, there's an enterprise which chews up a lot of taxpayer's money — we're talking about grants now and stuff like that, not just who wins philosophy arguments — that says, "Let's just not do it this way at all. Let's just do the biochemistry of the neurons." No intentional states at all, just biochemistry. Now, we're pretty sure that the physicalization of biochemistry, in this loose sense that I was talking about before, is possible. Unless we're pretty sure that the physicalization of the intentional is also possible, people who want to do intentional psychology seriously, as real science, are at a very serious disadvantage because they may be playing a game that can't be played. If it turns out that the physicalization — naturalization — of intentional science, in this rather loose sense, is impossible, if it turns out that there really are conceptual reasons why you can't give a physicalistic account of these states, then it seems to me that what you ought to do is do your science some other way. I mean, I don't think you ought to say like Dan Dennett, "Well, it's OK that there aren't any of these things, I'll keep talking as though they were." I don't think you should be an instrumentalist or an opportunist. If there really are conceptual problems about intentional states, if you really can't give an account of the role of the intentional in the physical world, then we should stop doing intentional science, and that counts for a lot more than some philosopher being worried. That's a matter of theoretical honesty. We're in a position in the development of these sciences, which makes it more important than whether philosophers are comfortable. The position we're in is, can we give a coherent account of this research project because, by Christ, if we can't we should stop spending the taxpayers' money.

Brooks: I'm glad we don't have to wait for the answer!

Perry: I can't help noticing you've just described a motivation for doing what you sometimes criticize people for: looking at some very simple cases and trying to understand them in some detail, such as why would it be useful, even though it's not mandatory, to give an intentional description of the pencil sharpener that you described. In other words, you've given a motivation for that — some kind of reductive motivation — that's different than the motivation you sometimes assume that in all the more complicated interesting cases of intentionality, this person must think. The true motivation may be just the one you described: to guess a couple of very clear examples of where intentional description though not mandatory is perfectly kosher and helpful.

Fodor: I don't think that's right, John. For one thing, I think it's not so clear what's a simple case. I mean, usually you're halfway through the battle in science when you know what is a simple case. My intuitions are that frogs and pencil sharpeners are not, simple or otherwise, good cases of intentionality. In fact, my guess is, in the case of pencil sharpeners, it's just false that the thing's got any intentional states. I don't think thermostats have beliefs and desires, I don't think pencil sharpeners have goals. You want to say, "OK, here's a state which, if I could understand it, then I'd be able to sleep like an honest citizen. When I do my intentional psychology, I would be convinced that the problem about naturalizing the intentional is sort of like the problem about naturalizing mountains: I see the kind of way it's going to go, although of course I can't do as many details." But I wouldn't choose pencil sharpeners as such a case. Being able to tell a story about pencil sharpeners doesn't stop me from being awake at night because I don't even think the intentional apparatus applies to them. I want to know about somebody believing that *p*, because that's the kind of intentional apparatus I use over and over again when I do my psychology.

Israel: I disagree.

Dretske: I agree with Jerry about the drive towards naturalization; there is always a kind of tension. We want to say it's some kind of physical way that we understand, that what intentional states amount to and how they figure in the explanation of behaviour is physical. But if that's true, then it looks like when you finish the job, if you ever finish it, then you'll have given over psychology to neurobiology; that is, to the people who are experts on what's going on there physically. How could you actually naturalize the method without giving away psychology in the bargain; that is, the science which says that we're going to appeal to competence or meanings to explain the behaviour of systems, not to the chemistry or the physics? Now one way of naturalizing content or intentionality without giving the explanation of behaviour over to the neurobiologists is to say that content or intentionality have to do not with what's going on inside the skin, but with the way the stuff inside the skin is related to the stuff outside the skin. It's those relations that make the system an intentional system. That's the kind of basic good idea behind information-based semantics. And that's why the neurobiologists won't have the last word, at least, the only word on the explanation of behaviour, because some of the explanation of behaviour has to appeal to the way the stuff on the inside is related to the stuff on the outside.

Pylyshyn: That's everybody's view...

Israel: Then we agree!

Fodor: Well, it's not the Churchlands' view.

Pylyshyn: That's because they don't believe in content, but for anybody who does believe in content, that is their view.

Perry: But if the definition of "intentional" is just "beliefs and desires"? I always thought the definition of "intentional" was "when a false proposition can be used to characterize a state that a thing actually had."

Fodor: Right. I don't think you have to have a theory of beliefs and desires. I think they are special cases.

Perry: But if the latter is intentionality, then the pencil sharpener is intentional.

Fodor: I don't see that.

Israel: But at least it could be. You agree that it could be, as John said, not mandatory, but useful and true to come up with an attitude — don't think of it as an attitude context, think of it as a sentence-embedded context — that you would apply to the pencil sharpener.

Fodor: The reason the pencil sharpener is no good isn't that it's so, you know, cheap or something. Let me admit this among friends. I actually think John Searle is right: it's very important to distinguish between derived and original intentionality. I think the reason the pencil sharpener is no damn good is its intentionality is utterly derived.

Israel: So do we, but you can't say it has derived intentionality if it doesn't have intentionality!

Fodor: But remember that the idea was to allow me to sleep the sleep of the just, right? There isn't any problem about explaining how something could have *derived* intentionality. Books have that. The problem is to explain how something could have non-derived intentionality, and the pencil sharpener, for example, doesn't take me any distance at all in the direction of doing that.

Perry: Well, not unless you pay careful attention to the details. Maybe that helps.

Fodor: I think if you cancel the talk about the goals, John, if you just drop that out, there's no intention — to put it very crudely, there's no smidgeon of intentionality left. The intentionality comes in by postulating God. The reason the pencil sharpener is intentional is because it's got a creator who has intentions. That I don't call naturalizing.

Dretske: I agree with Jerry, but isn't it true that if we're not worried about the derived/original distinction, gas gauges misrepresent sometimes when the battery gets low and they can misrepresent how much gas you've got left. We all talk this way. And if misrepresentation is intentionality, then gas gauges, simple electromechanical devices, have it. Their misrepresentation derives from the way we use them and our intentions of using them that way.

Perry: But there's no mathematical theory, even for those simple cases, that allows one to say carefully what you can say in English. That's what we're looking for — it's not this metaphysical stuff. Maybe there are two projects. But the one project, the project we're talking about is to get the mathematical framework that allows you to say very clearly exactly how a false proposition can be embedded in a true predicate, that helpfully characterizes in some way a system for some purpose. We claim no one has it, and that's what we're looking for. And until you have it, it seems a bit premature to get all worked up and believe in consciousness as posing some kind of insurmountable problem.

Pylyshyn: But why do you want a theory of intentionality for a system that doesn't have it?

Perry: So I can understand the mathematical nature of it.

Pylyshyn: That doesn't seem to me like the right way to go about it. Look, I think that there are two problems here. First, in virtue of what does a system have particular intentional contents? Second, what semantic contents do states of a system have? And I think that if you can separate these, then you can deal with the second set of questions, which we do all the time in cognitive science. The way you deal with the second set is that you ask yourself what contents you have to postulate in order to account for the generalizations that capture what's systematic about behaviour. So I find myself having to postulate that I have the belief that someone's going to reimburse my expenses, and a lot of generalizations about what I would do follow from that. And we build up a whole system that way. That's how we assign content. Now in virtue of what properties of my brain and its relationship to the world, does it have those contents? I haven't a clue and I leave it up to you guys to figure it out, but whenever I see you trying to do that, I'm glad I'm not in that business! I think it ought to be done, but it's just that it's factorable from the other part of the problem.

Dretske: John, I think you're after the same thing I am. I don't particularly care about the mathematics of intentionality, how that comes out. Maybe I misunderstand why you're interested in the mathematics. I thought the philosophical problem was to understand how a false proposition truly characterizes a physical system and can figure in the explanation of what that system is doing, why that should be a useful or possible part of the explanation of what a system is doing. Now how that works out formally or how you might symbolize it mathematically doesn't seem that central.

Perry: By "mathematically," I just mean "done with a great deal of care." So we need this constraint, this constraint, and this constraint, and this state, and if you have all those things, you have this, and you can say what those are.

Dretske: What's wrong then with just saying, look, we have examples about which we can be fairly rigorous. The thermostat turned the furnace on because it misrepresented the temperature of the room and it misrepresented the temperature of the room because someone was holding this ice cube up. Now, there we have a case where the thermostat, as it were, "says" the room is very cold and does something because it "says" that falsely, and does something because it has that false proposition characterizing it. There we've got what some people regard as the interesting beginning of intentionality. And I guess Jerry was saying that he doesn't find that terribly interesting — and I don't either — because all that's true of it in a metaphoric or extended or derived way.

Israel: What is this "metaphorical"?

Perry: Suppose somebody came along and said measurement has got to go. In fact, somebody did say that. Some philosophers are losing sleep about numbers, so I'm supposed to lose sleep about them? Measurement is very similar. You embed an abstract object in a predicate that some things have and some things don't, and somehow get an empirical predicate out of an abstract object. Matter of fact, you get empirical predicates out of abstract objects, some of whom are infinite, some of whom have probably never characterized anything. There are a lot of philosophical problems there. The first step, it seems to me, to understand those philosophical problems is to look carefully at the construction of measurement predicates, rather than starting with the assumption that there's something that has to be reduced. Instead of seeing how you're going to get rid of measurement predicates, maybe the thing to do is to study them, understand them, and see how they work. That's a perfectly good strategy.

Fodor: John, doesn't it bother you in choosing that case, that if all the people died, it would no longer be true that the thermostat says the room is too cold?

Perry: I can show you texts in the history of philosophy where people have worried about measurement, that if everybody died, there wouldn't be any numbers.

Fodor: They're wrong, but are you inclined to say that that's wrong in this case too?

Perry: Yes. If you're interested in grants and not philosophy, just try to understand the details of intentional attribution, keep at the details, and maybe slowly things will fall in place.

Israel: We're less ambitious than you guys, you might say. We just want to do a little bit of science and understand it.

Perry: One distinction is between "instrinsic" intentionality — that's what you need God for — and "natural" intentionality. Now, I agree that people shouldn't go around saying they have a theory of natural intentionality because they have a theory of evolution, and all that stuff, but it seems to me that it might be alright to say, well, I'm not going to lose too much sleep about natural intentionality. If I think that intentional attribution can go forward, then I think that there's two ways things can happen naturally: by being created for certain purposes by other things or by evolving. Then if I can work on getting the notion of the intentional content throughout the system, relative to constraints, relative to my sense of goals, if I can get that all straight, why would I lose sleep over the subject?

Fodor: Because I suspect that the way in which non-natural goals support intentionality is just *radically* different from the way in which natural goals could support intentionality, and that you could know everything you want to know about how people's goals support intentionality and have no idea at all of what the intentional content of the fly's visual state is when it's next to the frog. You just can't transfer the theory of one case to the other.

Perry: Let me suggest a very vague strategy. There are two objective 'pillars' in the world. One is information. It is an objective fact about whether a statement carries certain information, relative to true constraints; and the other is something like 'needs.' It is just a fact that there are certain needs. So the hunch is that you don't need intrinsic intentionality but you do need natural intentionality, and the starting point of natural intentionality, which is still probably far short of beliefs and desires, which are very complicated devices having to do with language and communication and people's intercommunicating, is probably that relation where the content of a state is appropriately related both to needs of the system whose state it is, and to information. So you have those two objective touchstones. That's not a reduction; I'm embarrassed to even say it, except that you laid the groundwork today and said, well, all I ask for is a little handwaving. That's the step we see between the uninteresting — except maybe as a way of getting started — notion of intentional content of a device relative to the goal that isn't its natural goal but just of something that comes from the outside, and the application of that, that claims that in some sense we have natural goals. Maybe it won't work, but it doesn't seem to be silly.

Fodor: I don't think it's silly. In fact, it's a very natural place to look. I just don't see any way of making it work. People have been beating around in that area for several years now, and at least the proposals I've seen all fail in the same way. What you need is a notion of goal which is determinate enough to give you an appropriately determinate notion of content. In fact, though, in *natural* technology those two indeterminacies go together. I claim that for every case where there's an indeterminacy of intentional content, there'll be a corresponding indeterminacy in the way you should describe the goal.

Israel: That's close to a theorem of our theory.

Fodor: But then it's going to be a theorem of your theory that in all sorts of cases where you want to appeal to the intentional content of the state, there will be no matter of fact about the intentional content of the state. But then you can't do the science.

Perry: That's like saying that there's no matter of fact because there's centigrade and all these other ways of measuring.

Fodor: This really comes down to the question that Fred's been asking and that I think I want to answer a different way than the way Fred does. I want there to be determinate matters of fact about intentional states because I want it to be in virtue of their possession of intentional states that organisms satisfy the laws of psychology. If I don't have a determinate notion of intentional state, then I will have no determinate notion of instantiating an intentional law, and when you don't have any determinate notion of instantiating an intentional law, you don't have intentional science. That's how these things connect. The one thing you really have to have for a science is cases where there's a matter of fact that the antecedent of some causal law has been satisfied.

Dretske: Could I make one remark? A question Hanson asked, which nobody addressed, concerned the connection between the meaning of something and the information carried. It has always seemed to me that's fairly obvious, but let me just say what I think the answer is, and see if anyone wants to talk about it. Obviously, when you say something false, you can nonetheless convey a lot of information, not necessarily about what you said, but about you and whether you've got a cold or not. All those facts are, I guess, obvious enough to everybody who worries about communication — that you often convey much more than you intend by what you said. And, often, what you say doesn't carry any information, although how you say it, and when you say it does. Only incidentally is the information that *p* communicated when you say that *p*. Very often that information is not communicated even when *p* happens to be true when you say it. Supposing I know that *p*, nonetheless, if I'm a notorious liar, I can say to you that *p* and assure you that *p* is true without communicating the information that *p*. That is, I can't bring you to know that *p* when I know that *p* if I'm a notorious liar because I'm not a channel through which the information can flow. Even though it got to me, I can't, as it were, communicate it. Nonetheless, in saying that *p*, I communicate a great deal of information about other matters unrelated to the content of what I say.

Asher: I have one question about the truism, and that is the use of the word "incidentally," insofar as it does not seem to me that it's a purely random track.

Dretske: That was a bad choice of words, you're right.

Asher: You'd want to say that the meaning of the statement plays some causal role in the communication of information and sure, it depends on a lot of other factors, in particular on how reliable the recipient takes the speaker to be and other things. Otherwise, I would think that the truism is right. If we are going to talk about that though, since that issue came up, I would like to just ask one question of David

Israel and John Perry. It seems as though one thing that you guys were doing — and I think it's an interesting task — is investigating this common sense theory of information, the matter-of-fact theory of information. You had these principles, A through J, that seemed to be right, and Fred Dretske also has lots of these principles A through J. They need an explanation and you'd like to have a theory of this. Now you could do a theory in situation semantics. You could also try writing out such a theory so as to reduce the notion of 'carries the information that ϕ', where that's a relation understood to hold between a fact and a proposition, to some kind of more belief-based notion of information, though it would seem that the problem that Fodor was raising would still remain. He would still want to know how to reduce the content of beliefs to something non-mental. But in any case, it seems like the two projects, yours and Dretske's, are just quite different.

Israel: The project of reducing or analyzing intentional content to informational content looks hopeless from the beginning, if you assume, as we do and as Fred does, that the informational content carried by a fact must be the case, that it's truth-involving. Because intentional content isn't. So it just looks like it's going to be really hard. This is not ordinary language, let me tell you!

Fodor: Well, that says they can't be identified. It doesn't say that a reduction can't be carried out.

Israel: No, that's right, I agree.

Asher: Why not an additional constraint that it be veridical and causally sustained? And actually, if you think about what facts are, veridicality should drop out for free, because any time you have that the fact that ϕ conveys the information that p, if you've got the fact that ϕ you know that ϕ automatically obtains.

Israel: No, what's of interest is the thing on the other side, namely conveying the information that p. In principle, I suppose, as you say, one could think of trying to analyse informational content in terms of intentional content plus some other stuff.

Asher: A notion of fact, a notion of relation between fact and proposition, which is what you guys have.

Israel: Right. I guess I'm a little at a loss to see what's at issue here.

Asher: That would give you the common sense theory of facts and information.

Israel: I see. Look, it is part of our account that we are absolutely happy about that we are awash with information. For us, it's good to start out with what is, in a certain sense, a very weak notion. There's the notion of information or of a state carrying information, both very weak, not terribly interesting notions. The notion of a state carrying information relative to a constraint is more interesting. It's still weak, in a certain way. That is, it's not terribly explanatory, if you like. That, by us, is good, that you should start out with such a notion like that.

Smith: I'd like to offer a sleeping pill to Jerry. You were saying that its seems like there are two possible ways to pursue psychology. One is actually to try to explain behaviour in terms of intentional content, and another is neurophysiology. What's striking to me is that there's sometimes thought to be a third possibility, which is to talk about computational states. Now, I guess what interests me is that computational states have not been mentioned at this conference really at all. I think that's actually proper but striking. The reason that I think it's proper is that I actually think all the current theories we have of computation are completely false. What computers actually are are just devices that have content. So I don't think there are any worries about the possibility of giving accounts that yield intentional content. In that respect, Jerry can sleep at night. However, I would add a cautionary note. I don't think we understand what it is to build computers. Nor do we understand what it is to be a computational explanation. I think that what's going on this weekend is an absolutely wonderful exercise in the foundations of computation. I'm not saying I think any of the problems that you think aren't solved, are solved. But I don't think there's any reason to worry that the project is not going to work. I think we're awash in the project's working! We just don't understand what the hell the project is.

Fodor: The trouble is, it's not unreasonable to believe, in fact I really do believe, not as a matter of principle, just as a matter of fact — that all those gadgets out in Silicon Valley are just pencil sharpeners; they're only intentional exactly the same way the pencil sharpeners are intentional. In fact, even the programs have content only relative to certain decisions about interpretation made by the programmer.

Smith: As it happens, I think they're not like pencil sharpeners. I also don't think there's any good argument that their intentional traits are all derivative on ours.

Fodor: But I think that's an interesting question. I think the reasons that have been given to the contrary have involved in a certain sense functionalist theories of content, that is, some notion that content emerges from, roughly speaking, computational causal role, and I think it's just inconceivable that such theories are correct. Now I could be wrong about that, but, but if I'm right about it then, contrary to what has been told to the Defense Department, they're really just pencil sharpeners. And what I'm losing sleep over is the guy who says the same thing about the computers that he says about us; that is, "Look, it's true that it's quite useful to talk about them this kind of way for certain purposes, I mean, for example, if you're selling them to the defense department. But in fact it's no more true of them that they've got intentional states or that their explanation is, requires, intentional thought, than it is about us. In fact, there *is* no such thing as an intentional state, *a fortiori* there are no intentional states."

Smith: That's not what they're saying.

Fodor: I agree that's not what *they're* saying. That's what they ought to be saying under certain conditions. The conditions are that it's impossible to give a naturalistic account of intentionality. Then all these ways of talking are just, you know, ways of talking — they don't mean anything. They're not science. They're not telling the truth about the world.

Smith: Something's been missed here. Pencil sharpeners were built before 1957. Computers weren't. Something happened. I actually think there is a crucial difference, a crucial property that computers have which pencil sharpeners do not. Now you're absolutely right about the defense department. There is in fact a funded project which is going to run the control room in the Cheyenne Mountains, for a full year following the demise of all people. If it is derived intentionality then on the day when we're all gone and the computers in the control room are just blinking away, according to your account, it may be that all of their content suddenly went away, as it were.

Fodor: No, look, Brian, I actually think that's wrong, and I think it's wrong for the following reason. It's true that computers answer a semantical question that we couldn't answer before. I mean, I'm very pro-computer. I think Turing was the only guy who ever had a *real* idea in Cognitive Science, I think we've just been living on his waste basket ever since. I'm very pro-computer but not for that reason. See, they do answer a semantical question that we couldn't answer before. The semantical question is roughly this: how could semantical properties be preserved by the transactions of a mechanism? That's an extremely interesting question. It's roughly the question of how you could have a mechanical theory of rationality. But it's a different question from the question of how you could have a mechanism have semantical properties. Computers answer the question 'given that this thing has semantical properties, how could its processes preserve it?' They don't answer the question any more than pencil sharpeners do, how they have semantical properties in the first place.

Smith: What if you propose to use computers in situations where the people who will actually buy them don't interact with them intentionally? I think there's something striking about that. Nowadays, people want to automate their houses and, you know, the computers involved actually interact with the weather and radio station and the like. They're out there participating in the world, not interacting with you under interpretation. Now, it's been suggested that they preserve semantic properties. But they don't interact with us semantically. So it seems to me very odd that they don't have semantic properties on their own.

Pylyshyn: How do they preserve the semantic properties if they don't have them? I think there's something right about what Brian is saying. There is something special about computational systems, and whether or not they have intrinsic intentionality remains to be seen. Your faith that they don't have it suggests that you have some criteria for what it is to have it, and if we had some criteria for what it is to have it, we'd really be in pretty good shape.

Fodor: I told you what it is to have it. Well, my account won't work, but I hope some lineal descendent of it in ten thousand years will. The point is, I'm not Searle. I'm not saying that you couldn't have a device made out of silicon which has bona fide intentional states. But it would have to do things — I don't know exactly which things, I'm trying to construct a theory about it like everybody else — but it would have to do things that none of those devices do. I suspect it would have to interact. Its internal state transitions wouldn't do it. They explain how it can preserve semantic properties, not how it can have them. It would have to interact with the world under certain conditions that I always thought the theory of information was trying to specify. But it's not, because you keep telling me, or I keep being told now, that the theory of information is not trying to specify conditions sufficient to having intentional states.

Smith: I'm not defending an informational theory of intentionality. All I'm saying is that all of the work in computer science at the moment has to do not really with internal state transitions but with connections with the world. That's what's going on in Silicon Valley, and I think that's interesting since a lot of smart people, some of whom are in this room, think two things. One is that ways of interacting with the world seem to be actually what's requisite in order to convey the real intentionality on to the source of determination. The other is that you probably don't have to have a theory of intentionality represented inside a machine in order to have intentionality. Current Silicon Valley machines don't have theories of intentionality. On the other hand, they are connected to the world so nothing that you guys have said makes me think that these machines may not actually be intentional.

Fodor: Oh, but that's funny, I mean there are all sorts of things that are connected to the world which clearly don't have intentional states, for example, this glass.

Pylyshyn: What we need is, we need a theory which tells us what the right way is. That's the foundational problem.

Smith: I'm agreeing. We don't have that theory and we need it. My only point is, I think there's good reason to believe that the project is going to work because we've got people constructing nonunderstood incidences.

Fodor: Let me just say one thing about that, Brian. Look, our sympathies are in exactly the same place. But it's a little question-begging. The question it's begging is that there are such states for the machine to have. Think of it another way. Think of it in the tone of voice of somebody who says 'Look, that's just fairy tales.' It's not that there are such states and we just have to get the machine to bang on the world in the way that's required to have them. There is no such thing as believing that *p*. There just isn't such a state, in which case, of course the machines aren't going to have them. Right? In fact, nothing has them. In fact, if you talk about believing that *p*, you're just not doing science.

Smith: Whoa, whoa, whoa ... you yourself said that computers are impressive because what they do is this — they have shown how it's possible to have mechanisms that appear to preserve and manipulate internal configurations which are in fact usefully classified under semantic interpretation. They have that fact which glasses don't. Furthermore, the advent of recognizing that you can build a chip like that caused an enormous revolution. So just from that point of view it seemed like a good idea, it had some force. Then the proposal is that what it is to have real content isn't just to have that stuff. It's also to be connected with the world. And those things *are* also connected with the world.

Fodor: But Brian, Turing showed us how to preserve content in case there is such a thing. That doesn't make it plausible all by itself that there is such a thing. I mean, it's true that we now have a theory of content-preserving transformations, in case there's some such stuff to preserve ...

Hanson: Time, gentlemen, please.

CHAPTER EIGHT

Belief and Mental Representation

Scott Soames*

In "Propositional Attitudes"[1] Jerry Fodor argues that beliefs and other propositional attitudes are "relations between organisms and internal representations" (177). However, this view is far from transparent, and Fodor fails to point out that it has several different interpretations. Most relevant are a strong interpretation, expressible as (1a) or (1b), and a weak interpretation, given as (2a) or (2b).[2]

(1a) For all declarative sentences S of English, there is a mental representation M such that for all individuals i (and times t), i satisfies (at t) ⌜x believes that S⌝, as used in a context of utterance C, iff i bears a certain relation R to M (at t), and the content of S in C is identical with the content of M, when taken as one of i's mental representations (at t).

(1b) For all propositions p, there is a mental representation M such that for all individuals i (and times t), i believes p (at t) iff i bears a certain relation R to M (at t), and p is the content of M, when taken as one of i's mental representations (at t).

(2a) For all declarative sentences S of English, individuals i (and times t), i satisfies (at t) ⌜x believes that S⌝, as used in a context C, iff there is (at t) a mental representation M such that i bears a certain relation R to M (at t), and the content of S in C is identical with the content of M, when taken as one of i's mental representations (at t).

(2b) For all propositions p, individuals i (and times t), i believes p (at t) iff there is (at t) a mental representation M such that i bears a certain relation R to M (at t), and p is the content of M, when taken as one of i's mental representations (at t).

In distinguishing these proposals it is important to bear in mind two limiting cases in which the differences between them are minimized. One case involves the assumption of a many-one correspondence between (believable) propositions and (a subset of) mental representations. Suppose that for every proposition that could be believed, there were exactly one mental representation that was capable of expressing it (in at least one context). Then, if x believed p in the sense of (2), it would follow that x bore R to a representation M that everyone who believed p had to bear R to. Hence the necessary conditions for believing p laid down by (1) would be satisfied. They would also be satisfied if R were so defined that it was impossible for an individual to bear R to a representation M without simultaneously bearing it to all other mental representations (including those of others) with the same content as M. In the first of these limiting cases the assumption of a many-one correspondence has the effect of strengthening (2) so that it encompasses (1). In the second limiting case a permissive definition of R weakens (1) to the point that it is subsumed by (2). Outside of these limiting cases the proposals are substantially different.

The second limiting case is ruled out by Fodor's insistence that the relation R be "syntactically," rather than semantically, defined. This is part and parcel of his view that any adequate philosophical analysis of belief must mesh with empirical accounts of mental processes — in particular, those of cognitive psychology. (This is his "condition V" on an adequate analysis of belief.) Thus, he says:

> Condition V, it will be remembered, permits us to choose among theories of PAs [propositional attitudes] in virtue of the lexico-syntactic form of the entities they assign as objects of the attitudes . . . it's not just cost-accounting [i.e., ranking beliefs by their psychological complexity] that is supposed to be determined by formal aspects of the objects of PAs; it's all the mental processes and properties that cognitive psychology explains. That's what it *means* to speak of a *computational* psychology. Computational principles are ones that apply in virtue of the form of entities in their domain. (201)

For Fodor, mental representations are syntactic objects on which individuals perform computational operations. For the belief relation R to hold between an individual i and a representation M is for computations involving M to play a certain kind of functional role in the mental life of i. It is not enough for M to have the same content as some other formula on which the computations are actually performed. Thus, in explaining why English sentences cannot play the role of objects of R he cites believers who do not know English, and hence do not use English sentences.

> Of course, relations are cheap; there must be *some* relation which a dog bears to 'it's raining' iff the dog believes that it's raining; albeit, perhaps, some not very interesting relation. So, why not choose *it* as the relation in virtue of which the belief-ascription holds of the dog? .... To put it generally if crudely, satisfying condition V depends on assuming that whatever the theory takes to be the object of a PA plays an appropriate role in the mental processes of the organism to which the atitude is ascribed. But English sentences play no role in the mental life of dogs. (192)

In light of this it is clear that R is to be understood in a way that does not result in the collapse of the theses in (1) into those in (2).

The other limiting case — in which the theses in (2) are strengthened to encompass those in (1) — is one in which a many-one correspondence between (believable) propositions and mental representations is assumed. Although Fodor does end up making this assumption, it does not provide a rationale for failing to distinguish (1) from (2). On the contrary, he makes the assumption because he runs (1) and (2) together, without exploring their differences.

Throughout most of the article, Fodor writes as if he were proposing (1). However, at various places in the discussion it seems as if (2) were at issue. For example, in defending his view against the objection that belief is a relation to a *proposition*, he indicates that as long as the relation is understood to be one that is *mediated* by mental representations, it need not conflict with his view.

> I am taking seriously the idea that the system of internal representations constitutes a (computational) language ... nothing stops us from specifying a semantics for the IRS [internal representational system] by saying (inter alia) that some of its formulae express propositions. If we do say this, then we can make sense of the notion that propositional attitudes are relations to propositions — viz. they are *mediated* relations to propositions, with internal representations doing the mediating.
>
> This is, quite generally, the way that representational theories of the mind work. So, in classical versions, thinking of John (construed opaquely) is a relation to an "idea" — viz., to an internal representation of John. But this is quite compatible with its also being (transparently) construable as a relation *to John*. In particu-

> lar, when Smith is thinking of John, he (normally) stands in relation to John, and does so *in virtue* of his standing in relation to an idea of John. Similarly, mutatis mutandis, if thinking that it will rain is standing in relation to a proposition, then, on the present account, you stand in that relation in virtue of your (functional/causal) relation to an internal formula, which expresses the proposition. (200-1)

This passage strongly suggests the theses in (2). To think of John one must have an idea that bears the right relation to him. However no one of the potentially many ideas of John is privileged over all others; there is no one idea such that to think of John one must entertain it. If, as Fodor claims, believing a proposition p is like this, then there should be no single mental representation such that to believe p one must stand in the appropriate mental relation to it. Rather, there should be potentially many mental representations that express p and are capable of mediating belief in it.

Nevertheless, it is evident that this is not the picture that Fodor has in mind. For example, in claiming that propositional attitudes are relations to internal representations, he is claiming that "in particular, the verb in a sentence like 'John believes it's raining' expresses a relation between John and something else, and a token of that sentence is true if [and only if][3] John stands in the belief relation to that thing" (178). Fodor could not maintain that beliefs are relations to mental representations *in this sense*, if he thought that (2) were true and that it were possible for different mental representations to have the same content (while being independently possible objects of the relation R).[4]

Fodor's commitment to (1) is most clearly illustrated by the way he motivates his view. After arguing that relations to linguistic representations play an important role in the analysis of propositional attitudes, he considers a "Neo-Carnapian" analysis, which serves as a prototype of his own theory. Fodor initially characterizes this analysis as one in which propositional attitudes are construed "as relations between people and sentences they are disposed to utter, e.g., between people and sentences of English" (187). He does not regard the behavioural characterization of the belief relation in terms of verbal dispositions as crucial to the account, and suggests that it might be functionally characterized instead. However he does indicate that the analysis is to be a strong one, in which to believe that Bill bit Mary is to bear the belief relation to "Bill bit Mary," and to believe that it is raining is to bear the belief relation to "it is raining" (rather than to any sentences that happen to have the same contents as these sentences)

(188). In short, Fodor's prototheory is to be understood along the lines of (3a) or (3b).[5]

(3a) For any sentence S of English, there is a sentence S′ of English such that for any individual i (and time t), i satisfies (at t) ⌜x believes that S⌝, as used in a context C, iff i bears a certain relation R to S′ (at t), and the content of S in C is identical with the content of S′ in an appropriately related context with agent i (and time t).

(3b) For any proposition p, there is an English sentence S′ such that for any individual i (and time t), i believes p (at t) iff i bears a certain relation R to S′ (at t), and p is the content of S′ in an appropriately related context with agent i (and time t).

Fodor uses this prototheory to motivate his own account. According to him the prototheory accommodates most of the observations that indicate that beliefs are relations to sentences, but fails to account for certain obvious facts. For example:

(i) the fact that it is possible to believe that it is raining without understanding "it is raining," or any other English sentence;
(ii) the fact that in certain cases different English sentences express the same belief (e.g., "John bit Mary" and "Mary was bitten by John");
(iii) the fact that some beliefs are inexpressible in English; and
(iv) the fact that some individuals hold beliefs without having learned any natural language.[6]

In summarizing these objections Fodor notes that (i) "would be without force if only everybody (viz. every subject of true propositional attitude ascriptions) talked English," that (ii) and (iii) "depend on the empirical likelihood that English sentences fail to correspond one to one to objects of propositional attitudes," and that (iv) "would be met if only English were innate" (197).

His strategy is simply to posit an internal language of mental representations with precisely the properties needed to avoid these objections, and to substitute its formulae for sentences of English as objects of the belief relation R in the prototheory.

> Indeed, I suppose an ultra hard-line Neo-Carnapian might consider saving the bacon by claiming that — appearances to the contrary notwithstanding — English *is* innate, universal, just rich enough, etc. My point is that this is the right *kind* of move to make; all we have against it is its palpable untruth.
>
> Whereas, it's part of the charm of the internal language story that, since practically nothing is known about the details of cognitive processes, we can make the corresponding assumptions about the internal representational system risking no more than gross implausibility at the very worst.

> So, let's assume — what we don't, at any event, *know* to be false — that the internal language is innate, that its formulae correspond one to one with the contents of propositional attitudes (e.g., that "John bit Mary" and "Mary was bitten by John" correspond to the same "internal sentences"), and that it is *as* universal as human psychology; viz., that to the extent that an organism shares our mental processes, it also shares our system of internal representations. On these assumptions, everything works. (197)

The end result is Fodor's theory, (1), of propositional attitudes.

This result was not inevitable. Fodor's objections to the prototheory are handled just as well by (2) as they are by (1). In fact, had he started with Carnap's actual theory,[7]

(4) An individual i satisfies ⌜x believes that S⌝ in English iff there is a sentence S′ and language L′ such that i is disposed to assertively utter (bears R to) S′ as a sentence of L′, and S′ in L′ is intentionally isomorphic to (has the same content as) S in English

several objections to the prototheory would not have arisen in the first place.[8] At most, objection (iv) might prompt a move from (4) to (2). However, this leaves the choice between (1) and (2) wide open. To make this choice we need to look closely at the ways in which these theses differ.

THEORETICALLY SIGNIFICANT DIFFERENCES BETWEEN (1) AND (2)

The first point to notice is that the proposals in (1) carry the strong presumption that each proposition is expressed by at most a single mental representation. For if a proposition p were expressed by more than one such representation — M, M′, M* — (in the same or different contexts), then there would be no evident reason why bearing R to *any* of them shouldn't count as believing p. Since this is allowed by the proposals in (2), they do not carry any presumption that every proposition is expressed by only one mental representation.

The proposals in (1) also entail that the propositions expressed by mental representations can vary from one context to another only if those propositions are cognitively inaccessible outside of certain privileged contexts. For suppose that M expresses p in context C and fails to express p in other contexts (either because it expresses other propositions or because it expresses no propositions in those contexts). Suppose further that the agent i of C believes p (at the time and place of C) by virtue of bearing R to M. It will then follow from (1) that for all C′ in which M does *not* express p, the agent i′ of C′ does *not* believe p (at the time and place of C') — *no matter what other mental representations i' may bear R to (at that time and place).*

The implausibility of this result can be illustrated by applying it to the prototheory (3b). For example, sentence

(5a) Classes start today

expresses different propositions on different days. It follows from (3b) that if a person believes p by virtue of accepting (bearing R to) (5a) on day d, then p cannot be believed on any other day — even if one accepts (5b) on day d-1, (5c) on day d + 1, or (5e) on an arbitrary day (where "day d" is a proper name of a day).

(5b) Classes start tomorrow.
(c) Classes started yesterday.
(d) Classes start on day d.

In short, the proposition expressed by (5a) on day d is cognitively inaccessible on other days. Analogous results hold for other context sensitive sentences. For instance, if I believe that I live in New Jersey by virtue of accepting

(6a) I live in New Jersey,

and Ruth believes that she does not live in New Jersey by virtue of accepting

(6b) I do not live in New Jersey,

then no one but me can believe that I live in New Jersey, and no one but Ruth can believe that she doesn't live in New Jersey, and no one at all can believe the conjunction of our two beliefs — namely that I live in New Jersey and she doesn't.

Such results may well seem absurd. However if mental representations are allowed to express different propositions in different contexts, then they will apply to Fodor's theory (1) as well. In light of this it is extremely tempting for a proponent of (1) to deny that mental representations are context sensitive — a proposal implicit in Fodor's assumption that the formulae of our mental language "correspond one to one with the contents of propositional attitudes" (197).

However, tempting or not, this strategy is unacceptable. One immediate consequence of it is to exclude indexicals from the system of mental representations. How then would one handle cases in which an individual expressed a belief p about an object o using a natural language indexical to refer to o? Presumably, the use of the indexical would have to be mapped onto a nonindexical representation that differed from the representation for every other object, while being identical with the one used by all other agents to believe the proposition p, about o. Thus, every moment of time about which one could express a belief using the indexical "now" would have to have its own unique name — the same for all believers — in the system of mental representations; similarly for every person and every grain of sand about which beliefs could be expressed using 'I' or 'that.'[9] The implausibility of this is an indictment of the strategy of trying to do away with indexicality in mental representations.

Moreover, the problem goes beyond implausibility. Like most proponents of mental representations, Fodor believes that they are "in the head." Thus, intrinsic properties of an individual's brain (at a given time) should determine which representations he entertains (at that time). One consequence of this is that physically indistinguishable individuals — e.g., molecule for molecule duplicates — must entertain the same representations.[10] But then the nonindexical strategy just sketched will fail to handle indexical variants of familiar "Twin Earth" cases.[11]

One such variant is David Kaplan's example of identical twins, Castor and Pollux, who are molecular and behavioural duplicates.[12] Each expresses a belief by sincerely uttering "I am older than my brother." Since one of the twins thereby believes a true proposition (about Pollux), whereas the other believes a false proposition (about Castor), the propositions believed are different. However, since the twins are duplicates, the beliefs must arise from the same mental representation. Thus, the strategy of mapping noncoreferential uses of natural language indexicals onto different nonindexical mental representations cannot succeed.

This leaves the proponent of (1) with no successful way of avoiding the seemingly absurd result that propositions expressed by indexical sentences, like those in (5) and (6), cannot be believed (or even apprehended) except in contexts in which they are expressed by those very sentences. Standard Twin Earth cases involving names and natural kind terms show that the same sort of absurdity arises even when indexicals are not involved.

For example, suppose that Oscar and Oscar′ are molecular duplicates who use the same natural kind term N to name different but observationally indistinguishable natural kinds. Both express beliefs by uttering ⌜N is F⌝. However the propositions believed are different; Oscar believes the proposition that kind K is F; Oscar′ believes that K′ is F. Since they are duplicates, their beliefs arise from the same representation M, which must have one propositional content when entertained by Oscar and a different content when entertained by Oscar′. It follows from (1) that each of these propositions p is such that an individual can believe p only if he bears R to M in a context in which p is the content of M. This means that as long as Oscar retains his belief that K is F, it will be impossible for him to acquire the belief that K′ is F. No matter that he may do what any other believer that K′ is F might do — namely encounter K′, use a natural kind term N* to name it, and sincerely assent to ⌜N* is F⌝.[13] As long as he continues to believe that K is F, the contentof M, when entertained by him, must be that proposition — rather than the proposition that K′ is F. Thus, (1) will tell us that he cannot believe that K′ is F. In fact, as long as M cannot simultaneously have different contents when entertained in the same context by the same individual, (1) will tell us that no one can believe both propositions. Surely this is incorrect.

In sum, the proposals in (1) cannot accommodate the following elementary facts:

(a) The same formula may express different propositions when interpreted from the point of view of different contexts, and different representational systems.

(b) The same proposition may be expressed by different formulae of the same system, or of different systems.

Faced with obvious instances of (a) in natural language, the proponent of (1) has two choices. He may attempt to explain away the relativity of linguistic content as a superficial feature of natural language which disappears at the more fundamental explanatory level of mental representation, or he may allow such relativity in mental representations at the cost of making their contents cognitively inaccessible in a great many cases. Both of these alternatives are dead ends. The first requires the incredible assumption that each object about which one can have a belief p has a nonindexical mentalese representation, unique to it, the same for all believers of p — an assumption refuted by the various Twin Earth cases. The second alternative founders from not being able to accommodate (b). Two individuals — Ruth and I, Castor and Pollux, Oscar and Oscar′ — may share beliefs by accepting different sentences that express the same proposition. Since the proposals in (1) cannot account for these facts, they should be rejected.

PHILOSOPHICAL ADVANTAGES OF (2)

In addition to giving a more accurate account of the fundamental facts about propositional attitudes, the proposals in (2) avoid a number of undesirable philosophical implications of those in (1). One such implication involves the relationship between mental representations and natural languages. Whereas (2) comfortably allows the possibility that an individual's system of mental representations may contain elements of the natural language(s) he speaks, (1) does not. For example, (1) requires everyone who believes that many people suffer from Acquired Immune Deficiency Syndrome to do so by virtue of being related to the same mental representation. Presumably it is possible for non-English speakers to believe this proposition even though their systems of mental representations do not include English terms like "Acquired Immune Deficiency Syndrome." But then it follows from (1) that even the mental representations in virtue of which monolingual speakers of English believe that many people suffer from Acquired Immune Deficiency Syndrome do not include any English terms. In short (1) implicitly denies that the natural language expressions we learn are directly involved in the propositional attitudes we hold. It is an advantage of (2) that it does not prejudge the issue in this way.

A further point, closely related to this, involves Fodor's startling claim that mentalese is entirely innate, and that natural language expressions are learned by connecting them with already understood mentalese counterparts.[14] If, as (2) allows, the system of mental representations of a given person includes elements of his natural language, then, since natural language expressions are not innate, it will follow that a person's system of mental representations is not entirely innate either — a welcome result.

It will also follow that it is possible for the explanation of *some* of an individual's beliefs to depend on the content of certain of his mental representations, *which in turn may depend on the content of expressions in a public language*. This picture fits well with accounts that see the contents of, for example, names and natural kind terms as arising from social, historical, and causal connections relating a speaker's use of these terms to other speakers, and ultimately to objects in the world. On such an account there is a social process that is crucially involved in determining the content of some expressions in a public language, which in turn are responsible for the contents of some beliefs. It should be noted that this picture is at variance with Fodor's reductionist conception in which the intentionality of natural language is reduced to the intentionality of propositional attitudes, which is then supposed to be reduced to the intentionality of mentalese. Although the proposals in (1) do not by themselves dictate this reductionist picture, they reinforce it by conferring an undeserved primacy on mentalese at the expense of natural language.

Finally, there is the question of the epistemological and metaphysical status of (1). Fodor's intention in offering his theory of propositional attitudes as relations to representations is to present an empirical, a posteriori thesis that nevertheless gives the essence of propositional attitudes, and supports counterfactual generalizations.

This emphasis on specifying the essence of a phenomenon, though natural in a philosophical theory, suggests an assertion of necessity that is difficult to incorporate into the theses in (1).

For example, consider the following attempts to strengthen (1b) to cover counterfactual circumstances.

(7a) Necessarily for all propositions p, there is a mental representation M such that for all individuals i (and times t), i believes p (at t) iff i bears R to M (at t), and p is the content of M, when taken as one of i's mental representations (at t).

(b) for all propositions p, there is a mental representation M such that necessarily for all individuals i (and times t), i believes p (at t) iff i bears R to M (at t), and p is the content of M, when taken as one of i's mental representations (at t).

Neither of these is attractive. The former implies that in each world there is one representation to which individuals must bear R in order to believe a given proposition p. However, it does not require representations carrying belief in p to remain invariant across worlds. Thus, it allows an individual to believe p by bearing R to M_1 in W_1, M_2 in W_2, and so on ($M_1 \neq M_2 \ldots$). But if one allows variation *across* worlds, it is hard to find grounds for denying it *within* worlds. In particular, if different representations can carry the same belief under different conditions, then either extensive empirical investigation or further conceptual analysis is needed to establish that these conditions are not jointly satisfied in the actual world. Since neither has been (or is likely to be) forthcoming, the proponent of (1) is pushed inexorably to (7b). But this just makes things worse by compounding the problems of the previous section to include cross-world cases in addition to the intra-world cases already discussed.[15]

Other Interpretations Of Fodor's Thesis

There is, then, ample reason to regard the proposals in (2) as more acceptable versions of the thesis that beliefs are relations to mental representations than those in (1). However, before settling on (2), we would do well to note that there are interpretations of Fodor's thesis that are intermediate between (1) and (2).

(1.5a) For all propositions p, individuals i (and times t), there is a mental representation M such that i believes p (at t) iff i bears a certain relation R to M (at t), and p is the content of M, when taken as one of i's mental representations (at t).

(b) For all propositions p, individuals i (and times t), there is a mental representation M with content p, when taken as one of i's mental representations (at t), such that i believes p (at t) iff i bears R to M (at t).

These proposals do not have all the obvious difficulties of (1). Thus, the question arises as to whether or not they are acceptable alternatives to (2).

They are not. All (1.5a) tells us is that for any proposition p, individual i, (and time t) there will be at least one representation M that makes the following biconditional true: *i believes p (at t) iff i bears a certain relation R to M (at t), and p is the content of M, when taken as one of i's mental representations (at t)*. One way for a biconditional to be true is for both sides to be false; and one way for the right-hand side of this biconditional to be false is for p not to be the content of M. Thus, (1.5a) leaves open the bizarre possibility that i might fail to believe p (e.g., that i lives in New Jersey) simply by virtue of the fact that some mental representation (e.g. one underlying '2 + 2 + 4') does not have p as content (when taken as one of i's mental representations at a given time). This possibility could be discounted if it could be guaranteed that bearing R to a representation expressing p were sufficient for believing p. However, (1.5a) doesn't guarantee this. It *fails* to predict that if (at t) i bears R to some, or even all, mental representations of i that have p as content, then i believes p (at t). Thus, (1.5a) is too weak. This doesn't mean that it is untrue; since it is a consequence of (2b) it had better be true. It does mean that (1.5a) cannot be an adequate theory of belief.

Thesis (1.5b) suffers from different, but related problems. According to it, bearing R to *all* one's representations that have p as content is *sufficient* for believing p, and bearing R to *at least one* such representation is *necessary* for believing p. However, (1.5b) is silent about cases in which one bears R to some but not all of one's representations that have p as content.[16] Moreover, it requires something that is obviously too strong — namely, that each believer have (at any given time) mental representations sufficient to express all propositions. Thus, it too is inadequate.[17]

Propositions, Mental Representations, And The Explanation Of Behaviour

In light of this, it appears that (2) is the most promising version of the thesis that beliefs are relations to mental representations. However, if one does adopt (2), one must be careful to note that the objects of belief — the things believed — are contents or propositions expressed by mental representations, rather than mental representations themselves. Moreover, one must build into these propositions much of the syntactic structure of the representations that express them; otherwise one will lose benefits crucial to Fodor's appeal to representations in the first place.

Such benefits are illustrated by the following remarks.

> It's plausible to claim that there is a fairly general parallelism between the complexity of beliefs and the complexity of sentences that express them. So, for example, I take it that "the Second Punic War was fought under conditions which neither of the combatants could have desired or forseen" is a more complex sentence then, e.g., "it's raining"; and, correspondingly, I take it that the thought that the Second Punic War was fought under conditions which neither of the combatants could have desired or forseen is a more complicated thought than the thought that it's raining. (188-9)

The view that beliefs are relations to mental representations is meant to explain this parallelism.

> A theory of propositional attitudes specifies a construal of the objects of the attitudes. It tells for such a theory if it can be shown to mesh with an independently plausible story about the "cost accounting" for mental processes. A cost accounting function is just a (partial) ordering of mental states by their relative complexity. Such an ordering is, in turn, responsive to a variety of types of empirical data, both intuitive and experimental. Roughly, one has a "mesh" between an empirically warranted cost accounting and a theory of the objects of PAs when one can predict the relative complexity of a mental state (or process) from the relative complexity of whatever the theory assigns as its object (or domain) . . . . Again, roughly: to require that the complexity of the putative objects of PAs predict the cost accounting for the attitudes is to impose empirical constraints on the *notation* of (canonical) belief-ascribing sentences. So, for example, we would clearly get different predictions about the relative complexity of beliefs if we take the object of a PA to be the correspondent [complement] of the belief ascribing sentence than if we take it to be, e.g., the correspondent [complement] transformed into disjunctive form. (189-90)

In short, Fodor suggests that (8a) and (8b) differ in psychological complexity, and that this difference mirrors the difference in syntactic complexity between (8c) and (8d).

(8a) the belief that $P \supset Q$
(b) the belief that $(P \& Q) \vee (-P \& Q) \vee (-P \& -Q)$
(c) $P \supset Q$
(d) $(P \& Q) \vee (-P \& Q) \vee (-P \& -Q)$.

This suggestion is plausible. Moreover, if it is correct, then the propositions expressed by (8c) and (8d) cannot be identical. They cannot be identical, since the state of believing one of them places fewer demands on an individual's psychological system, in a sense that Fodor takes to be measurable, than does the state of believing the other. This means that propositions with the same truth conditions may nevertheless be distinct. In short, propositions cannot be what I have elsewhere called sets of truth-supporting circumstances.[18]

Let us suppose, with Fodor, that syntactic complexity is a good measure of psychological complexity, which in turn is reflected in the propositions believed. It is then appropriate to view propositions as themselves syntactically structured objects constructed out of the contents of the constituents that make up the representations that express them.[19] This conception of content guarantees that although many different representations may express the same proposition, such representations will in general share the same syntactic structure, and hence have the same psychological complexity. This is what is required by the combination of (2) with Fodor's account of the psychological complexity of beliefs.[20]

In what sort of cases, then, can we expect different mental representations to express the same proposition? Presumably in cases of the sort illustrated by the following pairs.

(9a) Today is F. (Said at 11:55 p.m.)
(b) Yesterday was F. (Said 10 minutes later.)
(10a) This (pointing at o) is F.
(b) That (pointing at o from a slightly different perspective) is F.
(11a) London is pretty.
(b) Londres est jolie.
(12a) Catsup is a tomato-based condiment.
(b) Ketchup is a tomato-based condiment.
(13a) Everyone who has heard of catsup believes that catsup is catsup.
(b) Everyone who has heard of catsup believes that catsup is ketchup.

I take it that in the case of each of these pairs the two sentences express the same proposition, in the relevant contexts. The question then arises as to whether the two sentences are mapped onto the same or different mental representations (in the contexts). A good reason for assuming the latter is that competent speakers who understand them may nevertheless be unaware that they express the same proposition (in the relevant contexts).[21] If such speakers are unaware that the sentences express the same proposition, then they may be expected to react differently to them. Given the role of mental representations in explaining the linguistic and other behaviour of speakers, one will thus be led to posit different mental representations underlying the sentences in the relevant contexts.[22]

This conclusion has important consequences for cognitive explanations of behaviour. According to Fodor, the paradigm for such explanations is one

> where propositional attitudes interact causally and do so *in virtue of* their content. And the paradigm of this paradigm is the practical syllogism.... *John believes that it will rain if he washes his car. John wants it to rain. So John acts in a manner intended to be a car-washing.* (183)

According to Fodor, this account

> might be counterfactual-supporting in at least the following sense: John wouldn't have car-washed had the content of his beliefs, utilities, and intentions been other than they were .... To say that John's mental states interact causally *in virtue of* their content is, in part, to say that such counterfactuals hold. (183)

Fodor continues:

> If there are true, contingent counterfactuals which relate mental state *tokens* in virtue of their contents, that is presumably because there are true, contingent generalizations which relate mental state *types* in virtue of their contents. So ... we can schematize etiologies like the one above to get the underlying generalizaton: if x believes that A is an action x can perform; and if x believes that a performance of A is sufficient to bring it about that Q; and if x wants it to be the case that Q; then x acts in a fashion intended to be a performance of A. (183)

According to Fodor, explanation in cognitive science requires generalizations of this basic form. In his view,

> we can't state the theory-relevant generalization that is instan-

> tiated by the relations among John's mental state unless we allow reference to beliefs of the form *if X then Y*; desires of the form *that Y*; intentions of the form *that X should come about*; and so forth. Viewed one way (material mode), the recurrent schematic letters require identities of content among propositional attitudes. Viewed the other way (linguistically), they require formal identities among the complements [or better, the mental representations of the complements — my addition] of the PA-ascribing sentences which instantiate the generalizations of the theory that explains John's behavior. (184)

But if (2) is correct, and if different mental representations may express the same proposition without the agent realizing that they do, then Fodor's conception of psychological explanation cannot be accepted as stated. More precisely, claims of the form (14) will *not* require formal identity of mental representations corresponding to different occurrences of the same schematic letter in specifications of the agent's beliefs and desires.

> (14) If x believes that A is an action that x can perform, and if x believes that performing A will bring it about that Q; and if x wants it to be the case that Q; then x will act in a fashion intended to be a performance of A.

Because of this, claims of the form (14) cannot be expected to be true, exceptionless, universal generalizations.

The reason for this is easy to see. One can believe that doing A will bring it about that Q by bearing the belief relation R to a mental representation (15a).

> (15a) IF I DO A, THEN M_1.

One can desire to bring it about that Q by bearing the desire relation to a mental representation (15b).

> (15b) M_2

In this example M_1 and M_2 share the same content, *that Q*. However, if the agent fails to realize this, then he may well lack any inclination to do A. In such a case the agent has the requisite beliefs and desires, described in terms of content; however, he lacks the expected behavioural disposition because he holds the beliefs and desires in an unusual way — namely by virtue of being related to distinct mental representations that appear to him to be unrelated. Since cases like this can be expected to occur on at least some occasions, the pre-established harmony that Fodor imagines holding between explanations of behaviour that appeal to contents of propositional attitudes and explanations that appeal to internal computational operations breaks down.

In my view, this partial breakdown should not be cause for alarm. Generalizations like (14) require *ceteris paribus* clauses in any case, independent of any difficulties noted here. It is hard to see why we should be overly concerned about including in such clauses some indication that, for example, the agent apprehends the content *that Q* in the same way (i.e., via the same representations) in the relevant beliefs and desires. Moreover, whatever questions there may be regarding the details of such generalizations, such questions provide no grounds for a generalized doubt about the correctness of particular pretheoretic explanations of actions in terms of beliefs and desires. John may do A because he wants it to be the case that Q and believes that doing A will bring this about, even if not everyone with those beliefs and desires would do A. This is no more mysterious than the observation that John may fall and break his leg because he steps on a banana peel, even though not everyone who steps on a banana peel suffers a similar fate. Thus, the partial breakdown of Fodor's pre-established harmony need not threaten most ordinary explanations of actions in terms of beliefs and desires.

Nor does it threaten (2). Indeed, I can see no reason to doubt that the analysis of propositional attitudes given in (2) is superior to that given in (1). This is not to say that (2) has been established. Important questions remain about the exact nature and status of mental representations. Although I will not try to answer them here, I do hope to have shown that if such representations play a role in the analysis of propositional attitudes, then they do so along the lines of (2) rather than (1).

APPENDIX: SCIENTIFIC PSYCHOLOGY AND THE NOTION OF NARROW CONTENT

In chapter 1 of his recent book *Psychosemantics*, Fodor sums up his main point about the relationship between psychology and common sense as follows: "An explicit psychology that vindicates common-sense belief/desire explanations must permit the assignment of content to causally efficacious mental states and must recognize behavioral explanations in which covering generalizations refer to (or quantify over) the contents of the mental states they subsume" (14-15). The contents he has in mind are propositions that specify that which is believed or desired, and are the referents of "that"-clauses (*that P*, *that Q*, etc.) in standard propositional attitude ascriptions (p. 11). The content-involving causal generalizatons of the commonsense-vindicating psychology are supposed to parallel familiar commonsense generalizations of the sort illustrated by (14) above.

For Fodor, the key feature of this account is the correspondence between the causal powers of mental states and their (propositional) contents (12). Mental representations are seen as providing the link between the two (16-19). In particular, beliefs and desires with the same (different) causal powers are thought to be relations to the same (different) representations. If, in addition, beliefs and desires with the same (different) propositional contents involve the same (different) representations, then the contents of beliefs and desires will match up 1-1 with their causal powers. On such a picture it is no wonder that causal explanations of behaviour in terms of propositions believed and desired should be effective.

The difficulty, of course, is that this picture is inaccurate. I have argued that mental states involving different mental representations, with different causal powers, sometimes share the same propositional content, thereby creating exceptions to explanatory generalizations like (14). Twin-Earth type cases (both indexical and nonindexical) illustrate a related shortcoming. In such cases mental states involving the same representations, with the same causal powers, are assigned different propositional contents, due to differences in the surrounding environment. Although these cases do not threaten the truth of Fodor's commonsense explanatory generalizations, they do raise the specter of different psychological explanations — in terms of different propositions believed and desired — of identical behaviour of molecule for molecule twins. Fodor doesn't like this (37).

In Chapter 2 of *Psychosemantics*, he argues that the notion of content needed for scientific psychology must be individuated in terms of the causal powers of mental states, which in turn supervene on brain states. Twin-Earth cases show that commonsense (propositional) contents of beliefs and desires are not so individuated and do not so supervene. Thus, Fodor concludes that scientific psychology requires a notion of "narrow content" different from propositional content (44-5).

His proposal is that narrow contents should be identified with functions from "contexts" to propositional contents.[23]

> What, if anything does that mean? Well it's presumably common ground that there's something about the relation between Twin-Earth and Twin-Me in virtue of which his "water"-thoughts are about XYZ even though my water-thoughts are not. Call this condition that's satisfied by {Twin-Me, Twin-Earth} condition C (because it determines the *Context* of his 'water'-thoughts). Simi-

> larly, there must be something about the relation between me and Earth in virtue of which my water-thoughts are about H_2O even though my Twin's "water"-thoughts are not. . . .
>
> But now we have an extensional identity criterion for mental contents: Two thought contents are identical only if they effect the same mapping of thoughts and contexts onto truth conditions. Specifically, your thought is content-identical to mine only if in every context in which your thought has truth condition T, mine has truth condition T, and vice versa.
>
> It's worth reemphasizing that, by this criterion, my Twin's 'water'-thoughts are intentionally identical to my water-thoughts; they have the same contents even though, since their contexts are de facto different, they differ, de facto, in their truth conditions. In effect, what we have here is an extensional criterion for "narrow" content. The "broad content" of a thought, by contrast, is what you can semantically evaluate; it's what you get when you specify a narrow content *and fix a context*. (48)

Fodor's distinction between narrow and broad content resembles David Kaplan's distinction between character and content. However, there are important differences. For example, Kaplan observed that whereas speakers of a language use indexicals to express different propositions in different contexts of utterance, nonindexicals are not used in this way. In particular, the propositional contents expressed by English speakers using the word "water" do not vary from one context to another.[24] This is reflected in the fact that the characters of nonindexical expressions in a language — names, natural kind terms, other general terms, etc. — are always constant functions (that take the same propositional content as value in each context of utterance). Thus, for all contexts of utterance C and C′, and all nonindexical expressions N and N′ of a language, if the (propositional) content of N in C is identical with the (propositional) content of N′ in C′, then the characters of N and N′ will be identical.

This is not true for Fodorian narrow contents. The narrow contents of all representations (expressions) for which Twin-Earth type cases can be constructed — which includes virtually all representations (expressions)[25] — will be nonconstant functions (that assign different propositional contents as values to different Fodorian contexts).[26] As a result, representations E and E′ will share the same narrow content only if they have the same propositional content in all Fodorian contexts, including all variations of Twin-Earth type cases. But this condition will almost never be met. For nearly every pair of distinct representations, we can imagine a single context (Twin-Earth type environment) in which they stand for different things.[27] Thus, we can typically expect E and E′ to have the same Fodorian narrow contents if and only if E is the same representation as E′.

This means that, unlike the case with propositional content, one cannot drive a wedge between representations and their narrow contents. In criticizing the analysis of belief given in (1a) and (1b), I noted that they could not accommodate the following elementary facts:

(a) The same formula may express different propositions when interpreted from the point of view of different contexts, and different representational systems.

(b) The same proposition may be expressed by different formulae of the same system, or of different systems.

If reference to propositions in (a) and (b) is replaced by reference to Fodorian narrow contents, then this criticism will not apply. However, the same replacement will not save the proposals in (1).

The resulting principle in the case of (1a) is just like the original except for the addition of "narrow" before each of the two occurrences of 'content.' The falsity of this principle is easily shown. Imagine that I believe that I live in New Jersey by virtue of bearing R to representation M. Now imagine a different individual J in an epistemic circumstance similar to mine. Like me, J believes that he lives in New Jersey. J also says to himself "I live in New Jersey," and J also bears R to M. According to the modified principle (la-narrow), J satisfies "x believes that I live in New Jersey" when said by *me* in a context C with me as agent. But this is wrong. J doesn't believe that *I* live in New Jersey; J believes that *he* lives in New Jersey. Thus, the appeal to narrow content won't save (1a).

Nor will it save (1b). If reference to narrow contents replaces reference to propositions, then (1b-narrow) will identify narrow contents as the objects of belief. Since *what we believe* are propositions, this is incorrect. Fodor is well aware of this.

> the content that an English sentence expresses is ipso facto *anchored* content, hence ipso facto *not* narrow.
>
> So, in particular, qua expression of English "water is wet" is anchored to the wetness of water (i.e. of H_2O) just as, qua expression of Tw-English, "water2 is wet" is anchored to the wetness of water2 (i.e. of XYZ). And of course, since it is anchored to water, "water is wet" doesn't — can't — express the narrow content that my water-thoughts share with my Twin's. Indeed, if you mean by content what can be semantically evaluated, then what my water-thoughts share with Twin "water"-thoughts *isn't* content. Narrow content is radically inexpressible, because it's only content *poten-*

> *tially*; it's what gets to *be* content when — and only when — it gets to be anchored. We can't — to put it in a nutshell — say what Twin thoughts have in common. This is because *what can be said* [my emphasis] is ipso facto semantically evaluable; and what Twin-thoughts have in common is ipso facto not.
>
> . . . Looked at the other way around, when we use the content of a sentence to specify the content of a mental state (viz., by embedding the sentence to a verb of propositional attitude), the best we can do — in principle, *all* we can do — is avail ourselves of the content of the sentence qua anchored; for it's only anchored that sentences *have* content . . . . (p. 50)

Since (1a) and (1b) cannot be saved by appealing to narrow content, they are unsalvagable — as are (1.5a) and (1.5b). The proper interpretation of the thesis that propositional attitudes are relations to representations remains that given in (2). Although narrow content is not explicitly mentioned in (2), there is no problem accommodating it. If we assume that propositions result from applying narrow contents to Fodorian contexts, and that the propositional content of a mental representation M of an individual i in a context C with i as agent is the result of applying the narrow content of M to C, then (2b) will be true iff (2c) is true.[28]

(2c) for all propositions p, individuals i (and times t), i believes p (at t) iff there is (at t) a mental representation M, with narrow content N, and context C, with i as agent (and t as time), such that i bears a certain relation R to M (at t), and N(C) = p.

The end result is not an alternative to the analysis given in (2b), but an elaboration of it.

I emphasize this because it has been suggested to me[29] that appeal to narrow content provides "an obvious way" of avoiding both (2) and my critique of Fodor. In particular, (1c) has been suggested as "the obvious alternative" to my (2b).

(1c) For all narrow contents N, there is a mental representation M such that for all individuals i, and contexts C, i believes the proposition p that results from applying N to C iff i bears R to M, and N is the narrow content of M, and i is agent of C.

This proposal is hopeless. Since it fails to give necessary and sufficient conditions for believing p it won't do as an analysis of belief — for reasons analogous to those given in connection with (1.5a). Even worse, (1c) is plainly false. For example, let C be a context with Bill as agent, let N be such that N(C) = the proposition that Bill lives in New Jersey, and let Mary be distinct from Bill. Since Mary is not the agent of C, the right-hand side of the biconditional "Mary believes the proposition that Bill lives in New Jersey iff Mary bears R to M, and N is the narrow content of M, and Mary is agent of C," obtainable from (1c), is false thereby requiring the falsity of the left hand side to preserve the truth of the whole. Thus, (1c) has the absurd consequence that since Mary is not identical with Bill, she doesn't believe that Bill lives in New Jersey.

Perhaps the idea behind (1c) can be sympathetically recast as (1d).

(1d) For all narrow contents N, there is a mental representation M with N as narrow content such that for every individual i and context C with i as agent, i believes the proposition p that results from applying N to C iff i bears R to M.

Although this avoids the absurdity of (1c), it is still plainly inadequate. I have already indicated that different mental representations can be expected to have different Fodorian narrow contents — which means that, in general, an individual narow content will be associated with just one mental representation. Using this assumption plus a fact about propositional attitudes, we can derive a contradiction from (1d). I argued in the text (in the discussion of examples 9-13) that sometimes an agent may have different mental representations that express the same proposition. Such an agent may bear R to one of these representations while not bearing R to others. For example, i might bear R to M_1, with narrow content N_1, but not bear R to M_2, with narrow content N_2 (distinct from N_1), in a context C with i as agent, where $N_1(C) = N_2(C) = p$. It now follows from (1d) that if M_1 is the only mental representation with narrow content N_1 and M_2 is the only representation with narrow content N_2, then i both believes p and does not believe p. Since this is uncceptable, someone who adopts Fodor's notion of narrow content ought to reject (1d) as false.[30] As a result, the introduction of narrow content does not affect the analysis of belief given in (2).

However, it does raise a question about Fodor's conception of the relationship between scientific psychology and commonsense belief/desire explanations of behaviour. In Chapter 1 of *Psychosemantics*, he claims that the proper way to do scientific psychology is one that will "vindicate" commonsense generalizations like (14), which achieve their explanatory effect by generalizing over propositional contents. However, in Chapter 2, he takes Twin-Earth cases to show that the notion of content required by scientific psychology is narrow, rather than propositional. What, then, becomes of the "vindication" of commonsense?

Clearly, generalizations framed in terms of propositional contents will not appear in Fodor's imagined science. Nor, since narrow contents are neither believed nor desired, will generalizations that simply substitute narrow for propositional contents in principles like (14). At a minimum, if we find new objects to replace propositions, we will also need new relations to replace (commonsense) believing and desiring, which are relations to propositions. Perhaps we should call these new relations 'narrowly-believing' and 'narrowly-desiring.'

But now one must ask whether any such new theoretical inventions are needed. Having framed narrow content so as to match up 1-1 with mental representations, Fodor has no need to burden a properly austere scientific psychology with both. As far as the imagined explanation of behaviour is concerned, one might as well omit appeals to narrow content altogether, and rely entirely on computational relations to representations. Nothing would be lost, since representations and narrow contents have been designed to mirror one another anyway.

Suppose that some such view of scientific psychology were shown to be correct. We would then have a serious science of the mental which gave true explanations of behaviour in terms of the mental syntax of agents. It is worth noting that the truth of such explanations needn't conflict with the truth of typical content-involving commonsense explanations of behavior. Thus, such a science needn't be thought of as refuting particular commonsense belief/desire explanations of behaviour. However, vindication is more than the absence of refutation, and one might wonder whether such a science had provided any vindication of our commonsense explanations at all.

This issue, I think, depends on the relative priority of mental representations and narrow contents. If one thinks of mental representations as being somehow given in advance of any considerations of their narrow contents, then the invocation of such contents standing in a 1-1 relation to representations will naturally seem to be a theoretically superfluous excrescence. However, if mental representations are themselves individuated (in large part) by appeal to narrow contents —i.e., by appeal to prior commonsense judgments about the propositions believed by agents in different contexts[31] — then a representation-based psychology of the sort imagined by Fodor might well depend on, and even vindicate, our commonsense conception of propositional attitudes.

Although this would be a welcome result, it is not my intention to endorse it, or any other detailed conception, as a prescription for how scientific psychology ought to proceed. While I expect our commonsense conception of the attitudes to play a significant role in the development of scientific psychology, I cannot predict just what that role will be. This uncertainty does not infect my attachment to the attitudes themselves. In my opinion, our commonsense conception of propositional attitudes has an authority that is largely independent of its ultimate role in future science; and I am confident that we will find room in the world for beliefs and desires, no matter what surprises science may hold in store. My concern has been with how to explicate our commonsense conception. I agree with Fodor that mental representations are important for this. However, I have insisted that such an explication should be understood along the lines of (2) rather than (1). It has been the burden of this appendix to show that this result stands, whether or not Fodorian narrow content is incorporated into scientific psychology.

NOTES

* A version of this paper was read at the 19 August 1988 Philosophy Department colloquim at the University of Washington. The material in it was first presented in a seminar there that I taught jointly with Charles Marks during the summer of 1987. I am indebted to Professor Marks for illuminating discussions of Fodor's position, to other participants in the seminar and colloquim, and to Ali Akhtar Kazmi for their helpful comments.

1 *The Monist* 61:4 (1978); reprinted in Jerry A. Fodor, *Representations* (Cambridge MA: MIT Press 1981), 177-203. Page references in the text are to the reprinted version.

2 The quantifier phrase "a certain relation R" is to be understood throughout as having wide scope over everything else; thus R is invariant across individuals, times, etc. The difference between the interpretations in (1) and those in (2) involves the scope of existential quantification over mental representations. Other interpretations of Fodor's thesis involving intermediate scope for this quantification will be considered below.

In what follows, I will not focus on the differences between the (a) and (b) versions of theses (1) and (2). Since Fodor's primary intention is to give an analysis not just of propositional attitude ascriptions in English, but of propositional attitudes generally, I believe the (b) versions to be closer to his intent. However, he seems to prefer to talk in terms of "content" rather than "propositions," and so might prefer the (a) versions after all. (He ignores the need to relativize content to context, and does not formulate his proposals in the explicit manner of either (a) or (b)). Fortunately, any ambivalence on this score will make no difference to us, since my main points are independent of the differences between the (a) and (b) versions of the theses.

3 It is clear from the context that Fodor intends to give necessary as well as sufficient conditions for the truth of belief ascriptions. From time to time he informally expresses such truth conditions using conditionals, when strictly speaking biconditionals are in order. Another clear example of this occurs on page 180 in numbered paragraph 5.

4 Under these assumptions representations are not objects of the belief relation in the sense indicated in the quoted passage. For if M_1 and M_2 are different representations with the same propositional content, then (2) will allow ⌜John believes that S⌝ to be true either in virtue of John's bearing R to M_1 (but not M_2) or in virtue of John's bearing R to M_2 (but not M_1). But then neither M_1 nor M_2 is such that the belief ascription is true iff John bears R to it.

The relation Fodor is talking about in the quoted passage is clearly the semantic extension of the belief predicate. According to (2) the extension of "believe" is a relation between individuals and propositions, not mental representations. Nevertheless, it is a relation which holds between an individual and a proposition in virtue of another, psychological, relation holding between the individual and a mental representation. If (2) is correct, and if different representations can express the same proposition, then it is only in this latter psychological sense that beliefs are relations to representations.

5 Fodor's own discussion ignores context sensitivity and assimilates S′ to S. The formulation in the text is meant to correct this, and to bring out the parallels with (1).

6 It should be noted that (iii) is an objection only to (3b), and that in general the objections presuppose a strict, impermissive characterization of the relation R. For example, (ii) is an objection only if there are English sentences with the same content such that an individual can bear R to one without bearing R to the others. I have used (iv) to cover two Fodorian objections — the fact that humans must have prior beliefs in order to learn an initial natural language, and the fact that some organisms have beliefs even though they never learn languages.

7 Rudolf Carnap, *Meaning and Necessity*, 2nd edition (Chicago: University of Chicago Press 1956) 62. In Appendix C, Carnap modified his view slightly to make it less behaviouristic. In the modified view, having a disposition to utter S′ is no longer treated as a necessary or sufficient condition for bearing the relation R in the analysis to 'S', but rather is seen as providing strong inductive evidence that one bears R to S′. It should also be noted that Carnap's model of English was highly simplified, and did not include context sensitivity.

8 In discussing objection (ii) above Fodor says the following:

> The natural way to read the Carnap theory is to take type identity of the correspondents of belief-ascribing sentences [i.e. S in ⌜x believes that S⌝] as necessary and sufficient for type identity of the ascribed beliefs: and it is at least arguable that this cuts the PAs [propositional attitudes] too thin . . . . A way to cope would be to allow that the objects of beliefs are, in effect, *translation sets* of sentences; something like this seems to be the impetus for Carnap's doctrine of intentional isomorphism. (191)

However, Fodor finds this appeal to sameness of content problematic.

> It may well be, for example, that the right way to characterize a translation relation for sentences is by referring to the communicative intentions of speaker/hearers of whatever language the sentences belong to. (S_1 translates S_2 if the two sentences are both standardly used with the same communicative intentions.) But, of course, we can't both identify translations by reference to intentions and individuate propositional attitudes (including, n.b., intentions) by reference to translations. (191)

Applied to (4), which doesn't mention translation, this objection appears to be based on a commitment to *reduce* semantic contents of natural language sentences to propositional attitudes of speakers — which in turn are to be explained in terms of semantic contents of sentences of some type. One who does not share this reductionist program need not share Fodor's objection to Carnap's actual proposal. Moreover, even if the reductionist program is accepted,the objection to appealing to different *natural language sentences* with the same content in an analysis of belief does not rule out appealing to different *mental representations* with the same content. Thus it provides no grounds for preferring (1) to (2).

9 This result requires each object to have at least one mentalese name (unique to it). It also requires that if an agent believes a proposition p about an object o by virtue of bearing R to a representation containing a mentalese name n (connected to a use of an indexical that refers to o), then every agent who believes p must do so in virtue of bearing R to M, and hence must use n to refer to o. Since the point of the strategy is to ensure that propositions expressed by indexical sentences of natural

language are cognitively accessible to believers generally, there is no avoiding the conclusion that different agents must use the same mentalese names for the same objects.

It is worth noting that these requirements do not depend on any particular semantic analysis of names or indexicals; in particular they take no stand on the question of whether or not the semantic contents of such terms, relative to contexts, are their referents, relative to those contexts. The strategy requires that each content (propositional constituent) expressed by a mentalese name be expressed by that name alone. So long as this is guaranteed, it is noncommital about the nature of the contents themselves. (If contents are referents, then an object can be named by only one mentalese name; if they are more fine grained than that, then an object might have more than one such name.)

I am indebted to Ali Akhtar Kazmi for a discussion of this point.

10 The reason for insisting that mental representations be "in the head" in this sense involves the theoretical role of such representations in cognitive theories, as understood by Fodor and others. Mental representations are meant to capture the contributions of an agent's internal cognitive states to explanations of his behaviour. Since molecular duplicates can be assumed to be internally the same, they must be assigned the same mental representations. This is reflected in Fodor's insistence that mental representations are *syntactically*, rather than semantically, individuated formulae in an internal language, and that psychological processes are *formal computational* operations on these syntactic objects.

11 The original Twin Earth case was presented by Hilary Putnam in "The Meaning of 'Meaning'," in K. Gunderson (ed.), *Language, Mind and Knowledge*, Minnesota Studies in the Philosophy of Science, VII, (Minneapolis: University of Minnesota Press 1975), 131-93. The characteristic feature of such cases is that of physically and psychologically indistinguishable individuals with different beliefs. Although Putnam's original case involved beliefs expressed using the natural kind term "water," the phenomenon generalizes to all terms whose contents are (partially) determined by factors "outside the head."

12 Section XVII, "Demonstratives," in J. Almog, J. Perry, and H. Wettstein (eds.), *Themes from Kaplan* (New York: Oxford University Press 1989), 481–563.

13 While rejecting, or suspending judgment on $\ulcorner N = N^{*} \urcorner$.

14 Fodor's argument for this is given in chapter 2 of his *The Language of Thought* (New York: Crowell 1975). A thorough critique of the argument can be found in Kim Sterelny, "Fodor's Nativism," forthcoming, *Philosophical Studies*.

15 On page 202, Fodor shows some reluctance to accept either (7a) or (7b). First he grants that it is *conceivable* that propositional attitudes are not relations to internal representations, by which he means that his thesis that propositional attitudes are relations to internal representations is *not a priori*. Next he considers the objection that it is *empirically possible* that propositional attitudes are not relations to internal representations.

> it may be *empirically possible* that there should be creatures that have the same propositional attitudes we do (e.g., the same beliefs) but *not* the same system of internal representations; creatures that, as it were, share our epistemic states but not our psychology. Suppose, for example, it turns out that Martians, or porpoises, believe what we do but have a very different sort of cost accounting. We might then want to say that there are translation relations among systems of internal representation (viz., that formally distinct representations can express the same proposition).... Whether we can actually make sense of this sort of view remains to be seen.... (p.202)

Fodor's mention of translation seems to be a way of indicating a potential retreat from (1) to (2), should the objection prove correct. But what would it be for it to be correct? If the possibility mentioned in the objection is the normal non-epistemic kind, the correctness of the objection depends not just on what turns out to be the case in the actual world, but also on what could have been the case (in counterfactual circumstances). Once it is admitted that there could have been creatures of the sort Fodor hypothesizes, (7b) must be rejected, as must (7a), if it is further acknowledged that these creatures could have co-existed with us in a world in which we retained our actual psychology.

16 In my view, it is possible for an individual i to believe p by virtue of bearing R to a representation M_1 that expresses p, even if he fails to bear R to another of his representations, M_2, that (unknown to him) also expresses p. This point is connected to the discussion of examples (9-13) in the next section, and to the literature cited in note 21.

17 On p. 17 of *Psychosemantics* (Cambridge, MA: MIT Press 1987), Fodor makes the following claim.

> Claim 1 (the nature of propositional attitudes): For any organism O, and any attitude A toward the proposition P, there is a ("computational"/"functional") relation R and a mental representation MP such that MP means that P, and O has A iff O bears R to MP.

Aside from the obviously undesirable feature of allowing the relation R to vary not just from attitude to attitude (belief, hope, desire, etc.) but also from organism to organism and proposition to proposition, this formulation is a straightforward generalization of (1.5b), and is therefore inadequate for the same reasons as (1.5b).

However, it is not clear that Fodor's formulation really captures his intention. For example, in explaining what he means by the claim, he says:

> A cruder but more intelligible way of putting claim 1 would be this: To believe that such and such is to have a mental symbol that means that such and such tokened in your head in a certain way; it's to have such a token 'in your belief box,' as I'll sometimes say. (17)

But this is an informal statement of (2), rather than (1.5b). Moreover, on page 20 Fodor says that claim 1 has the following consequences:

> For each tokening of a propositional attitude, there is a tokening of a corresponding relation between an organism and a mental representation;
> and
> For each tokening of that relation, there is a corresponding tokening of a propositional attitude.

Although both of these are consequences of (2), only the first is a consequence of claim 1 (or of 1.5b). So perhaps by the time of *Psychosemantics* Fodor really means to adopt (2). If so, this is a significant and unacknowledged change from his position in "Propositional Attitudes."

I am indebted to Ali Akhtar Kazmi for directing my attention to Fodor's comments on page 17 of *Psychosemantics*.

18 Scott Soames, "Lost Innocence," *Linguistics and Philosophy*, 8 (1985), 59-71; "Direct Reference, Propositional Attitudes and Semantic Content," *Philosophical Topics*, 15 (1987), 47-87, reprinted in Nathan Salmon and Scott Soames (eds.), *Propositions and Attitudes* (New York: Oxford University Press 1988); and "Direct Reference and Propositional Attitudes," in Joseph Almog, John Perry, and Howard Wettstein (eds.), *Themes from Kaplan*, (New York: Oxford University Press 1989).

19 See the articles mentioned in the previous note, plus Nathan Salmon, *Frege's Puzzle*, Cambridge, MA: MIT Press 1986).

20 The combination of (2) with Fodor's account of psychological complexity requires contents of sentences to be structured propositions; but this does not mean that if (2) were replaced by (1), the more familiar conception of contents as truth conditions, or sets of truth-supporting circumstances, could be maintained. On that

conception there is no way of preventing the construction of semantically equivalent sentences that differ greatly in syntactic complexity — something that would make Fodor's account of the psychological complexity of beliefs impossible.

21 For discussions supporting this characterization of these examples see: (i) in the case of (9) and (10), David Kaplan, "Demonstratives"; and John Perry, "The Problem of the Essential Indexical," in *Nous*, 13 (1979), 3-21, reprinted in *Propositions and Attitudes*; (ii) in the case of (11), Saul Kripke, "A Puzzle about Belief," in A. Margalit (ed.), *Meaning and Use* (Dordrecht): Reidel (1979), 239-83, reprinted in *Propositions and Attitudes*; and Nathan Salmon, Appendix A of *Frege's Puzzle*; (iii) in the case of (12), Nathan Salmon, "How to Become a Millian Heir," *Nous*, 23 (1989), 211–20; and "A Millian Heir Rejects the Wages of *Sinn*," in Anthony Anderson and Joseph Owens (eds.), *The Proceedings of the Minnesota Conference on Propositional Attitudes: The Role of Content in Logic, Language, and Mind*, forthcoming (Minnesota: University of Minnesota Press); (iv) in the case of (13), Alonzo Church, "Intensionality and the Identity of Belief," *Philosophical Studies*, 5 (1954), 65-73, reprinted in *Propositions and Attitudes*; Section IX of Scott Soames, "Substitutivity," in J.J. Thomson (ed.), *On Being and Saying: Essays for Richard Cartwright* (Cambridge, MA: MIT Press 1987), 99-132; and the introduction to *Propositions and Attitudes*.

22 Recall that (2) allows cases in which the mental representation of a sentence contains the words in the sentence, and may even be identical with the sentence itself.

23 Fodor introduces this notion on page 47, where he describes narrow contents as "functions from contexts and thoughts onto truth conditions." There, Fodor assumes for the sake of argument that propositional contents can be identified with truth conditions. He indicates that he would be content with a more general formulation in terms of functions from contexts and thoughts to propositional contents, if the identification were questioned. His inclusion of thoughts among the arguments of narrow contents (of thoughts) is more puzzling. It is clear that "thought" in this discussion stands for the mental state of believing a proposition, rather than the proposition believed. Since mental state tokens tend not to survive even tiny variations in contexts, a thought whose narrow content allows it to have different propositional contents in a variety of different contexts should presumably be a mental state type. But then, the narrow content of a mental state type will be a function from arbitrary contexts to propositions (it expresses in those contexts), not a function from arbitrary contexts and arbitrary mental state types to propositions (expressed by those types in those contexts). This is how I shall understand Fodor's proposal. (An alternative would be to take a narrow content as belonging to a mental state token, and as being defined over pairs of contexts and tokens of that type. The issues raised below are neutral between these alternatives, and could be expressed under either formulation.)

24. This will hold even if, descending the steps of my spacecraft after landing on Twin-Earth, I say "That [gesturing at Twin-water] is water." Even though the utterance occurs on Twin-Earth, the content of my utterance of 'water' will be the same as the content of 'water' utterances on Earth. Of course if I stay on Twin-Earth long enough I may, through interaction with the natives, lapse into Twin-English, in which case my 'water' utterances will come to share the content of those of my twins. However, as long as I continue to speak English — as it is spoken back home — my 'water' utterances will retain their normal English contents. Note the contrast with a genuine indexical like 'that.' My (English) utterance of 'that' on Twin-Earth stands for twin-water, even though no 'that' utterances on Earth have ever stood for twin-water.

In Kaplan's terminology, the content of the English word 'water' does not vary from context to context, and similarly for the content of the Twin-English word 'water.' Since the contents, and characters, of the two are different, the English word 'water' is not identical with the Twin-English word 'water'; instead, they are phonological and orthographic twins, or homonyms. A similar analysis is given to ordinary cases in which different individuals are said to "have the same name." In

Kaplan's system, such cases typically involve homonymy — phonologically similar names with different contents that remain fixed across contexts.

25 Fodor notes the generality of Twin-Earth type cases on page 29 of *Psychosemantics*.

26 Since even nonindexical expressions have nonconstant functions as narrow contents, Fodorian contexts must somehow provide such expressions with propositional contents, in addition to assigning contents to indexicals. Fodor's goal is to use this framework to give a naturalistic account of how expressions acquire representational content. The general idea seems to be that the propositional content of a representation arises from the causal properties of its tokens, including both relations to mental events entirely internal to the agent (encoded in narrow contents) and relations to other agents and objects in the world (encoded in contexts).

27 With indexicals, the case is transparent. Imagine a representational system with demonstratives 'that_1', 'that_2', etc., which can be used to refer to arbitrary objects (or arbitrary objects within a specified range or ranges). Even if each demonstrative behaves in accord with the same rule (e.g., one that determines its referent or content in a context by tracing its causal connections to objects in the environment), the existence of at least one context in which 'that_1' and 'that_2' have different referents/contents will be sufficient to guarantee that they (always) have different narrow contents. Since there will typically be such contexts, different demonstratives can be expected to have different narrow contents. Given the parallel between the narrow contents of indexicals and nonindexicals, as well as their similar roles in Fodor's account of Twin-Earth cases, one would expect the point to carry over to representations generally.

This can be illustrated with the help of Hilary Putnam's well known "elm"/"beech" example in "The Meaning of 'Meaning.'" Putnam says that he cannot tell the difference between a beech tree and an elm tree, and even that his "concept" of a beech is the same as his "concept" of an elm. Nevertheless the two words, as used by him, refer to different things — beeches and elms — and therefore have different propositional contents. In each case, Putnam uses the word to express the same propositions about the same things that other speakers do. Thus, each word has the same reference and propositional content for him as it does for other speakers in the linguistic community. In Fodor's scheme, Putnam would associate different mental representations with the two words and these representations would have different narrow contents, despite the fact that Putnam has no mental resources for distinguishing beeches from elms.

The point generalizes across the board. Imagine, for example, a related character, Putnam*, who is in an analogous position regarding 'ketchup' and 'catsup.' Putnam* lives in a community just like ours except for the fact that the red tomato-based condiment that comes in transparent bottles marked 'ketchup' is slightly different from the red tomato-based condiment that comes in transparent bottles marked 'catsup.' Though the difference between the two can be discerned by experts, most speakers, including Putnam*, can't distinguish them. He may even be unsure whether the two words refer to the same or different things. Still the words 'ketchup' and 'catsup,' as used by Putnam* and other speakers, would, in Fodor's system, be associated with different mental representations with different narrow contents. Finally, if our own community of English speakers contains one of Putnam*'s molecule for molecule twins, then his distinct 'ketchup' and 'catsup' representations will have different narrow contents, despite the fact that 'ketchup' and 'catsup' are fully synonymous in English. (This example is adapted from Nathan Salmon's "A Millian Heir Rejects the Wages of Sinn.")

The general point to be noted is that different representations will virtually always have different Fodorian narrow contents, no matter how similarly they function in the internal cognitive economies of their agents. The reason for this is that Fodorian narrow contents are, by definition, sensitive to everything capable of influencing the propositional content of a representation in a context. Since environmental factors are typically capable of this, and since they may interact with distinct representations in different ways, distinct representations will in general have different Fodorian narrow contents. Exceptions are possible only when factors entirely internal to the agent determine that a pair of different internal representations must have the same content, independent of external considerations. For example, it might be possible (even if pointless) to have two formally distinct internal representations that functioned as first person singular pronouns with the same narrow content. However, outside unusual cases like this the generalization should hold.

28 These principles could, of course, also be relativized to possible worlds. In the case of (2c) the result would be: For all propositions p, individuals i (times t and worlds w), i believes p (at t, w) iff there is (at t, w) a mental representation M, with narrow content N, and a context C, with i as agent (t as time and w as world), such that i bears a certain relation R to M (at t, w), and N(C) = p.

29 By an anonymous referee for this volume.

30 Of course, someone with views different from Fodor's might try to develop a different notion of narrow content, in which distinct mental representations are allowed to share the same narrow content. However, even then (1d) would not be an acceptable analysis of belief — as indicated by the way in which it would fail to provide necessary and sufficient conditions for believing p. Suppose, in the case of the above example, that N_1 and N_2 were identical, even though M_1 and M_2 were distinct. Although no contradiction would then be forthcoming, (1d) would make no prediction about whether i believed p or did not believe p. In general, where N is any narrow content and C is a context with i as agent and N(C) = p, the most that could be said would be that a necessary condition for i to believe p was for i to bear R to at least one mental representation with narrow content N, and a sufficient condition for i to believe p was for i to bear R to all mental representations (of others as well as of i himself) with narrow content N. Since this is not enough for an analysis of belief, (1d) would still pose no threat to the analysis of belief given in (2).

31 As illustrated in the discussion of the examples in note 27.

CHAPTER NINE

Partial Information, Modality, And Intentionality

Fred Landman*

It is almost a truism to say that a semantic theory for natural language, which relates linguistic expressions with nonlinguistic interpretations, has to take into account the fact that language users are in general only partially informed about the intended domain of interpretation. Several semantic phenomena can only be adequately dealt with within a theory that takes this partiality seriously; vice versa, the semantics equally constrains the theory of partial information.

In this paper I will discuss some issues concerning the form that a semantic theory of information should take. Several of the issues I will bring up here, I have discussed in previous work (especially, Landman 1986), but I will follow a different line of presentation here and put some accents differently.

The main topic of the present paper will be the semantics of epistemic modals and identity statements. I will take as my starting point Hans Kamp's theory of Discourse Representation Theory (Kamp 1981), discuss its status as a theory of information, and point out some (obvious) problems it may come into if it is to deal with epistemic modals and their interplay with identity statements. The discussion of how to remedy such problems will lead me to some general discussion of partial semantics, and a sketch of a particular proposal for a partial semantics for Discourse Representation Theory for the phenomena in question. The semantics I will present takes a rather nonstandard view on what objects are. I will spend some time discussing how the semantics should be interpreted and how it fits in a broader theory of conversation and intensionality.

DISCOURSE REPRESENTATION THEORY

Discourse Representation Theory (DRT) is a rather fruitful semantic theory, developed by Hans Kamp (Kamp 1981) and Irene Heim (Heim 1982), for representing information that can be extracted out of sentences uttered in discourse. In Kamp's version, DRT contains a level of discourse representation structures at which for instance anaphoric links between discourse anaphora and their antecedents are represented, i.e., the links between the underlined expressions in (1) and (2):

(1) *A man* comes in. *He* wears a hat.
(2) If *a man* comes in, *he* wears a hat.

These structures get a semantic interpretation through recursive embedding conditions in a model. So the structure of the theory is:

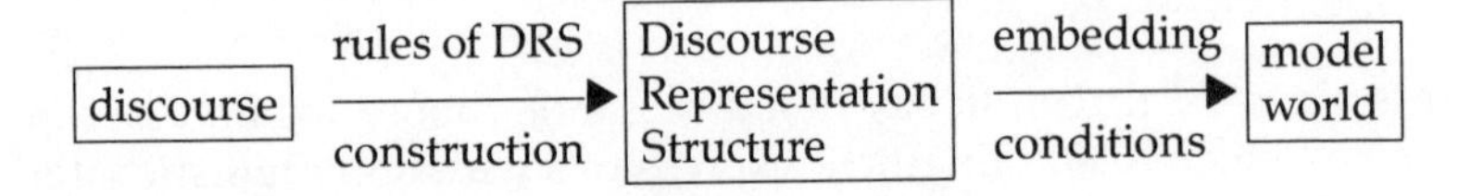

For a sentence like (1) a drs like D_1 is constructed:

D_1	x y man(x) come in(x) hat(y) wear(x, y)

x and y are discourse referents; statements like man(x) are called discourse conditions. The crux of the analysis of discourse anaphora is that indefinites like *a man, a hat* do not introduce existential quantifiers but discourse referents, **variables**, that are bound by explicit or implicit quantifiers in the discourse. Discourse anaphora like *he* in (1) can be interpreted coreferentially with antecedents like *a man* only if they are in the scope of the quantifier that binds the antecedent discourse referent (variable). The embedding conditions for D_1 give the drs its existential force: D_1 is true in a model M if **there is** a truthful embedding for D_1 in M, a function g assigning to x and y objects in M such that in M g(x) is a man that comes in, g(y) is a hat and g(x) wears g(y).

Here the two sentences of the discourse are interpreted in the same box. Other constructions, like conditionals, universal expressions, negation, etc. introduce subordinated structure in the drs. (2) roughly has the structure D_2:

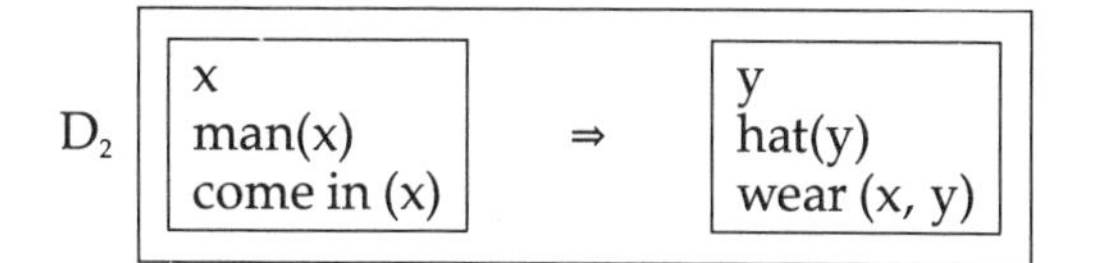

D_2 is true if every truthful embedding in M for the antecedent can be extended to a truthful embedding for the consequent in M (that is, every function g, assigning to x in M an object that is a man who comes in, can be extended to a function assigning to y a hat that g(x) wears. This is true in the model if for every man who comes in you can find a hat that he wears).

The interpretation of the conditional will give the sentence its universal force; moreover, *he* can be coreferential with *a man*, because (roughly) both are bound by the universal quantification linking the antecedent drs with the consequent drs. Note that we could not felicitously continue (2) as in (3):

(3) If *a man* comes in *he* wears a hat. **He* sits down.

The reason is that the *he* in the second sentence is not in the scope of the quantification that binds the discourse referent introduced in the antecedent of the conditional.

The drs, then, is a level of representation of information that is semantically interpreted through embedding conditions. The basic idea is that language users make a partial representation of the world, a "picture", on the basis of the sentences in the discourse that they hear and accept as true, and they try to match this picture with the world.

This interpretation of the theory brings out an important aspect: DRT is a theory of information with a **classical semantics**: it is based on a recursive specification of truth conditions (relative to a model, world) and hence based on an objective notion of truth. To use a metaphor, the semantics works like a telescope: in order to evaluate a discourse (representation structure) we point the telescope at the world and see whether we find the things that the discourse expresses in the world, we see whether it is true.

PARTIALITY

If DRT is to be used not only as a semantic theory, but also as a framework of representing information that language users do or do not have, it obviously has to have an opinion on partial information, on lack of information. That is, we have to be able to express in the theory what it means for a language user not to be informed about whether something is true or not. How does DRT deal with this?

Suppose it is undetermined for our language user whether John has property P. Since the semantics is classical, in the model John either has P or he doesn't. So it is not at that level that the lack of information can be expressed. Here is the crucial observation. If the discourse representation stucture is empty, then every embedding function will verify it in the world M. As more information is entered into the drs, less and less embedding functions will be able to verify it in M. In view of this, we can regard discourse representation structures as **restrictions** on embedding functions. Adding conditions to the drs will eliminate previously available truthful embeddings, and only by adding conditions to the drs can embeddings be eliminated.

This means that the fact that it is not yet determined whether John has P or not, can only mean that there is no condition **in the drs** concerning P(j): **neither** P(j) nor $\neg$P(j) are represented in the drs. The world, the model, has of course an opinion on these, but **as far as the DRS is concerned**, no opinion is expressed on P(j) or $\neg$P(j). What is not mentioned in the drs is not yet decided as far as the language user is concerned. The language user does not have an opinion on P(j) if the drs does not contain either P(j) or $\neg$P(j).

But a problem arises. Such an analysis of partial information may be plausible for the case that I discussed, but it comes into problems if the drs contains expressions whose semantic function it is to **express** lack of information, partial information. For such expressions, the theory of information does not describe, so to say, from the outside what information is or is not available, but enters into the semantics for the drses themselves, into the embedding conditions. Expressions that function in this way are for instance epistemic modals *must* and *may* and conditionals.

An example may help to make clear that this is a problem for the simple perspective I have sketched. Take a sentence like (4):

(4) Maybe John is coming in, maybe he isn't.

This sentence **expresses** that, for a language user who utters or accepts it, it is undetermined whether John comes in. The drs is something like D_3:

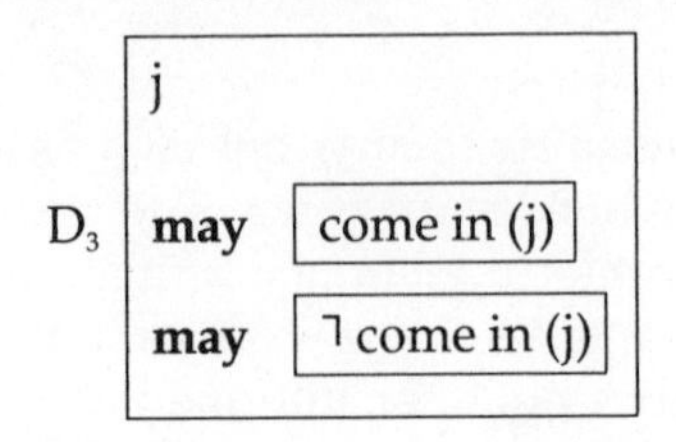

The question to be asked is: given that D_3 expresses the language user's lack of information concerning *John is coming in* (C(j)), and given the perspective on partial information discussed above, what should the embedding conditions for this drs be? In other words, what semantics should we give to **may**?

In the case of D_3, what **may** C(j) expresses is something like the following: even though I may not at the moment have the information that John is coming in, it is compatible with my present information that he is, it is possible that at a certain point I will have that information. Given the perspective on partial information that I mentioned, this 'at a certain point I will' has to mean that at a certain point C(j) will be explicitly in the drs (because that is, on this perspective, what it means to have that information).

So let us consider as a first guess (or strawman) the following semantics for **may**:

> **may** C(j) is true relative to g, a truthful embedding for D_3, if it is possible to extend D_3 to some D that explicitly contains C(j).

This analysis, of course, should be rejected out of hand. The problem is that it is an uninterpreted procedure that in no way captures the meaning of **may**. Any drs D_3 can be extended to some drs D containing C(j), in fact, it can be extended to a D′ containing both C(j) and $\neg$C(j) (of course, no embedding function will verify it). With the above rule a **may**-statement cannot fail to be true.

So let us try to turn this condition into a more semantic one, like the following; our second guess/strawman:

> **may** C(j) is true relative to g, a truthful embedding for D_3 iff D_3 can possibly be extended to some D, containing explicitly C(j), such that g can be extended to a truthful embedding for D.

The difference between this condition and the previous one is that this is a semantic one. We now say: D_3 itself doesn't yet contain and verify C(j), but some possible extension of it does. The problem with it is that if in D_3 we make **may** C(j) true in this way, we cannot possibly make **may** $\neg$C(j) true in D_3 at the same time. This follows from the classical semantics. g, the embedding function for D_3 maps j onto some object such that some extension of g verifies C(j). Since no other discourse referents are introduced, that simply means that g verifies C(j) in M, and then, of course, no extension of g can verify $\neg$C(j) in M. In other words, this semantics fails to provide a consistent interpretation to D_3, D_3 cannot be true on this semantics.

Rather than trying to modify the clause once more, let us try to diagnose what is wrong with it.

PARTIAL SEMANTICS

Let me repeat the problematic clause for convenience.

> (a) **may** C(j) is true relative to g, a truthful embedding for D_3 iff (b) D_3 can possibly be extended to some D, containing explicitly C(j), (c) such that g can be extended to a truthful embedding for D.

This clause was primarily an attempt to combine the notion of lack of information as being not yet represented in the drs with the idea that there is a truthful embedding for **may** C(j) in M if there can be a truthful embedding for C(j) in M. However, given the extensional semantics, we have to conclude that this *can* in the second clause is really misleading: it suggests that we are looking at possible extensions, it suggests a modality here, but the semantics does not provide this. Clearly we need a modal theory of some sort here. This is not the only problem, however. Suppose we do replace the last part (c) by a suitable modal theory (Like: (c′) 'such that there is a truthful embedding of C(j) in some other world'). It will then be that modal theory, i.e., the structure of available possible worlds, and not whether something is or is not represented in the drs that will determine whether the language user has the information that C(j) or not. In other words, a correct semantics for epistemic modals will provide us with a **semantic** theory of information, distinct from the "syntactic," representational one introduced earlier, and it is this semantic notion of information and not the syntactic one that will play a role in our semantics: once we get the semantics of **may** right, the condition that the drs containing **may** C(j) can be extended to a drs containing explicitly C(j) (b) is totally superfluous. (In other words, the definition could be: (a) **may** C(j) is true relative to g, a truthful embedding for D_3 iff (c′) g can be extended to a truthful embedding of C(j) in some other world.)

I have criticized the middle (b) and the end (c) of the above unfortunate clause, let's now turn to the beginning (a), that is the part 'g is a truthful embedding for **may** C(j) in M iff. . . .' I called the middle superfluous and the end wrong, the beginning I would characterize as nonsense.

Let me use the telescope metaphor once more. What we are doing, if we assume a semantics along the lines sketched is the following: we point the telescope at the world (M) and we ask: is it true in that world that maybe John is coming in? And this seems to be a nonsensical question to ask for modals like *maybe John is coming in*. In other words, it seems to be nonsense to try to embed *maybe John is coming in* into the world.

This is a serious point, because it affects not just a particular semantic clause but the whole semantic theory: apparently, when it comes to epistemic modals it won't do to take the notion "g is a truthful embedding for ϕ in M" as the basis of our semantic recursion.

Back to the telescope. Our telescope is a mechanism for inspecting semantic objects (like worlds). If we need a semantic notion of information in our theory then information states will be semantic objects, and hence our telescope can inspect them. So let us not point the telescope at the world, but at our information about the world. In other words, let us not try to match drses with the world, but with our information about the world. Now we again ask: is it true according to the information we are inspecting that John is coming in? The answer is yes if this information tells us so, no if it tells us that he doesn't, and undetermined if the information doesn't tell. Now ask: is it true according to the information that maybe John is coming in? This will be answered positively if the information can still possibly be extended to better information that tells us that John is coming in; negatively if such an extension is not possible.

Note that by changing the metaphor in this way, not only do the non-modal questions stay well formed, but now the modal questions make sense as well.

Let me here insert a warning about terminology. When I use the words "information (state)" and "true according to the information" here, I do so without intending to restrict myself to correct information, that is, information that is true of our world. Information states encode the things that language users accept in the course of conversation and there is no reason for not accepting non-factual information states, nor is there any reason why we cannot evaluate a sentence relative to non-factual information. Information states as I use them then correspond closely with Robert Stalnaker's common grounds and context sets (e.g., Stalnaker 1979), Frank Veltman's prejudice sets (Veltman 1976) and Angelika Kratzer's modal bases (for instance, Kratzer 1981).

The point I made above can be reformulated: modals are context dependent expressions: we do not ask whether **may** ϕ is true in the world, but only whether it is true given and in view of a contextually given information state. This context dependency of modals is well established in the literature (most strongly and systematically in Angelika Kratzer's work, see for instance, Kratzer 1981). So the very least we should do is extend the notion of context (already necessary for indexicals) with an information state (or a structure of information states, if we want to be able to deal with different modalities) and assume that the semantic interpretation is sensitive to this context.

Context sensitivity and indexicality

Typical examples of context sensitive expressions are, of course, indexicals, like *I* and *now*. Such expressions are "rigid": regardless of whether they are embedded under modals, tenses, or attitude verbs, indexicals are interpreted relative to the outside context of use (*I* denotes the speaker in all of *I am ill, maybe I am ill, I will be ill, John thinks I am ill*). In order to interpret these sentences we do need to relativize the semantic recursion to the context (ϕ is true in w relative to c), but unlike the possible world parameter it does not play a real role in the semantic recursion. The context gives us the parameters that have to be set before we can start the semantic recursion. Or, as Kaplan 1979 and Stalnaker 1979 formulate it: the context determines what proposition the sentence expresses, given that, the semantic recursion on the possible worlds will determine whether it is true.

Not all context sensitivity is indexicality. Partee 1987 discusses the need to distinguish the general context of use from the local context. For instance, spatial adverbials like *nearby* are context dependent in that they need a spatial location to which they are related, but this spatial location need not come from the general context of use ('near to us') but can come from the local context: cf.

(5) Mary lives nearby
(6) John lives on State Street and Mary lives nearby.

The second sentence is ambigious in that it can mean 'near to us' or 'near to John.' Let us think of the local context of *Mary lives nearby* in the second sentence, as derived from the outside context and the interpretation of the first part of that sentence. Then indeed the interpretation of *nearby* is context sensitive, but not indexical: it would be indexical if we could only interpret it relative to the place of the outside context, but in fact it can also be interpreted relative to the local context.

Given that modals are context sensitive, let us ask whether they are indexical.

I will discuss this question in some detail here, because answering it is important for the discussion of modality later in this paper. My purpose there will be to argue in favour of one of the two plausible perspectives one can take on modals. Those two analyses are both context dependent, but not indexical. The arguments I will give there, however, are not arguments against an indexical analysis. Their strength, then, would be undermined, if an indexical analysis is a viable alternative.

Concerning the indexical analysis, some of the discussion in Kratzer 1986 can be interpreted as a defense of this position, though it would be too strong to ascribe the claim that modals are nothing but indexicals to her there. The indexical position claims that just like *I am walking* can only be true in a world relative to a contextually provided speaker, *Maybe I am walking* can only be true relative to a contextually provided information state, contextually provided by the context of use.

This position differs drastically from most theories of conditionals and modals (including Kratzer 1981). Look at the following example:

(7) Maybe John killed the doctor. But if he was at home that night, he can't have killed her.

The *maybe* in the first sentence is interpreted relative to the contextually given information state, the sentence expresses that this information still allows for the possibility that John killed the doctor. If modals are indexical, then the *can't* in the second sentence (under the conditional) is interpreted relative to the same contextually given information state. But that would mean that the discourse can only be false or trivially true (if the antecedent of the conditional is false) with respect to that information state, because (8) is a contradiction:

(8) Maybe John killed her and he can't have killed her.

This example seems to suggest directly that modals are context sensitive, but not indexical, just like the mentioned spatial adverbials.

The only way we can get out of this problem is by assuming that all such problem cases, where we have a modal embedded in the consequent of a conditional, in fact have a **logical form** where the conditional is under the modal, more precisely *if* ϕ *then may* ψ has a logical form *may* (ϕ, ψ). This view fits, of course, very well with the idea of DRT (going back to Lewis 1975) that antecedents of conditionals only serve to **restrict** a quantifier (in this case *may*). This position is taken in Heim 1982 (for the examples she discusses there), where it is further assumed that this logical form is interpreted as: 'there is a possible world compatible with the contextual information where both ϕ and ψ are true.' This analysis fits with our tendency to read (9) as (10)

(9) If John comes to the party, Mary may come too
(10) Maybe John and Mary come to the party.

Heim's particular analysis is problematic for the following reason. Suppose I say to you:

(11) If I get my driver's licence next month, I may come and visit you.

What you don't know is that I know for sure that I won't have my driver's licence next month: I'm just being polite.

For sure, I did a bad thing. If you find out about it, you will be offended. Given my information, my statement was surely **incorrect**.

But was it false? I don't think so. I don't think that you can accuse me of **lying** to you (my response to such an accusation will be a treacherous, but nevertheless correct: "I wasn't lying, I said: **if** I get my licence . . .").

This makes sense if *may* is under the scope of the conditional, because then *maybe I get my driver's licence* is a quantity conversational implicature (rather than an entailment) of my statement (11), and violating a quantity implicature is not lying.

However, on the wide scope theory of **may** in Heim's version, my statement had the same meaning as (12):

(12) Maybe I get my driver's licence and I come and visit you.

This sentence can plausibly be argued to be a lie, because I know I won't get my driver's licence.

So Heim's analysis is not adequate. There is an alternative, however, that is more in line with Lewis' 1975 analysis of adverbs of quantification anyway: let us give the two-place **may** (ϕ,ψ) the following interpretation: 'of all the worlds allowed by the contextual information where ϕ holds, in some of those ψ holds.' We interpret a sentence where *may* seems to have only one argument, like *maybe John is coming*, as taking as implicit first argument the whole conversational information: **may** (S, John is coming), i.e., we take *Maybe John is coming* to mean: of all the possible worlds allowed by the context, in some of them John is coming. On the plausible pragmatic assumption that S is not empty, this will give us the correct interpretation.

We avoid now the problem with the example I gave, because the logical form **may** (ϕ,ψ) does not entail **may** (S,ϕ), but at most implicates that. In fact, what we are doing here is giving **may** (ϕ,ψ) an interpretation that is only subtly different from the interpretation that is in other theories given to $\phi \rightarrow$ **may** ψ.

The indexicality theory of modals then makes the following claim: whenever we see a modal that in surface form is embedded under other operations, like conditionals, in logical form it is not embedded and we don't find modals that we have to interpret **under the scope** of conditionals or other modals (because they don't have scope).

This simplifies our logical, semantical work considerably. For instance, we don't have to worry anymore about the semantics of conditionals that are not first degree, like $\phi \rightarrow (\psi \rightarrow \kappa)$: we simply won't interpret this directly, but give it the logical form: $(\phi \wedge \psi) \rightarrow \kappa$ (i.e., treating ϕ and ψ as restricting the same main quantifier): basically, we work all modals and conditionals up to the highest possible level).

Now, because this is not an approach that is systematically worked out and defended, I cannot argue that it is inadequate. But let me express some serious worries.

The disturbing aspect of this approach is the degree to which it has to rely on logical forms that are more and more remote from the surface forms of the expressions. In the above example, the use of logical form was still relatively mild (just move the **may** over the conditional). But when we look at what would have to happen with sentences where, say, the consequent of the conditional is itself logically complex and has modals embedded in it, the prospects of finding a systematic mechanism that will assign the right logical forms to sentences become highly dubious to me. For instance, in the following example we would have to work the modal up out of the conjunction, without it having scope over the first conjunct, and without losing the fact that the *it* in the second conjunct has *a hat*, introduced in the first conjunct, as its antecedent (and hence that should be accessible):

(13) If a man comes in, he wears a hat and he may give it to you.
$\phi \rightarrow (\psi \wedge \textbf{may } \kappa)$.

This means that we will have to give this sentence a logical form like: $(\phi \rightarrow \psi) \wedge \textbf{may } (\phi \wedge \psi, \kappa)$. Now, it may be possible to find such a rule and find a rationale for it, but at present it seems rather hopeless to me.

A discourse representation theory that is able to interpret the **may** in situ in this example does not need any logical reconstruction of this sort of the discourse and seems to be more attractive (to me at least) for that reason.

But such a theory has to deal with the logical problem of iterations of logical operations and modals and this is a highly non-trivial task (this is a very hard problem that I cannot go into here, see Landman 1986 for extensive discussion).

It is very hard to give knockdown empirical arguments in favor of either one position (modals as indexicals, vs. modals as context sensitive expressions, that can be sensitive to the local context) for various reasons. One is the vagueness of the present state of the indexical theory (if one logical form makes the wrong predictions, you can always draw back to another one). Another is the subtlety and vagueness of the data: it is just very hard to find strong intuitions about iterated modalities. Another point is the closeness of the approaches. For instance: the analysis of **may** (ϕ,ψ) that I sketched above differs only subtly from the analysis of $\phi \rightarrow$ **may** ψ that you would find on the other approach, basically because in the other theory, the distinction between $\phi \rightarrow$ **may** ψ, **may** $(\phi \rightarrow \psi)$ and **may** $(\phi \wedge \psi)$ is very slight. Finally, an approach that says that modals and conditionals can have scope with respect to each other, will tend to assume that sentences with embedded modals show scope ambiguities: the fact that the theory allows, say, the sentence 'if ϕ then may ψ' to mean either one of the above three almost equivalent formulas doesn't make the task of deciding what it means and what it doesn't mean easier.

Let me discuss two more examples that bear on the question of the scope of modals in conditionals.
Let us listen to the following story:

> "There was this boy in my hometown before the war. We were in love. I'm not sure whether I still want to marry him. I don't even know whether he is still alive. Moreover, I don't know whether I will ever go back. What's the point: maybe I'll go back and find out that he is dead. **But if I ever go back, I may marry him.**"

I tend to think that the last sentence, in the context set up, has a reading where it is false (to formulate it overly cautiously), a reading where it is incompatible with the one but last sentence. This is the narrow scope reading: the sentence is true if every possible situation (allowed by the context) where she goes back, still allows for her marrying him, but this is excluded in the possible situation where she goes back and finds out that he is dead. The indexical theory predicts that the only reading that the sentence has in this context is: "it is possible that I go back and marry him." Again, a scope theory does not exclude this reading, but tells you that it has the other as well.

A second example has to do with donkey-sentences. We hear the following description:

> "This guy is so paranoid. He sees Jack the Ripper everywhere. His methodology is: **If a prostitute is killed, then Jack may have killed her.**"

In this context, this sentence does not mean that out of all the possible cases where a prostitute is killed, in some (possible ones) Jack did it, but rather that for each possible case, the possibility that Jack killed her cannot be excluded. Again, the indexical theory seems to predict that this reading doesn't exist.

As I mentioned before, I won't in this paper go into the analysis of iterations, in fact, I will for simplicity use an analysis that for iterations makes blatantly wrong predictions (but see Landman 1986 for how it can be modified). Yet I hope to have given some motivation for a central assumption that I will make, an assumption that I share with most of the work on conditionals (especially the work closest to the present paper, Kratzer 1977, 1981; Veltman 1976, 1981, 1985, 1986; Landman 1986), that there are important differences between the context dependency of indexicals and of modals.

The conclusion, then, that can be drawn from all this is that information states play a more fundamental semantic role in the analysis of modals than the context of use (i.e., whatever deals with indexicals). The information state crucially enters into the semantic recursion (to give us the local context), while the context of use, so to say, stays on the outside. Maybe it clarifies the discussion to point out the difference with a more standard modal logic. In a standard modal logic the semantics determines, given the context, the possible worlds in which a complex sentence is true, given the possible worlds in which its parts are true. On the present perspective the semantics determines, given the context, the information states in virtue of which the complex sentence is true, given the information states in virtue of which its parts are true. Theories that have the latter form are given in the mentioned works by Veltman, Kratzer and myself.

The theory in Kratzer 1981 is in fact a mixed theory, where the semantics determines the world-information state pairs in virtue of which a complex sentence is true, given the world-information state pairs where its parts are true. In her theory the world part is used for non-modal expressions and the information part for modals. The distinction is important and leads to subtle differences with the approach followed by Frank Veltman and me. Since the semantics for non-modal expressions is classical (a standard possible world semantics) the notion of a speaker having partial information with respect to a non-modal expression is not given by the semantics directly, but has to be defined using the notions that the semantic theory provides (this can easily be done in the classical way: the information state as a set of alternatives provides the information that ϕ if ϕ is true in all alternatives). Such a definition is an essential part of the theory, because epistemic uncertainty is not a semantic borderline phenomenon that can be isolated in the grammar by saying 'there are a few expressions (*must* and *may*) that are information sensitive, they are context dependent.' Since the modals and conditionals interact strongly with the other connectives (like negation and disjunction), epistemic uncertainty and the use of devices to express among others what information we have or do not have is at the heart of semantics (although we might decide to call it pragmatics, rather than semantics). A more classical setup like Kratzer's 1981 theory makes predictions about the interplay between modal and non-modal expressions that are interestingly different from a theory that takes the notion of an information state as the basis for the semantic recursion for both modal and non-modal expressions. One such case (the relation between ϕ and **must** ϕ) I will come back to in the next section.

I said earlier, when I talked about the telescope metaphor, that it is sensible not just for modals but also for non-modal expressions to ask whether they are true or false relative to the information. Given what I just said about epistemic uncertainty being at the heart of semantics it becomes attractive to develop this metaphor (pointing at the information) into an alternative to the classical theory based on the classical metaphor (pointing at the world) (that is, the theory that brought us into problems to start with). This we could call a **partial semantics**: the semantics does not specify conditions for truth/falsity per se (though truth and falsity can be defined as the borderline case of total information), but **the semantics specifies recursive conditions for truth on the basis of the information/falsity on the basis of the information**. Again, the terminology is meant to be neutral, I don't care whether you want to call them truth-on-the-basis-of-the-information conditions, evidence conditions, verification conditions, or assertability conditions (although each choice of terminology has its own welcome and unwelcome associations; assertability (Dummett's term) is probably the most neutral), the important point is that (even though a particular semantic theory developed in this framework can be classical) the framework is non-classical in that it does not take the capacity to recognize whether a sentence is true or false in a situation to be at the heart of the semantic recursion, but the capacity to recognize whether an information state does or does not justify an assertion.

Before I continue to fill in some details, let me draw a general moral for discourse representation theory out of the preceding discussion. The work on modals shows that we need to recognize a semantic level of partial information. Besides the level of discourse syntax (drses), we need a level of discourse semantics as a level of partial information. If we incorporate this in the theory as a partial semantics, then the structure of the theory will be as follows:

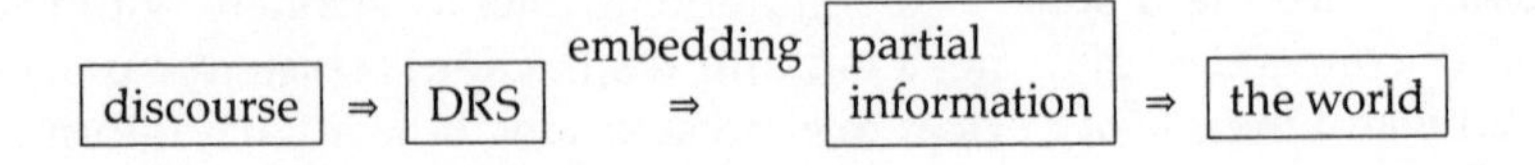

SOME ASPECTS OF DATA SEMANTICS

In this section I will do two things. I will first give a short sketch of some relevant aspects of data semantics. Apart from some remarks on discourse anaphora, the points I will discuss here are not new. Most of them can already be found in Veltman 1981, one example (15-18) comes from Veltman 1987, some more discussion from Landman 1986. Since I have to be brief here, I have to refer the reader to the works by Frank Veltman and by me in the references for details, motivation, etc. After this, I will compare the data semantic analysis of modals with the more well known possible world style analysis.

The models for the theory are familiar Kripke models (with some conditions on them that I won't go into here).
A model is a tuple $<S, \leq, D, i^+, i^->$.
S is a set of possible information states, partially ordered by a relation of information extension $\leq$.
For every information state s, D(s) is the domain of s. I will assume that domains grow as information grows: i.e., if $s \leq s'$ then $D(s) \subseteq D(s')$.
i^+ and i^- assign to every predicate and information state a positive and negative extension respectively, i.e., $i^+(P, s)$ is the set of objects that have P on the basis of s, $i^-(P, s)$ is the set of objects that (definitely) do not have P on the basis of s. Obvious conditions are:
$i^+(P, s) \cap i^-(P, s) = \phi$ (the positive and negative extension of a predicate do not overlap)
$i^+(P, s) \cup i^-(P, s) \subseteq D(s)$ (both the positive and the negative extension of P are subsets of the domain)
if $s \leq s'$ then $i^+(P, s) \subseteq i^+(P, s')$ and $i^-(P, s) \subseteq i^-(P, s')$ (not only the domains, but also the interpretations grow)
Several other things should be made explicit (concerning for instance total information states), but I will omit such discussion here.

The semantics defines recursively the following two relations:

$g \Vdash_s C$: embedding g verifies condition C **on the basis of** s
$g \dashv\vert_s C$: embedding g falsifies condition C **on the basis of** s.

Again I should be much more explicit about the semantics, in particular about how this fits in with a full DRT-semantics. I hope to do that in another paper. The cases that I will discuss do not particularly depend on the suppressed parts.

Partial and total information

Given the models introduced above, the clauses for atomic conditions and their negations are unsurprising:

$g \Vdash_s P(x)$ iff $g(x) \in i^+(P, s)$
$g \dashV_s P(x)$ iff $g(x) \in i^-(P, s)$
$g \Vdash_s \neg A$ iff $g \dashV_s A$
$g \dashV_s \neg A$ iff $g \Vdash_s A$

On partial information both P(x) and ¬P(x) can be undefined. On total information all epistemic uncertainty will be resolved, the logic becomes classical logic. So classical logic indeed is the limit case of total information.

Instability of may

$g \Vdash_s$ **may** A iff for some $s' \geq s$ and some $g_A \supseteq g$: $g_A \Vdash_{s'} A$.
(Here and in what follows I am assuming the formulation of DRT that is given in Landman 1987 as Theory 4.)

maybe John is coming in is unstable (non-monotonic) under information growth: it can be true on the basis of partial information (because there is still a possible extension where John is coming in), but false on the basis of better information (because the previous extension is no longer possible). This is different for atomic expressions: *John is coming in* is stable: once it is in the information, it will stay there (under information growth).

The unstability of **may** can be observed as well in (5):

(14) Maybe John killed the doctor. But if he was at home that night, he didn't kill her.

The contextual information state still allows an extension to an information state containing the possible fact that John killed the doctor, but the information states with respect to which the consequent of the conditional is evaluated (which are information states extending the contextual one with the antecedent information) cannot be extended to such an information state any more.

The indirectness of must

$g \Vdash_s$ **must** A iff for every $s' \geq s$ there is an $s'' \geq s'$ such that $g \Vdash_{s''} A$

(I am simplifying by ignoring the assignment extensions here.) This means: in whatever way the information may grow, in the end A comes in the information. The distinction between A and **must** A is the distinction between direct and indirect information: A is stronger than **must** A (at least for atomic A): A is directly true on the basis of s if A is one of the facts in s, it is indirectly true on the basis of s if there is no way of extending the information to an information state where $\neg$A holds. (If A is stable and directly true it is also indirectly, but A can be indirectly true without being yet directly true.)

Of course, one may wonder: isn't the notion of direct information a philosophically unfounded notion? Isn't all information when it comes to it indirect, based on inference? And if so, why try to incorporate a notion of direct information in the semantics?

The answer to these questions, which I have defended at length in Landman 1986, relies on the distinction between (semantic) information growth and (pragmatic) information refinement. Notions like partial information, total information, and with them direct and indirect information, are strongly context dependent. Information is only total or direct with respect to a standard of precision. What is a simple piece of information under a loose standard of precision may turn out to be a highly complicated structure of facts under a more precise standard; what is a total description of the world may turn out to be highly partial if on a more fine-grained standard of precision other facts are taken into account; moreover, pragmatic accommodation can introduce possibilities that were previously not considered (for accommodation, see Lewis 1979; the fact that pragmatic accommodation can introduce possibilities is a crucial aspect of Stalnaker's conception of possible worlds (see Stalnaker 1984) and the way they allow for partiality).

So we need to allow for the possibility that pragmatic accommodation changes our initial information state and/or its relation to other information states. This is not incompatible with a semantic notion of total information or direct information. The sense in which all information is indirect is pragmatic: we can always refine our information pragmatically in such a way that what was direct before turns out to be indirect according to the new standard of precision. The situation is similar to the analysis of vagueness (Kamp 1975): here too we want to semantically relate a vague predicate to all possible ways of making it precise. But 'precise' means: precise with respect to a certain standard of precision; a predicate is inherently vague if we can always redraw the standard of precision is such a way that the sharp extensions according to the old standard, turn out to be vague with respect to the new standard (cf. also Lewis 1979's discussion of the classical example *France is hexagonal*).

But having recognized the strong influence of context and accommodation, one should also recognize the semantic need for notions like total information and the direct/indirect distinction. For instance, though we realize that we can accommodate a possibility that was previously not considered, therewith making **may** A true, we should at the same time realize that it involved **correction**, that after we have accepted **may** A, we are no longer willing to accept **must** ¬A. This means that our semantic theory should capture the fact that, when evaluated with respect to one and the same information state **may** A and **must** ¬A are incompatible. And we can do that by distinguishing between information growth, the relation between an information state and its extensions, relative to a standard of precision (a background context), and information refinement, the relation between one system of information states and its refinements.

Similarly, there are semantic reasons for introducing a difference between direct and indirect information. That *John must be home* is a weaker statement than *John is home* has often been observed (for instance, Karttunen 1972) and the distinction is neatly shown in examples (15)-(18):

(15) I am hungry.
(16) ?I must be hungry.
(you can say (16), but it seems as if you infer it from external, indirect cues)

(17) You must be hungry.
(18) ?You are hungry.
(again, you can say (18), but it is normally impolite to inform someone about something which they normally have direct information about and you don't).

A semantic theory should represent these distinctions, and the proper way to do that, it seems to me, is to say that even though in the end the evidence on which my utterance of (15) is based may be indirect as well, still there is a clear sense in which, relative to the standard of precision of a normal context, (15) is direct in a way that (16) is not. This is what the data semantic clauses intend to capture. I will come back to the semantics of *must* later in this section.

Negation of conditionals

$g \Vdash_s A \rightarrow B$ iff for every $s' \geq s$ and every $g_A \supseteq g$ such that $g_A \Vdash_{s'} A$ there is an $s'' \geq s'$ and a $g_B \supseteq g_A$ such that $g_B \Vdash_{s''} B$.

The basic scheme of the conditional is: every extension verifying the antecedent can be extended to an extension verifying the consequent. In this paper I won't be concerned with the semantics of the conditional; for extensive discussion of this clause and its modifications and alternatives, see Landman 1986, Veltman 1985. I will make some remarks about the corresponding falsifying clause, though. The obvious data semantic clause (given the above verification clause) is the following:

$g \dashv\!\vdash_s A \rightarrow B$ iff for some $s' \geq s$ and some $g_A \supseteq g$: $g_A \Vdash_{s'} A$ and $g_A \dashv\!\vdash_{s'} B$.

This means that $\neg(A \rightarrow B)$ is going to be (weakly) equivalent to **may** $(A \wedge \neg B)$

A lot has been written on the proper analysis of negations of conditional sentences. I here want to indicate that donkey sentences and discourse anaphora may add some new arguments in favour of the above analysis. Take the following examples:

(19) If a farmer owns a donkey, he beats it.
(20) It's not the case that if a farmer owns a donkey, he beats it.

Now look at (21)-(24):

(21) It's not the case that if *a farmer* owns a donkey, he beats it. **He* owns a cow as well.
(22) *A farmer* owns a donkey and doesn't beat it. *He* owns a cow as well.
(23) If a farmer owns a donkey, he doesn't beat it.
(24) Maybe *a farmer* owns a donkey and doesn't beat it. **He* owns a cow as well.

(19) is the classical donkey sentence, which quantifies universally over farmer-donkey pairs. What does (20) mean? Classical logic tells us that (20) is equivalent to the first sentence in (22) ($\neg(A \rightarrow B)$ iff $A \wedge \neg B$), so (21) and (22) should have the same meaning. But they don't: apart from the fact that (22) seems to make a much stronger claim than (21), the anaphora is not possible in (21), while it is in (22).

Stalnaker 1968 tells us that (20) is equivalent to (23) ($\neg(A \rightarrow B)$ iff $A \rightarrow \neg B$). Since (23) is a donkey sentence as well, it means: for every farmer-donkey pair such that the farmer owns the donkey, the farmer doesn't beat the donkey. Clearly this is not right for donkey sentences, (20) certainly does not mean (23). The data semantics analysis tells us that (20) is equivalent to the first sentence in (24) (and hence (21) and (24) are equivalent). Not only does this intuitively sound quite plausible, but it makes the right predication about anaphora possibilities as well.

This connection seems to be even strengthened if we look at modal subordination cases. Modal subordination (see Roberts 1987) is the phenomenon that under a certain continuity of modality anaphora is possible, but stays "in the scope of the modals," as in:

(25) *A tiger* may come in. *It* would eat you first.

We have observed above that in general anaphora outside a modal or a conditional, or the negation of a conditional cannot take an antecedent inside those. But they can take such an antecedent on the modal subordination interpretation. The observation that I want to make is that for all three of the following examples (26)-(28) a modal subordination continuation seems possible and on that reading they all seem to be equivalent to (29):

(26) Maybe *some farmer* doesn't own a donkey. For instance, maybe *he* owns a cow instead.
(27) It's not the case that *every farmer* owns a donkey. For instance, maybe *he* owns a cow instead.
(28) It's not the case that if *someone* is a farmer, he owns a donkey. For instance, maybe *he* owns a cow instead.
(29) Maybe some farmer doesn't own a donkey but a cow.

Again, I think that the parallel suggests that the data semantic analysis is on the right track.

Note that this analysis of the negation of conditionals will bring the instability of **may** into the logic of the conditionals. A consequence of this is that the principle of *modus tollens* does not hold generally (see Veltman 1981, 1985; Landman 1986).

Modus Tollens

Take the following argument: Maybe there will be war, maybe there won't be war. But if B wins the elections there must be war, war is unavoidable.
With modus tollens we can conclude:
may W, **may** ¬W (= ¬ **must W**), B → **must W**, hence ¬B. In other words, if modus tollens is valid, the above argument implies: B doesn't win the elections. This is of course much too strong. In data logic we get the following dichotomy:

If ¬Y is stable, we have indeed: X → Y, ¬Y, hence ¬X
If ¬Y is instable, we only have: X → Y, ¬Y, hence **may** ¬X

Since ¬ **must A** is instable, the above argument only leads to the weaker (and correct) conclusion: *maybe B doesn't win*.

Indirectness of must *in data semantics and possible world semantics*

In the next section, I will argue that the data semantics analysis of modals which I have sketched here forces us to adopt a radically nonstandard analysis of identity statements. Although the above analysis differs on all the points mentioned from what we are used to in possible world semantics, we may wonder whether a rich framework like possible world semantics is not able to take over the desirable aspects of the data semantic analysis. This is an important question, because, if we can incorporate the important aspects of data semantics in a possible world-style theory, it is quite likely that we will be able to combine the analysis of modals with a standard, classical analysis of identity statements. Thus, the argument in the next section, that the analysis of modality unavoidably leads to a non-classical theory of identity, would be unfounded.

I will argue here that the distinctions on which data semantics is based — in particular the distinction between direct and indirect information — cannot be captured correctly in standard possible world semantics (of course it can be captured in a less standard possible world semantics: data semantics is a less standard possible world semantics); moreover I will defend the data semantics analysis of *must* against the analyses that **are** possible in possible world semantics.

Let us see how possible world semantics could take over a data semantic analysis.

In possible world semantics, sentences denote propositions, sets of possible worlds. Following Stalnaker's work (e.g., Stalnaker 1979), there are two notions of information state. The first is what Stalnaker calls the common ground and what Kratzer calls the modal base, a set of propositions which are regarded as common information by the speech participants. The second is the context set, a set of alternatives (i.e., possible worlds). This is the set of all worlds compatible with the information in the common ground, i.e., the set of all worlds in which every proposition in the common ground is true (technically, this is equivalent to the closure of the common ground under logical consequence and conjunction).

Let us ask the following question: how can we capture the distinction between direct and indirect information, i.e., between A and **must** A in such a theory?

Already in the oldest work on *must* and *may* (Karttunen 1972) it is assumed that the distinction can be analysed as that between what **is in** the information and what is not yet in the information, but follows from it. Let C be a common ground and let $\cap C$, the intersection of all the propositions in C (i.e., the set of worlds in which every proposition in C is true) be its corresponding context set. Then we can define: A is true relative to common ground C if $A \in C$; *must* A is true relative to common ground C if A is true in every world in $\cap C$.

So indeed A is true relative to C if A is in C; *must* A is true relative to C if A follows from C.

The common ground is not closed under logical consequence, the context set is. So something can follow from the context set, without being explicitly (directly) in the common ground. In that situation it would be indirectly true.

However, as I argue in Landman 1986, this leads to a notion of direct information that is devoid of any content. Which propositions can provide direct evidence for the truth of A? In the present analysis, hardly anything but A itself gives us sufficient information to conclude that A is directly true; any other information gives us at most **must** A (the reason is that C is not closed under logical consequence). To give an example, not even if we have the information that John is sixteen years old can we conclude that "John is older than three" is directly true (because the first proposition can be in C without the second being in C). So we can introduce a direct/indirect distinction in this way, but it gives us a useless notion of direct information.

So we cannot use the logical gap between the common ground and the context set to characterize the direct/indirect distinction. In this framework, then, the only thing we can possibly try to use to define the direct/indirect contrast in a non-void way, is the fact that the corresponding context set, in virtue of its being a set of alternatives, is a partial information state. With this I mean the following: the context set is the set of all worlds compatible with the information. This set is constrained by the context, so it doesn't contain all possible worlds, but since our information is not total, it doesn't eliminate all possible worlds (except one) either: different alternatives for what the facts in fact are still open in the context set. In this sense, the notion of a context set comes very close to the data semantic notion of an information state. So the option that we can try to work out is to take the data semantic clauses for A and *must* A and impose them on the notion of context set. For the truth conditions of A, we have no choice but to follow the standard possible world practice: A is true relative to $\cap C$ iff $\forall w \in \cap C$: A is true relative to w. For *must* A, we take the data semantics clause: *must* A is true relative to $\cap C$ iff for every context set X extending $\cap C$, there is a context set Y, extending X such that A is true relative to Y.

What does it mean that X extends $\cap C$? It means that X contains better information than $\cap C$, which in possible world semantics means that X has eliminated certain alternatives that $\cap C$ hasn't, i.e. X is a subset of $\cap C$. Given this, it is easy to see that these two notions of direct and indirect information collapse: A is true relative to $\cap C$ iff *must* A is true relative to $\cap C$.

This exhausts the possibilities of the framework as it is, so we see that in the framework as I have presented it above, there is no sensible way of drawing the distinction between direct and indirect information.

In the possible world framework, what I think is the only way out is the way that Kratzer 1981 takes.

Modals, in her theory, are not just interpreted relative to one information state, the modal base, but relative to two: the modal base and the ordering source. The modal base C (as before a set of propositions) determines what kind of modality is involved (epistemic, deontic, etc.). The ordering source O, which is also a set of propositions, orders the worlds in $\cap$C. Leaving aside certain complications that are irrelevant for our purposes, O determines within $\cap$C a subset $\cap$O.

Here is an example of the usefulness of this notion. Suppose C is a deontic modal base. Then $\cap$C is the set of all deontically possible worlds. C determines the truth of sentences like: *I must not steal*: in view of what C allows and forbids, I must not steal, and this means that the proposition that I steal is incompatible with C. However, certain modals seem to make a less categorical statement than the previous one, like *I should give to the poor*. It's just too strong to claim that *I don't give to the poor* is incompatible with C: the above modal statement is more an 'adhortative' than an 'imperative.' For this reason, Kratzer assumes that within the set of deontically possible worlds, the ordering source selects out a subset of which we can think as those deontically possible worlds in which the deontic ideal that the ordering source contains (like 'the ideal of a good life') is realized. *I should give to the poor* is true relative to C and O if in all those permissible worlds (worlds in C) which are worlds in which I realize the ideals of a good life, over and above the fact that I don't do anything that isn't forbidden, if in all those worlds (and they are the worlds in O) I give to the poor.

I have no problem with the notion of ordering source as such. In fact, I think Kratzer makes a compelling case for them in her paper. I will here only be concerned with the question whether they are the right instrument to get us out of our problems.

This is the idea:
we keep the definition of A being directly true relative to C if A is true in every world in $\cap$C. We now redefine 'A is indirectly true (or **must** A is true) relative to C′ as: A is true in every world in $\cap$O. Presumably we should think of $\cap$C as the set of epistemic alternatives and $\cap$O as the set of epistemically trustworthy alternatives.

In this theory A is stronger than **must** A (if A is true throughout $\cap$C, it is certainly true in $\cap$O), and also, **must** A can be true, without

A being yet true (relative to C and O). Hence, we have been able to create a distinction in possible world semantics after all.

However, there is a crucial difference between this theory, and the data semantic one I introduced above.

Let us consider a situation where we are willing to assert **must** A, but not yet A. For instance, you and I are walking over Cornell campus, you're figuring your way out with a map and finally you say: 'This must be Morrill Hall.' I am in a position to confirm this, I know the building, so I say: 'Indeed, this is Morrill Hall.' Let us assume that your information state C before my utterance was one where **must** A is true, but not yet A. The situation according to data semantics is that A is true in every total extension of C, but not yet in C itself. The situation according to Kratzer is that A is true in every world in $\cap O$, but not in every world in $\cap C$, which means that in some world in $\cap C$ A is false: in other words, there is an epistemic alternative, but not one very close to the ideal O, in which A is false. In Kratzer's terminology: someone who accepts **must** A in this situation allows for a slight possibility that A is false (in fact, there is a modal operator which we can call SP (for slight possibility) that expresses this: SP(A) is true relative to C and O if there is a world in $\cap C - \cap O$ where A is true).

Both theories claim the **must** A is accepted on the basis of indirect evidence. The difference between the two theories lies in what they think indirect evidence is. Data semantics claims that a language user, although she distinguishes between direct and indirect evidence, still treats indirect evidence as **good** evidence: someone who accepts **must** A does not leave open the possibility that $\neg$A any more than someone who accepts A does. Kratzer's theory, on the other hand claims that a language user treats indirect evidence as **bad** evidence (or relatively bad evidence): it is not possible to have indirect evidence that A without leaving open at least the slight possibility that $\neg$A, while direct evidence that A is inconsistent with even the slightest possibility that $\neg$A.

We observe that the theories make different predictions. Kratzer predicts that in the situation described in fact **must** A is equivalent to SP($\neg$A) $\wedge$ **must** A. Data semantics predicts that SP($\neg$A) and **must** A should be inconsistent. Now look at the following sentence:

(30) There is a slight possibility that this is not Morrill Hall, but it must be Morrill Hall.

Kratzer predicts that this sentence should be perfectly all right, and in fact, in the situation described equivalent to (31)

(31) This must be Morrill Hall.

To me, this seems to be wrong: (30) is not equivalent to (31), but feels as inconsistent as (32):

(32) There is a slight possibility that this is not Morrill Hall, but it is Morrill Hall.

We should be careful, though. What about (33):

(33) Although it is logically possible that John is not the murderer, all the evidence we have gathered tells us that he must be the murderer.

This example (which I owe to Angelika Kratzer) does not seem to be inconsistent and seems to have the structure SP(¬A) ∧ **must** A. However, I don't think that this is a convincing example. The second sentence (the **must**) is evaluated relative to an **epistemic** modal base. But the first modal is not related to the **same** epistemic modal base, but to a (less informative) **logical** modal base (specifying what is logically, rather than epistemically possible). Of course, the data semantics' claim is only that SP(¬A) and **must** A are incompatible if related to the same modal base, so this example is not a counterexample to it. On the contrary, I would even say, we can take the fact that we make explicit that in the first sentence a different modality is involved than in the second as an indication that we want to avoid a contradictory interpretation as in (30).

However, what should we say about the following example? You have guided me to Morrill Hall by use of the map and you say:

(34) This must be Morrill Hall. Well, of course there's a possibility that I made a mistake, that the map is inaccurate, and that this isn't Morrill Hall.

This also doesn't seem to be inconsistent in the way (30) is, and it doesn't necessarily have to involve the notion of **logical** possibility. So this would be a more convincing example in favour of Kratzer's theory. Another example that seems perfect I owe to one of the anonymous referees:

(35) John must not have heard what I said, but if he did, he can tell you.

However, I don't think that these examples are arguments for Kratzer's theory.

Let us look at (35) first. The case is slightly different from the other case, because the second sentence, if evaluated relative to the modal base relative to which the first is true would not be a contradiction but a tautology. However, the *but* indicates a contrast and what it does in this context is indicate that a slightly different modal base may be relevant. That this is so is shown by example (36):

(36) John **didn't** hear what I said, but if he did, he can tell you.

Here the first sentence is not a modal at all, yet the sentence is perfect. If so, then we have every reason to assume that whatever makes (36) well-formed makes (35) well-formed as well.

Concerning (34), I have the strong impression that what is going on here is accommodation. The speaker makes a strong statement: this must be Morrill Hall. Then she wants to be more cautious, she **weakens** her statement slightly by bringing certain possibilities into the discourse. Consider (37):

(37) This must be Morrill Hall. Well, there is a slight possibility that it isn't Morrill Hall. But still, this must be Morrill Hall.

Unlike (34), (37) sounds to me as contradictory as (30). The explanation that I gave for (34) and a data semantics explain this. The second sentence would be inconsistent with the first if interpreted on the same information state. So instead the information state is (pragmatically) changed: a possibility is accommodated. But as soon as the second sentence is accepted as true, the first is no longer acceptable, because they are inconsistent. So adding the first sentence after the second has been accommodated results in a contradiction indeed.

On Kratzer's theory, the only thing that is wrong with (37) is that it is repetitive: given the first sentence, the second and the third do not add any new information. To me, that seems to be inadequate.

Summarizing: (37) shows that (34) is a case of accommodation or correction, that in the second part of (34) the claim made in the first part is weakened, in fact, that on accepting the second part, the first part is no longer acceptable, i.e., that the second part **corrects** the first part. The only reasonable explanation for this is, I think, that after all SP(¬A) and **must** A are semantically inconsistent if evaluated on the same information state. In other words: language users treat indirect information that A as good information, information that is not more compatible with the possibility of ¬A than the direct information A itself is: the seeming compatibility is in fact a case of pragmatic correction.

This argument, if plausible, has some strong consequences. Kratzer's theory is, as I indicated, in fact the only available option in a theory that builds information states out of possible worlds, more precisely, a theory that reduces truth on the basis of an information state to truth in the worlds in that information state, i.e., a theory that tries to define assertability conditions in terms of truth conditions. I have argued that the distinction that underlies the difference between A and **must** A is not a distinction between good and less good evidence, but a distinction between good direct and good indirect evidence. As I have indicated here and have argued at length in Landman 1986, this is a distinction that you cannot make in a classical theory like possible world semantics; the fact that you can make the distinction in data semantics depends crucially on the fact that the partiality enters into the semantic recursion there. But then we have a strong argument in favour of the data semantic approach. The central notions of Stalnaker's and Kratzer's theories can be formulated as easily on information states as they can be formulated on sets of possible worlds. In fact, possible world semantics itself can be formulated in data semantics, if wanted, because total information states can be taken as possible worlds and accessibility relations can be defined on them (also, there is no principled reason to take $\leq$ to be the only accessibility relation on information states). But data semantics gives us more than possible world semantics, because it allows for a better account of the direct/indirect contrast.

IDENTITY STATEMENTS AND EPISTEMIC MODALS

Let me now turn to identify statements. Since identity is a two-place relation, the clause for atomic formulas that I have given earlier provides the basics of the semantics for identity statements:

$$g \Vdash_s a = b \text{ iff } \langle g(a), g(b)\rangle \in i^+(=,s)$$
$$g \dashv\!\mid_s a = b \text{ iff } \langle g(a), g(b)\rangle \in i^-(=,s)$$

g(a) and g(b) are objects in the model (in the domain of s). But what are objects? Are they classical, 'real' objects? What are real objects? Well, whatever a real object is, it is clear that if something is a real object, it is identical to itself and to nothing else. This means that if the objects in our models are real objects, the positive and negative extension of the identity are fixed in the following way:

$$i^+(=,s) = \{\langle d,d\rangle : d \in D(s)\}$$
$$i^-(=,s) = \{\langle d,d'\rangle : d,d' \in D(s) \;\&\; d \neq d'\}$$

The problem with this is that this makes = a **total predicate** on the domain of every information state. That is, for every information state s, it holds that if a and b are defined on s (that is if g(a), g(b) $\in$ D(s)), then:

$$g \Vdash_s a = b \text{ or } g \dashv\!\mid_s a = b.$$

In other words, there is no way of **not** knowing whether a and b are the same or not.

Let us take the following drs, representing a Babylonian conversation concerning Hesperus and Phosporus: (this could be the drs of: *Something shines in the morning, something shines in the evening. Maybe they're the same, maybe not.*

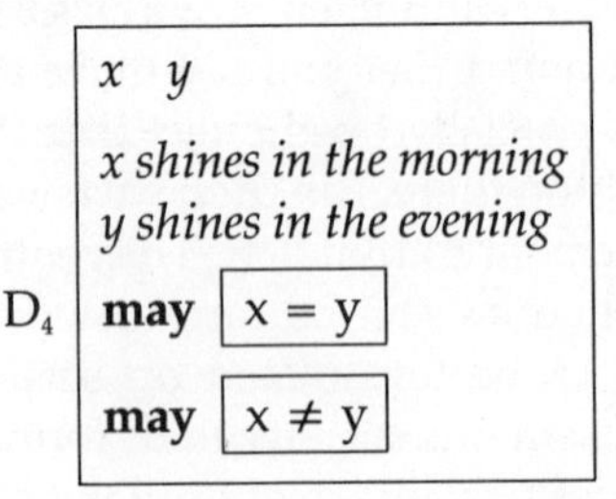

If an embedding g verifies the first two conditions relative to some information state s, then g(x), g(y) ϵ D(s). But that means that either $<g(x),g(y)> \in i^+(=,s)$, or $<g(x),g(y)> \in i^-(=,s)$. But if that is the case, then that is the case for every extension of s as well:

If $<g(x),g(y)> \in i^+(=,s)$ then $\forall s' \geq s$: $<g(x),g(y)> \in i^+(=,s)$
If $<g(x),g(y)> \in i^-(=,s)$ then $\forall s' \geq s$: $<g(x),g(y)> \in i^-(=,s)$

This just follows from the fact that our interpretations are growing (and given the interpretation of objects as classical objects, there is certainly not a way around that). But this means that if x and y are defined, **may** x = y is logically equivalent to x = y, and similarly **may** x $\neq$ y is logically equivalent to x $\neq$ y. Namely: clearly x = y implies **may** x = y. If x and y are defined on s then either g $\Vdash$s x = y or g $\Vdash$s x $\neq$ y. If g $\Vdash$s **may** x = y then for some extension s' of s: g $\Vdash$s' x = y. But then, given the above it is impossible that g $\Vdash$s x $\neq$ y, hence g $\Vdash$s x = y. The same argument for x $\neq$ y and **may** x $\neq$ y.

This shows that, if identity is a total predicate on every information state, the above drs D_4 is a contradiction. But clearly it is not, D_4 represents the information of someone who wonders about whether Hesperus and Phosphorus are one and the same object or not, and there is of course nothing incoherent in that kind of wonder.

There is only one way out of this problem: we have to make = a partial predicate. That is, we have to **weaken** the classical conditions:

$i^+(=,s) = \{<d,d>: d \in D(s)\}$
$i^-(=,s) = \{<d,d'>: d,d' \in D(s) \;\&\; d \neq d'\}$

Since I see no reason to remove the ordered pairs $<d,d>$ from the positive extension of =, this means that:

$i^+(=,s) \supseteq \{<d,d>: d \in D(s)\}$
$i^-(=,s) \subseteq \{<d,d'>: d,d' \in D(s) \;\&\; d \neq d'\}$

with the normal condition that the positive and the negative extension of = on s do not overlap.

This means two things. In the first place, if d ϵ D(s) then $<d,d> \epsilon\, i^+(=,s)$. So much is as before. But now there are objects d_1, $d_2 \epsilon$ D(s) of which we may find out that they are identical and of which we may find out that they are not identical, there are the objects in the gap of = on s. In other words, we cannot find out of d that it is not identical to d, but if d and d′ are in the gap of = on s, then there may be an extension of s, s′ such that $<d,d'> \epsilon\, i^+(=,s')$, and there may be an extension s″ of s where $<d,d'> \epsilon\, i^-(=,s'')$.

As I said before, of classical objects we know what they are identical to (themselves). So the things in our models are not classical objects.

(All this doesn't come as a surprise, of course, for readers familiar with intuitionistic logic. But note, for the interpretation that I will give in the next section, and its semantic consequences it is crucial that partial objects are the objects in D(s), and not their equivalence classes under identity on s, as in intuitionistic logic.)

We have to put some more restrictions on the identity predicate (for a more detailed account, see Landman 1986). First a definition:

> d and d′ are **indiscernible on the basis of** s iff no property tells them apart on the basis of s (i.e., s does not put d in the positive extension of some predicate and d′ in the negative extension of that predicate or vice versa). Else d and d′ are **discernible on the basis of** s.

When are d and d′ in the **negative** extension of = on s? Obviously, if s can tell d and d′ apart:

> $<d,d'> \epsilon\, i^-(=,s)$ iff d and d′ are **discernible** on the basis of s.

When are d and d′ in the **positive** extension of = on s? Not if d and d′ are indiscernible on the basis of s, not if s cannot tell them apart. Maybe some extension of s can. I take them to be identical on the basis of s if it is not possible to extend s to an information state that tells d and d′ apart:

> $<d,d'> \epsilon\, i^+(=,s)$ iff for all $s' \geq s$: d and d′ are indiscernible on the basis of s′.

Assuming, as I do, that every information state can be extended to a total one, that is, an information state where all predicates are totally defined (i.e., on every possible way of getting better informed, in the end you reach a state where all predicates are total), it now follows that = is an equivalence relation on every information state.

If the objects in our model are not real objects, what then are real objects? Possible worlds, I have already said, correspond to total information states. Let w be a total information state (say, the information state corresponding to the real world). The real **objects** in w are the equivalence classes of the objects in D(w) under = on w.

In this way we reconstruct objects as their informational approximations (with respect to a certain standard of precision).

If one feels dissatisfied with this, one can of course live one's realist convictions by adding to the structures that I have described, domains of possible worlds and their individuals with a correspondence relation between the informational approximations and (some of) the real worlds and objects. (The situation is similar to that of event logic. The program of constructing moments of time out of events (see for instance Kamp 1979) does not forbid some (e.g., Hinrichs 1985) to add to the event structures a primitive domain of (real) moments of time).

Such an addition may satisfy some who feel that in the theory sketched here the solid ground of the real world is somehow disappearing under their feet. Whether there are compelling **semantic** reasons for such a modification remains to be seen, however.

Given this reinterpretation of the identity predicate, we can give a consistent interpretation to drs D_4:

D_4 is true on the basis of s iff for some g:
$g(x) \in i^{+}$(shine in the morning,s) and
$g(y) \in i^{+}$(shine in the evening,s) and
for some $s' \geq s$: $<g(x),g(y)> \in i^{+}(=,s')$ and
for some $s'' \geq s$: $<g(x),g(y)> \in i^{-}(=,s'')$

and this is perfectly consistent.

We are, then, forced to the conclusion that objects at the level of discourse semantics, the level of information, are not classical objects, but partial objects, objects only partially determined by their identity conditions. How should we interpret such non-classical objects?

PEGS

What are pegs? Pegs are things to hang coats on and take them off again. They don't change themselves if you do that. The objects in our domains are pegs, informational pegs. Information states hang properties on pegs (and relations on tuples of pegs) in the process of information growth (and take them off again in correction) (in other words, information states ascribe properties to pegs). Since identity is a relation, identity conditions are also hung upon pegs in this way (not just upon **a** peg, but upon pegs).

Pegs are informational objects, they are the trick that we use to overcome the problem of communication in a situation of partial information. We are commonly too badly informed to distinguish one real object from another that we know as much about (for instance, when someone reports to us: 'there's a man coming down the street. He wears a hat.'). **Pegs are objects we postulate in conversation as stand-in's for real objects. They are means of keeping track of what we talk about in information exchange**. As such they are postulated in shared information: we share the pegs we are talking about (because that is what they are for).

We cannot reach real objects. So we postulate pegs as stand-ins for real objects. When one person postulates a peg, other people can continue, and ascribe properties to the same peg. We talk about them as if they have independent existence, existence outside of us, like real objects.

It should be clear that I am taking a strongly pragmatic view on the content of the model-theoretic information structures that I am using. Information structures are pragmatically postulated, intersubjective structures "in between" mental representations and the world. The key to their existence is agreement between language users. Neither a strongly mentalistic, nor a strongly realistic view on those structures is particularly suited. Pegs are not mental objects, but agreed upon **interpretations** of those, and their nature as postulated objects does not make them very real. This is not to say that the theory can do without either mental representations or the world. Though it is a pragmatic choice **which** pegs are actually postulated in a conversation, and pegs are conventional in that way, this does not mean that the **structure** of the level of information is. In other words, though information structures form a medium in which we can make relatively arbitrary choices, this does not mean that the medium is arbitrary in any way, in fact we would expect the medium to be strongly determined both by our cognitive apparatus and the world.

But pegs are pragmatic objects. It is the context in a conversation, and its standard of precision that determines what distinctions are made, i.e., that determines the domain of pegs, and that manipulates domains of pegs: context change can make things that were discernable before indiscernible by adding new pegs (for examples, see Landman 1986).

Pegs are postulated objects, whose function is to keep track of objects in information states. When, in an introduction to semantics class, we talk about Hesperus and Phosphorus, **we** talk **as if there are**

two objects. We use the plural **they**, even when we say that **they** are identical, and when we say that they are not, on the basis of the Babylonian information. It is, I think, important to realize this. For instance, when Kripke (Kripke 1979) presents his puzzle about belief, he argues that the puzzle does not rely on substitution, but only on the uncontroversial assumption that in representing a French sentence in English we can translate 'Londres' by 'London.' This assumption is uncontroversial indeed in normal circumstances, but precisely **not** in the case that Kripke discusses. In order to translate a French name into its English equivalent it does not suffice to assume that they stand for the same object, but we need the further assumption that the original and its translation stand for the same peg, the same conversational object. And this is not the case in Kripke's case; indeed, the case is constructed in such a way that precisely that condition is violated: 'Londres' and 'London' stand for two pegs, referring to the same real object. When Frege talks about 'Venus,' 'der Morgenstern,' 'der Abendstern,' he talks about them as if they are three things, the text postulates three pegs. And, of course, a translator cannot say: 'Oh, it's so confusing, those different expressions, let me all translate them with Venus.' The situation is different if an astronomy book uses those three names as stylistic variants. In that case there is only one peg, and no problem arises if a translator decides to eliminate that variation. (For more discussion of Kripke's puzzle, see Landman 1986.)

Still, when we talk about the world-view of the Babylonians and introduce two pegs for Hesperus and Phosphorus, we use **the same** pegs when we describe their information, as we do when we compare it with ours. This is crucial for correction. Only in this way can we hope to convince the Babylonians that they have wrong information, in other words, that they have attributed the wrong properties to these same pegs. In this sense **pegs are objective** (or conversationally uniform). In a picture as seen on next page.

We see indeed that pegs perform their primary function: we keep track of Hesperus and Phosphorus through different information states.

The theory as developed here is meant to combine three seemingly incompatible theories:

(1) Pegs are like (Fregean) senses, **conceptual** objects. Though they are postulated by us, they are postulated as independent of us. They are more intensional, conceptual, though, than intensions in possible world semantics. If $\Box(h = p)$, then h and p have the same intension. However, as I have indicated above, even if $\Box(h = p)$ is true, we can

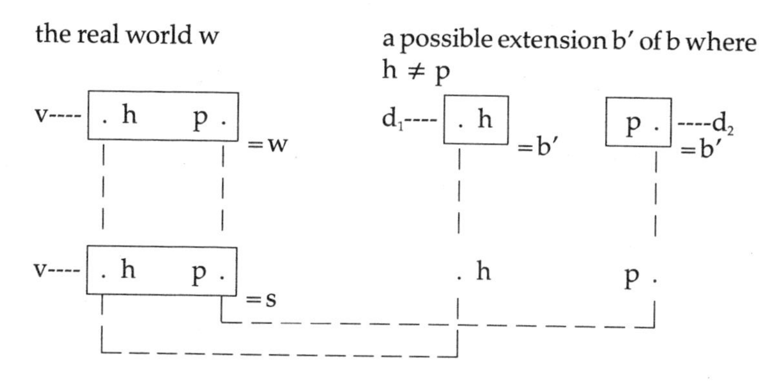

assume that h and p are interpreted as different pegs. This is so, precisely because pegs are conversationally postulated objects.

(2) For the same reason, pegs are like (Husserlian) intentional objects, objects of postulation, the objects as we talk about them, as we intend them, the objects we aim at.

It will be clear that given this, pegs have a lot in common both with Castaneda's guises and with Kaplan's vivid names, to mention a few. There is a third aspect, however, that I'd like to stress:

(3) Pegs are like (Kripkean) rigid designators. Consider the following quotation from Kripke 1972: "A possible world isn't a distant country that we are coming across, or viewing through a telescope. . . . A possible world is **given by the descriptive conditions we associate with it**. . . . Why can't it be part of the **description** of a possible world that it contains **Nixon** and that in that world **Nixon** did not win the elections? . . . There is no reason why we cannot **stipulate** that, in talking about what would have happened to Nixon in a certain counterfactual situation, we are talking about what would have happened to **him**" (Kripke 1972:267).

The view on counterfactuals that seems to fit this passage very well is that we take Nixon in our world and bring **him** to different worlds, by changing the facts. The fact that the name 'Nixon' denotes in those worlds the same person as it does in ours, then, is a direct consequence of this interpretation of the counterfactual: we keep track of Nixon in the process that brings us from our world to counterfactual ones.

The analyses of counterfactuals that are most faithful to this idea are those of Veltman 1976 and Kratzer 1977. The basic idea is that, for a counterfactual 'If Nixon had lost in 1968, then. . . ,' you first go back in information till "Nixon lost in 1968" becomes compatible, and then you evaluate the conditional as an indicative conditional (i.e., you check whether if you add the antecedent to those information states, that commits you to the consequent). Pegs, as I have stressed, are primarily means of keeping track of objects through information change. In my perspective, if, in evaluating the counterfactual, you go to other information states, by removing properties that our present information state ascribes to the peg Nixon, it is still **that peg** that you take to different information states. In this sense, you keep track of Nixon in the course of information change, and in this sense *Nixon* is a rigid designator.

One might say that I am perverting the notion of rigid designation. One might say the following: 'Okay, you're keeping track of Nixon, by taking that peg to different information states. But your theory allows us to take the peg n(ixon) to an information state on the basis of which n = j(ane fonda) is true. You can't possibly do that on Kripke's view. So there is clearly a sense in which your notion of rigid designation differs from Kripke's. Or to say it differently, Kripke's notion of rigid designator has philosophical bite, while yours does not."

I tend to agree with this criticism, but I also tend to prefer my interpretation. The judgments on the truth and falsity of counterfactuals in as far as they bear on rigid designators are a consequence of two aspects of the semantics (and/or pragmatics) of counterfactuals. The first is **minimality**. The assumption shared by most theories of counterfactuals is that in order to evaluate a counterfactual you go to information states (worlds) that are **minimally** different from the present one in that the negation of the antecedent doesn't hold. Although technically identity is a relation like others, this does not mean that language users are not aware that its semantic nature differs from a relation like, say, **love**. It is not at all implausible to assume that in general you stay closer to this world if you change the extension of relations like **love** than if you start tinkering with the ontology. It seems to be a plausible constraint on the minimality involved in counterfactual change that counterfactuals will not remove identity conditions **unless** they are explicitly forced to (I will come back to this).

It is this minimality that makes us expect that in discussing whether Nixon could have been a movie star, we will try to change Nixon's career decisions at various times of his life, to see whether we end up in Hollywood, rather than trying to make him identical to Jane Fonda.

Still, this doesn't answer the question: Could Nixon possibly have been Jane Fonda? This is the real question at issue: Kripke's answer, the answer that we have learned to accept, is a strong no. This is where the second aspect of the semantics of counterfactuals comes in. This second aspect is the context dependency we have seen before. Kripke is very clear about what he is concerned with: he is asking a metaphysical question: given the way the world is and the fact that Nixon is not Jane Fonda, could that fact have been different? Given that the answer to that metaphysical question is no, it is clear that if names are rigid designators the counterfactual process that leads us through the sphere of metaphysically possible worlds will never identify the two.

The modal base, then, with respect to which these counterfactuals are evaluated is one that contains our metaphysical assumptions. I have no problem with the assumption that for a "metaphysical modal base" (that is, if the information state and the accessibility relation tell us what is possible in view of our metaphysical assumptions) there is a constraint that the identity relations are the same in all accessible information states (or at least in all total ones). But a metaphysical interpretation is not the only one that counterfactuals can get. It is at least as natural (and quite a bit more common, I would say) to give counterfactuals an epistemic interpretation, relating them to an epistemic modal base and, as Kripke has taught us, we have to account for the fact that statements that may be metaphysically necessary can be 'epistemically contingent.' The whole point of the information semantics for identity statements that I have given is to account for this epistemic contingency, to account for the fact that statements like *maybe Hesperus is Phosphorus, maybe not* and questions like *Is Hesperus Phosphorus?* are **contentful**, meaningful, and that it makes sense to raise them, and similarly that one can be surprised about the answer. Above, I formulated a minimality condition on counterfactuals that I repeat here: counterfactuals will not remove identity conditions **unless** they are explicitly forced to. My claim is that for counterfactuals, we cannot understand their meaningfulness if we do not allow for this 'unless we are explicitly forced to,' if they are related to an epistemic modal base.

Consider the following sentence:

(38) If Hesperus had not been Phosphorus, then Babylonian science would be more respected nowadays.

If you would say this sentence at a party to me, I would disagree with you, because I think that Babylonian science is quite respected nowadays, and the Hesperus/Phosphorus business would not have made a difference.

In disagreeing with you, I take your sentence as contentful (as contentful as (39):

(39) Suppose we would find out tomorrow that Hesperus is not Phosphorus after all. Would Babylonian science be more respected?

We can perform such thought experiments and quarrel about their outcome. All that would be impossible in possible world semantics. (38) is vacuously true, because $h \neq p$ is neccessarily false, false in all possible worlds.

If we assume that Hesperus and Phosphorus are pegs that are identical on the basis of our present information, then (38) and (39) instruct us to go back in information to states where their identity conditions are no longer certain and see what happens if we add the information that they are not identical. Although we may not be able to tinker with the ontological furniture of the world, we are certainly able to imagine that we wouldn't have certain information that in fact we do have, and we are able to imagine what we would do if our information were different. It is this capacity that makes counterfactuals like (38) meaningful, and our semantic theory has to be rich enough to account for this.

What we need, then, is a more intentional theory of meaning (both in the philosophical sense and in the technical sense of Thomason 1980), one that does not lose aboutness, but has a place for epistemic contingency. This will necessarily be a theory that has a less direct relation between meanings and extensions than the notion of an intension, as a function from possible worlds to extensions, gives us. In other words, this has to be a theory that in fact does not equate meaning and necessity.

It seems to me that the information perspective that I have sketched can form an interesting guide line for such a theory and that the theory of pegs that I have outlined here can form a natural part of it.

ACKNOWLEDGMENTS

*I would like to thank Nirit Kadmon for stimulating discussions and comments that made me write the present paper and write it in its present form. Further, discussions at various times with Gennaro Chierchia, Angelika Kratzer, Barbara Partee, Craige Roberts, and Ray Turner were very helpful. This paper has further benefitted from the comments of one of the anonymous referees.

Parts of the present paper were presented in various talks that I gave at the University of Massachusetts at Amherst and at Cornell University. I would like to thank the audiences of those talks for their comments.

Part of the research for this paper was done while I held a research position at the University of Massachusetts, financially supported by Barbara Partee's grant "Foundations of Semantics" from the System Development Foundation (SDF). This support is gratefully acknowledged.

REFERENCES

Heim, I. (1982). *The Semantics of Definite and Indefinite Noun Phrases*, Diss., GLSA, Amherst: University of Massachusetts

Hinrichs, E. (1985). *A Compositional Semantics for Aktionsarten and NP Reference in English*, diss., Ohio State University

Kamp, H. (1975). Two theories about adjectives. In Keenan (ed.), *Formal Semantics for Natural Language*. Cambridge, Eng.: Cambridge University Press

— (1979). Events, instants, and temporal reference. In Bauerle, Egli, and von Stechow (eds.), *Semantics from Different Points of View*. Berlin: Springer

— (1981). A theory of truth and semantic representation. In Groenendijk, Janssen, and Stokhof (eds.), *Formal Methods in the Study of Language*. Amsterdam: Mathematical Centre Tracts. Reprinted in Groenendijk, Janssen, and Stokhof. (eds.) (1983), *Truth, Interpretation, Information*, GRASS 2. Dordrecht: Foris

Kaplan, D. (1979). On the logic of demonstratives. *Journal of Philosophical Logic* 8

Karttunen, L. (1972). Possible and must. In Kimball (ed.), *Syntax and Semantics 1*. New York: Academic Press

Kratzer, A. (1977). What 'must' and 'can' can and must mean. *Linguistics and Philosophy* 1

— (1981). The notional category of modality. In Eikmeyer and Rieser (eds.), *Words, Worlds and Contexts*. Berlin: de Gruyter

— (1986). On conditionals. In Farley, Farley, and McCullough (eds.), *Papers from the Parasession on Pragmatics and Grammatical Theory*, CLS 22

Kripke, S. (1972). Naming and necessity. In Davidson and Harman (eds.), *Semantics of Natural Language*, Dordrecht: Reidel

— (1979). A puzzle about belief. In Margalit (ed.), *Meaning and Use*. Dordrecht: Reidel

Landman, F. (1986). *Towards a Theory of Information*, GRASS 6. Dordrecht: Foris

Landman, F. (1987). A handful of versions of discourse representation theory, ms, Cornell University

Lewis, D. (1975). Adverbs of quantification. In Keenan (ed.), *Formal Semantics for Natural Language*. Cambridge, Eng.: Cambridge University Press

— (1979). Scorekeeping in a language game. In Bauerle, Egli, and von Stechow (eds.), *Semantics from Different Points of View*. Berlin: Springer

Partee, B. (1987). Quantifying over contexts, paper presented at the Sixth Amsterdam Colloquium, April

Roberts, C. (1987). *Modal Subordination, Anaphora, and Distributivity*, diss., GLSA, Amherst: University of Massachusetts

Stalnaker, R. (1968). A theory of conditionals. In Rescher (ed.), *Studies in Logical Theory*. Oxford: Blackwell

— (1974). Pragmatic presuppositions. In Munitz and Unger (eds.), *Semantics and Philosophy*. New York: New York University Press

— (1979). Assertion. In Cole (ed.), *Syntax and Semantics* 9. New York: Academic Press

— (1984). *Inquiry*. Cambridge, MA: MIT press

Thomason, R. (1980). A model theory for propositional attitudes. *Linguistics and Philosophy* 4

Veltman, F. (1976). Prejudices, presuppositions and the theory of Conditionals. In Groenendijk and Stokhof (eds.), *Proceedings of the First Amsterdam Colloquium*, Amsterdam Papers in Formal Grammar 1. University of Amsterdam: Dept. of Philosophy

— (1981). Data semantics. In Groenendijk, Janssen, and Stokhof (eds.), *Formal Methods in the Study of Language*. Amsterdam: Mathematical Centre Tracts. Reprinted in Groenendijk, Janssen, and Stokhof (eds.), (1983). *Truth, Interpretation, Information*, GRASS 2. Dordrecht: Foris

— (1985). *Two Logics for Conditionals*, diss., University of Amsterdam

— (1986). Data semantics and the pragmatics of indicative conditional sentences. In Traugott et al. (eds.), *On Conditionals*. Cambridge, Eng.: University Press

CHAPTER TEN

Unifying Partial Descriptions of Sets

Carl J. Pollard and M. Drew Moshier

INTRODUCTION AND LINGUISTIC MOTIVATION

In unification-based grammatical theories and formalisms (see, e.g., Shieber et al. 1983; Fenstad et al. 1987; Pollard & Sag 1987; Zeevat, Klein, & Calder 1987), feature structures are used to model (depict, represent, describe) linguistic objects of various types, be they phonological shapes, syntactic categories, functional structures, semantic contents, components of the utterance context, or linguistic signs.[1] In this paper, we presuppose a general familiarity with the use of feature structures in linguistic theory;[2] our purpose is to show how the standard feature structure apparatus can be augmented with a linguistically motivated notion of *set value* in a mathematically precise and computationally feasible way. Before turning to the technical development, however, we attempt in the remainder of this section to provide some general background and linguistic motivation.

The attribute-value matrix (AVM) in Figure 1 is a slightly simplified picture of the feature structure used in head-driven phrase structure grammar (HPSG), (Pollard & Sag 1987) to model a certain linguistic sign, the English third-person singular verb *walks*. The same feature structure is depicted graphically in Figure 2.

An understanding of the linguistic details is not necessary to follow the subsequent mathematical development, but a little explanation may be in order to put the motivating examples in context. In the HPSG framework, we take the subject-matter to be linguistic objects of various types, such as signs (which include as subtypes both words and phrases), phonological shapes, syntactic categories, phrase-structure trees, and various types of semantic objects. In general, linguistic objects have complex internal structures; that is, they consist of several components or substructures, each of which may in turn have complex internal structure of its own. To reflect such structures, we model each linguistic object as a feature structure. The *sort* of the feature structure tells what type of object is being modelled, and the values of the attributes in the feature structure model the components of the object. And characteristically, for each sort of feature structure, the value of each attribute will be of a certain other sort; this reflects the fact that each component of a linguistic object of a certain type must itself be of a certain type. An HPSG grammar will then be a simultaneous-recursive specification of the linguistically well-formed feature structures of every sort (and indirectly, via the modelling relationship, a specification of the well-formed linguistic objects of every type).

For example, all signs are assumed to have phonological, syntactic, and semantic components; and phrases (phrasal signs) but not words (lexical signs) are assumed to have a fourth component called constituent structure as well.[3] Thus the feature structure in Figures 1 and 2 has the sort *sign* and the attributes PHONOLOGY, SYNTAX, and SEMANTICS. Here the value of the PHONOLOGY attribute is given as a string of phonemic symbols, which models a certain phonological shape. The values of the SYNTAX and SEMANTICS attributes are feature structures of sorts *syntactic-category* and *semantic-object* respectively; let us briefly consider both of these in turn.[4]

The syntactic category of a sign — the value of its SYNTAX attribute — is an object with the two attributes HEAD and SUBCATEGORIZATION (abbreviated SUBCAT). The value of the HEAD attribute in turn is a feature structure (of sort *head*) consisting of specifications for certain syntactic features which are often referred to by linguists as "head features." Head features include such things as part of speech (here modelled by the MAJOR feature), i.e., whether the sign in question is (or is a phrasal "projection" of) a noun, a verb, an adjective, or a preposition; inflectional features, such as the case (CASE) of a noun (in English, whether it is nominative or accusative) or the form (VFORM) of a verb (whether it is finite, infinite, gerundive, past participial, etc.); and the "auxiliariness" (AUX) of a verb, i.e., whether it is an auxiliary verb or a main verb.[5] The head features of a sign are the syntactic features that it shares with its projections (the phrases that have that sign as its grammatical head). For example, in the sentence *Kim walks fast*, the head features [MAJOR *verb*, VFORM *finite*, AUX *minus*] of the verb *walks* are shared by both the verb phrase (VP) *walks fast* (whose head is the verb itself) and the whole sentence (whose head is the verb phrase).

The subcategorization (SUBCAT) component in the syntactic cate-

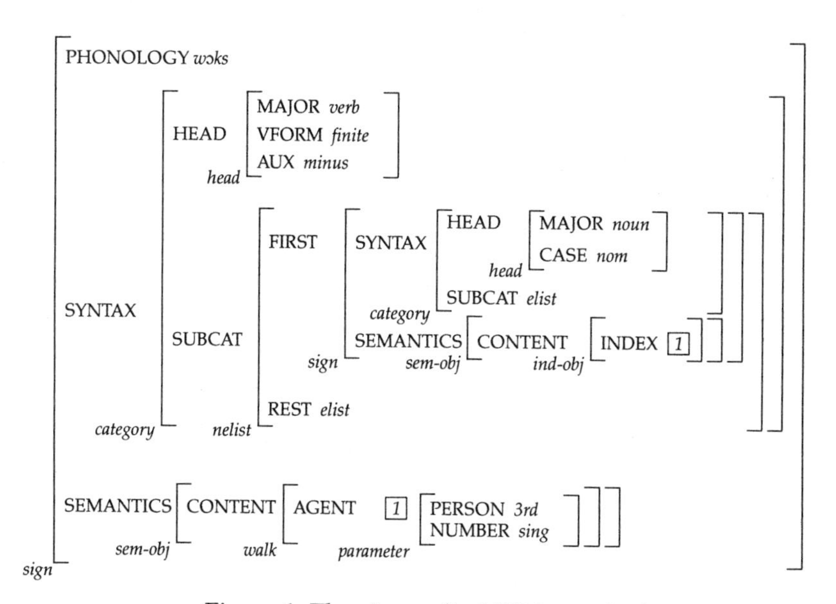

Figure 1: The sign *walks* (AVM notation)

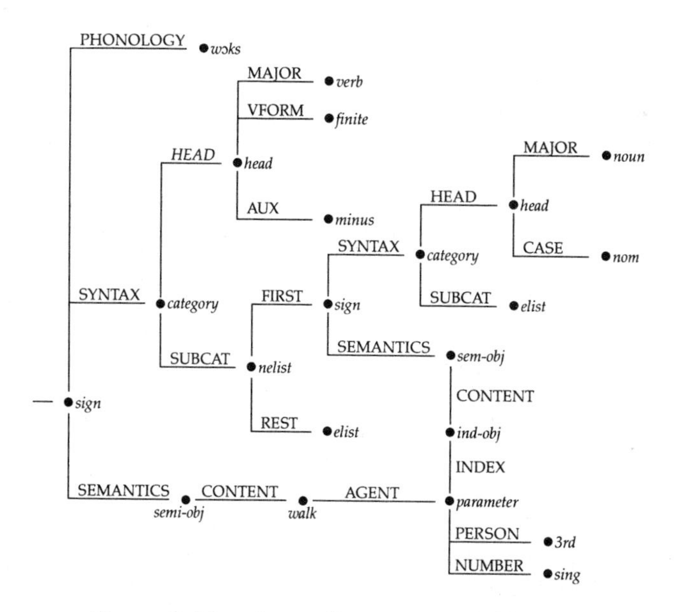

Figure 2: The sign *walks* (graph notation)

gory of a sign is a list of specifications which characterizes the sign in terms of its grammatical "valence," i.e., the number and kind of other signs that the sign in question must be combined with in order to become "complete" or "saturated."[6] In the case of the intransitive verb *walks*, the SUBCAT value contains only one sign (the FIRST value), corresponding to the grammatical subject, which is required to be a third person singular nominative noun phrase (NP); that is, *walks* has to be combined with a third-singular nominative NP in order to form a saturated verbal projection (a sentence). But a transitive verb like *sees* would have two signs on its SUBCAT list, corresponding to the subject and the object; and a double-object verb like *gives* would have a SUBCAT list of length three.[7]

We complete our discussion of the sign shown in Figures 1 and 2 by considering the SEMANTICS attribute, whose value is a feature structure of sort *semantic-object*. Such objects have two attributes, CONTENT and CONTEXT, which correspond roughly to the notions "described situation" and "utterance situation" in situation semantics (Barwise & Perry 1983). For present purposes the described situation can be thought of as similar to the literal meaning, and the utterance situation can be identified with pragmatic features of an utterance standardly discussed under such rubrics as presupposition and conventional implicature. For simplicity, only the CONTENT attribute is shown in Figures 1 and 2.[8] An example of the CONTEXT attribute appears in (2b) below.

For a verb, the value of the CONTENT attribute corresponds to the type of object known in situation theory as a *parametrized state-of-affairs;*[9] the intuition is that this parametrized state-of-affairs obtains, or holds, in the described situation. A *state-of-affairs* is a situation-theoretic object characterized by a *relation* together with an appropriate assignment of objects to the *roles* of that relation. (States-of-affairs also have a *polarity*, either positive or negative, but for simplicity we disregard that here.) We model the relation by the feature structure sort (in this case *walk*) and the roles of the relation by the attributes (in this case the single attribute AGENT). Thus the attributes in the semantic content of a verb correspond roughly to what linguists variously call "semantic roles," "theta-roles," or "thematic relations." A *parametric* state-of-affairs is similar to a state-of-affairs, except that some of the roles have assigned to them not full-fledged situation-theoretic objects but rather *parameters* (or, more generally, objects which in turn contain parameters). The precise nature of parameters in situation theory is unresolved, but for present purposes they can be thought of as something like variables in classical function theory (not in modern mathematical logic): a parametric object yields a full-fledged object only when each of its parameters is either replaced by (in situation jargon, "anchored to") a full-fledged object, or else bound (in situation jargon "absorbed") by some kind of binding operation (e.g., quantification or property abstraction).

Unlike verbs and their projections, which have a parametrized state-of-affairs as semantic content, we assume that nouns and their projections have as their semantic content objects of a different type, which (for lack of a better name) we call *indexed objects*. Just what indexed objects are need not concern us here; what is important is that they have an attribute — the INDEX attribute — whose value is a parameter. For example, in the case of the proper name *Bill*, this parameter will function rather like a free variable which, in any token use, will be anchored to some object named "Bill" in the utterance context.[10] In the case of a quantified noun phrase like *every student*, the indexed object will correspond to what is often called a "generalized quantifier," with the parameter corresponding to the variable bound by the quantification.

A parameter which is the value of the INDEX attribute in the semantic content of an NP (briefly, an NP parameter) plays a role in HPSG analogous to that of syntactic indices in government-binding theory (GB) (Chomsky 1981) and reference markers in discourse representation theory (DRT, Kamp 1981). Notice that we consider NP parameters to have attributes of their own, namely the *agreement* features of person, number, and gender.[11] For example, as shown in Figure 2, the parameter corresponding to the subject of *walks* has third person and singular number; indeed, a third-singular verb is simply a verb that subcategorizes for a subject that introduces a third-singular parameter. One last point to be noted in connection with Figures 1 and 2 is that the co-indexing (or structure-sharing) of this parameter, indicated in Figure 1 by the two occurrences of the tag [1] , is what sets up the correspondence between the grammatical subject of *walks* and the AGENT role in the state-of-affairs of walking; loosely speaking, the verb *walks* "assigns" the AGENT role to its subject.

We now move the focus of discussion from the linguistic objects modelled by feature structures, to the modelling objects themselves, which are the real concern of this paper. Feature structures of this general character are standard in unification-based grammar (though the attribute labels employed are specific to HPSG), except that we follow the practice, common in knowledge representation (see, e.g., Ait-Kaci 1984) of allowing any node (not just the terminal nodes) to be assigned an atomic value, or, as we are calling it, a *sort*. As noted by Pollard and Sag (1987), for truly robust linguistic knowledge representation, it seems necessary to build a notion of *inherit-*

ance into the feature structure formalism. That is, each feature structure has a sort, and which attributes are appropriate depends on the sort; e.g., the attributes FIRST and REST are appropriate for feature structures of sort *nelist* (nonempty list); but in addition the sorts are partially ordered. For example, *list* subsumes (is a supersort of) both *elist* (empty list) and *nelist*; both *plus* and *minus* are subsumed by (are subsorts of) *boolean*, and any sort has at least those attributes which are appropriate for its supersorts. Moreover the values of a given attribute for a given sort may be required to be of a certain sort, and such restrictions are also inherited by sorts from their supersorts; e.g., binary attributes such as AUX are "typed" to *boolean*, and REST is typed to *list*. Because issues of inheritance are orthogonal to the matters discussed here, we will make the simplifying assumptions that (1) the signature (i.e., the ordering on the sorts) is the trivial one; (2) all attributes are appropriate for all sorts; and (3) there are no sortal restrictions on any attributes. But we will make some sketchy indications from time to time as to how our results can be generalized to nontrivial signatures.[12]

It will be important for our purposes to make explicit some assumptions (that are generally adopted without comment by practitioners of feature-structure based theories) about the connection between feature structures themselves and the objects in the empirical domain of study which they model. Two of these are as follows:

Modelling convention. A feature structure A appropriately models an object X of the empirical domain provided:

(1) if the root of A is labelled by the sort symbol a then X is of the type conventionally denoted by a;
(2) if A has an arc from the root that bears the attribute label l, then X is of a type that has the attribute conventionally denoted by l, and that attribute of X is appropriately modelled by the feature structure A/l (the *shift* of A along l, i.e., the feature structure reached by following the arc labelled l from the root of A).

Thus, theoretical frameworks which model objects of the empirical domain by feature structure theories are committed to the reality of certain types (properties) of objects, and to the existence of certain attributes for objects of given types, though they may differ with respect to what types and attributes are posited, or with respect to what ontological status is accorded to types, e.g., types as sets, Fregean properties, or "realistic properties." Theories may also differ with respect to what ontological status is accorded to the notion of attribute, e.g., functions from the objects of one type to the objects of another, or unanalysed primitives such as the roles of situation theory. At the level of generality of the present discussion, it will be safe to abstract away from such metatheoretical issues.

Now it often happens in linguistic theory that a linguistic object that we wish to model is a set. For example, the "gappiness" of a sign — roughly, its capacity to enter into long-distance syntactic dependencies of the kind usually discussed under such rubrics as "wh-movement," "extraction," or "filler-gap relations" — can be thought of in this way.[13] In HPSG, the gappiness of a sign is modelled by the value of an additional syntactic attribute SLASH.[14] Thus, the verb phrase *get to __ on __* in the sentence (1a) can be (partially) modelled by the feature structure shown in (1b):[15]

(1a) $[_s$[This bus]$_1$, I don't think [Palo Alto]$_2$ is very easy to $[_{vp}$get to t_2 on t_1]]?

(1b) $\left[_{sign}\ \text{SYNTAX}\ \left[_{category}\ \text{SLASH}\ \{.\ \text{NP, NP}\ .\}\right]\right]$

Here each occurrence of the symbol "NP" stands for an instance of the feature structure:

$$\left[_{sign}\ \text{SYNTAX}\ \left[_{category}\ \begin{array}{ll}\text{HEAD} & \left[\text{MAJOR}\ \textit{noun}\right]\\ \text{SUBCAT} & \textit{elist}\end{array}\right]\right]$$

Intuitively speaking, the notation {.NP, NP.} means something like the type of set that contains two NP's and nothing else; a complication, as we shall see below, is that the "two" NP's might actually turn out to be the same. The reasons that the gappiness of a phrase is suitably modelled by something like a set are that (1) in principle a phrase can contain arbitrarily many gaps, and (2) there is no real sense in which the gaps in a phrase can be considered to be ordered with respect to each other. The example in (1a) has only two gaps, but examples with three gaps are not unknown in some Scandinavian dialects, and some examples from Hebrew with up to five gaps within a single phrase have been cited in the linguistic literature. Although the rough representation of the sentence in (1a) seems to indicate that one of the gaps precedes the other (so that perhaps a list, rather than a set, might be appropriate to model gappiness), this appearance is deceptive. Traces actually are part of the constituent structure of a phrase, not part of its phonological shape; and in theoretical frameworks like HPSG, the constituent structure of a phrase is considered to include the immediate dominance relations that hold between a sign and its constituents, but crucially *not* the relations of temporal ordering that obtain between their phonological realizations (the values of their PHONOLOGY attributes). Indeed, one of the most distinctive properties of a gap is that it is phonologically null; in terms of the phonological shape of a phrase containing it, the gap has no location. This point is a little difficult to appreciate with respect to a relatively fixed-constituent-order language like English where we sense that we can tell where the gap must be by finding the spot in the sentence where something appears to be missing; but in free-constituent-order languages (like Japanese), even this appearance is no longer present; there simply is no such thing as the location of the gap in the phonological structure of the gap that contains it.

Another point worth noting here is that the notation in (1b) appears to be at variance with standard set-theoretic notation, wherein, for example $\{a, a\}$ and $\{a\}$ denotes one and the same thing, viz. the singleton set whose member is a.[16] This appearance arises from the different interpretation of the symbols inside the curly brackets. In the standard set-theoretic notation, each symbol type inside the set braces denotes a set-theoretic object. Hence multiple tokens of the same symbol type denote one and the same object; that is, each symbol type corresponds to a token object, either a set or an urelement. But in (1b), each symbol between dotted curly brackets denotes a certain type of sign (the noun phrase type), not a particular token of that type. Thus we are not at liberty to collapse multiple occurrences of the same symbol down to one occurrence, unless the symbol denotes a type which in principle can have only one token (for example, under appropriate set-theoretic assumptions, the empty type of set). We will return to this point in the section on feature structures with set-valued nodes (pp. 306-17).

As a second example, consider the sentence (2a), which contains two pronouns *she* and *him*. To describe the semantics of this sentence, it is necessary to characterize not only its content, but also the context, which for present purposes we will consider to consist of the set of NP parameters that a sign (or, more generally, a discourse) introduces. This characterization is indicated in (2b) on the following page.[17]

Again speaking intuitively, the value of CONTEXT here is something like the type of set that contains a third-singular-feminine parameter and a third-singular-masculine parameter (and nothing else).

Again, there is no upper bound on the possible size of the CONTEXT value, since a sentence can contain arbitrarily many NP's. And again, there is no natural sense in which the parameters in the CONTEXT value can be viewed as ordered with respect to each other. This latter point can be brought out by considering languages such as Japanese

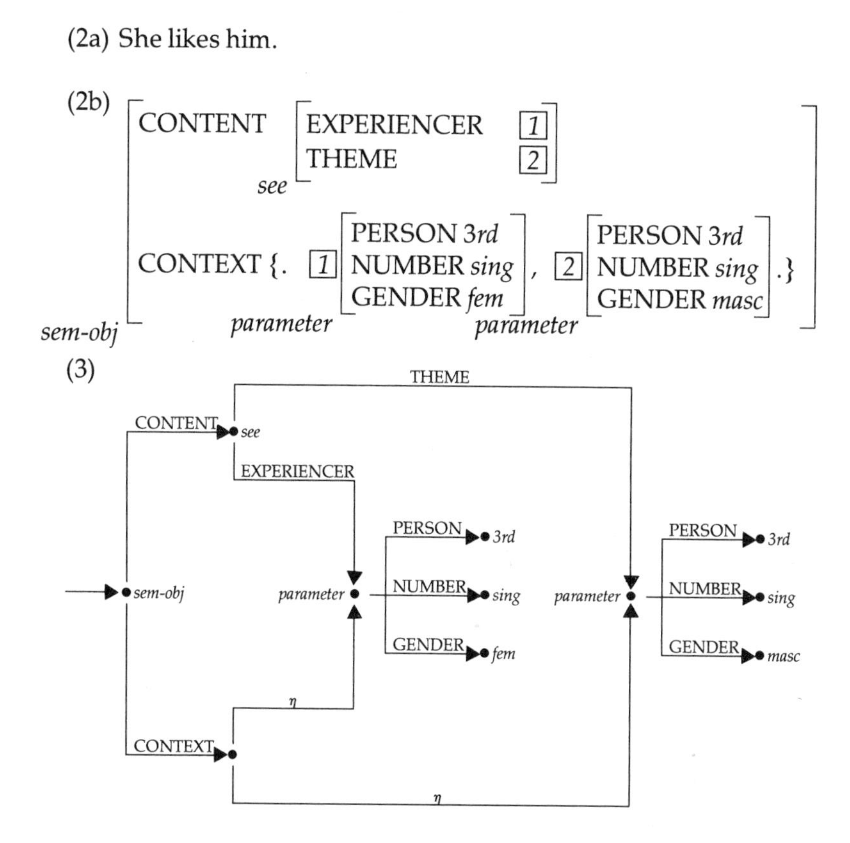

which have both free constituent order and zero pronouns (NP's without phonological shape, functioning just as overt pronouns function in English). Like a trace, a zero pronoun has no phonological location, and in free constituent order languages (again, just as with traces), there is no sense in which we can pick out a place in the phonological form where it "would have been if it had been overt."

But what do these notations inside dotted braces mean? This question really has two parts. First, what kinds of mathematical objects should we use to model types of sets? And second, what should be the modelling convention for determining whether a given such object appropriately models a given set?

The first of these questions is easily answered, at least in a preliminary fashion, as follows: feature structures with "set values" are just graphs with a new kind of arc; for example, (2b) can be alternatively described graphically as in (3). In such a graph containing arcs labelled "η," a node with outgoing η-arcs is regarded as modelling a set, with the nodes at the ends of those η-arcs modelling the elements of the set. We return to a mathematical characterization of such graphs below (pp. 30n).

To approach the second question, we consider first the following special case: what should be the modelling convention between a graph with η-arcs out of the root node and a set that this graph appropriately describes? Obviously part of the convention should be that the feature structure at the end of each η-arc should appropriately describe one of the members of the set. Are there any further desiderata? To answer this question, we examine a couple of additional linguistic examples.

Consider first the verb phrase (4a), which appears to contain two gaps:

(4a) [$_{VP}$ sell pictures of t_2 to t_1]

As in (1), the fact that this phrase stands in need of two NP fillers can be indicated by assigning it the SLASH value {.NP, NP.}. Now there are two distinct types of linguistic situations in which this VP might occur. In the first type, exemplified by (4b), the apparent two gaps are really one, not only insofar as they are coindexed, but in the much stronger sense that they are filled (or bound) by one and the same sign token *which indiscreet politician*:

(4b) [Which indiscreet politician]$_{1=2}$ did you sell pictures of t_2 to t_1?

In (4c), by contrast, the two gaps are really distinct, and they are filled by different NP's:

(4c) [Someone that repressed]$_1$ even [the most beautiful woman in the world]$_2$ would be difficult to sell pictures of t_2 to t_1.

On the other hand, the VP (4a) certainly cannot have *more* than two gaps in it; one and two are the only possibilities.

Second, consider the sentence in (5a):

(5a) He$_1$ thinks he$_2$ is smart.

Here presumably two referential parameters are introduced, as described in (5b):

(5b)

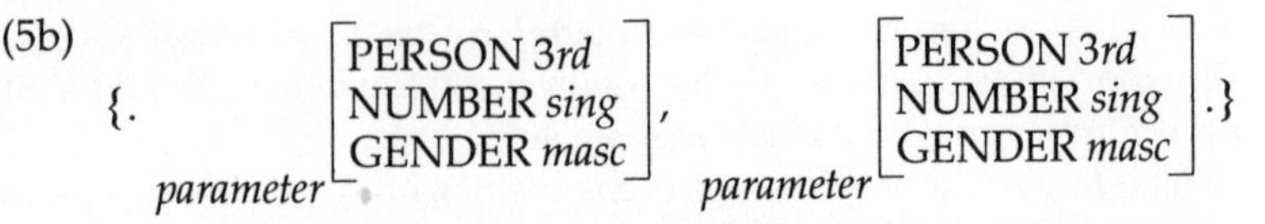

Now one possibility for (5a) is that the pronouns are coindexed, i.e., they introduce the same referential parameters, as in (5c):

(5c) John$_{1=2}$ is an idiot. But he$_1$ thinks he$_2$ is smart.

But the parameters may also be distinct, as in (5d):

(5d) John_1 doubts that Bill_2 is honest. But he_1 thinks he_2 is smart.

On the other hand, there is no linguistic situation in which (5a) introduces more than two third-singular-masculine parameters.

Considerations such as these motivate an augmentation of the notion of feature structure to allow the possibility of η-arcs, together with the following addition to the foregoing modelling convention:

3. if the root of A is a set node, then X is a set, and there is a function μ from the children of A onto X such that each child B appropriately models $\mu(B)$.

The key idea here is that the nodes of our augmented feature structures come in two flavours: individual nodes and set nodes. Individual nodes, which describe nonsets, can have only arcs bearing attribute labels, while set nodes can have only η-arcs; for a feature structure rooted at a set node to describe a set, there must be an association of children of the root (a *child* of a given node is a node which lies at the end of an η-arc originating from the given node) to members of the set, such that the feature structure rooted at each child appropriately describes the element associated with that child and such that every element of the set is associated with at least one child.

Crucially, however, although such an association must be onto, there is no requirement that it be one-to-one. As a result, just as ordinary (i.e., without η-arcs) feature structures are often thought of as partial models of the (nonset) objects that they describe, so an augmented feature structure A can be thought of as a partial model of a set X that it describes, and indeed the modelling is partial in two distinct ways. First, of course, each child of A's root is a partial model of a member of X; consider, for example, an augmented feature structure such that all the children of the root are ordinary feature structures. But second, the modelling is partial with respect to the *individuation* of the members of X: it does not necessarily furnish total information with respect to the question of which elements are distinct from each other. Thus the augmented feature structure shown in (6) is a partial model in the sense that the elements are only partially described (we are told only that they are cats). But it is also partial in the sense that we lack the information as to whether the two cats are indeed the same cat. Singleton and doubleton sets of cats exhaust the possibilities for sets appropriately modelled by (6), but that is as much as we can say.

(6)

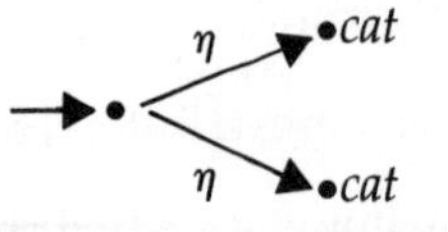

In the remaining sections, we will make mathematically precise the notion of feature structures with set values, and the sense in which they are partial, concluding with a formal characterization of the *unification* of augmented feature structures.

REVIEW OF ORDINARY FEATURE STRUCTURES

Subsumption, generalization, and unification

Let *Sorts* and *Labels* be finite disjoint sets of *sort symbols* and *attribute labels*.

Definition 1. An *ordinary feature structure* over $< Sorts, Labels >$ is a tuple $< Q, q_0, \delta, \alpha >$ such that:

(1) Q is a nonempty finite set of *nodes*;
(2) q_0 is a distinguished element of Q called the *root* node;
(3) δ is a partial function from $Q \times Labels$ to Q called the *transition* function;
(4) α is a partial function from Q to *Sorts* called the *sort assignment* function;
(5) (*connectivity*) for every node q, there exists a sequence of nodes $< q_0, \ldots, q_n >$ with $q_n = q$ and a sequence of labels $< l_1, \ldots, l_n >$ such that $\delta(q_{i-1}, l_i) = q_i$ for each $i = 1, \ldots, n$.

The set of all ordinary feature structures is denoted by *OF*(*Sorts*, *Labels*), or simply *OF* when the context is clear (the same convention applies to the other spaces we will define); it is obviously countable. The transition function δ is extended in the obvious way to a partial function from $Q \times Labels^*$ to Q; elements of $Labels^*$ are called *paths*. For notational ease the extended function is also written δ. It should be observed that there is no prohibition on cyclicity (i.e., nodes q such that $q = \delta(q, u)$ for some non-null path u). In addition, a node q can be in the domain of α even if it is a nonterminal (i.e., $< q, l >$ is in the domain of δ for some l), and there can be "copies" of sort symbols (i.e., α need not be injective).

Ordinary feature structures are partial in nature, or, to put it another way, one feature structure may be at least as complete a description as another. This notion is captured via the *subsumption* relation $\sqsubseteq_{OF}$, which is defined as follows:

Definition 2. For A and A' in *OF*, $A \sqsubseteq_{OF} A'$ iff there is a mapping $h : Q \rightarrow Q'$ such that:

(1) $h(q_0) = q'_0$;
(2) if $\delta(q, l)$ is defined, then $\delta'(h(q), l)$ is defined and equals $h(\delta(q, l))$;
(3) if $\alpha(q)$ is defined, then $\alpha'(h(q))$ is defined and equals $\alpha(q)$.

The mappings h are called *OF-morphisms*. Intuitively, A subsumes A' just in case the collection of things that it appropriately describes includes the collection of things appropriately described by A'; equivalently, A' is at least as complete, or at least as informative, as A. If $A \sqsubseteq_{OF} A'$, then h is easily seen to be unique. It is trivially verified that $\sqsubseteq_{OF}$ is a preorder (i.e., transitive and reflexive).

If A and A' subsume each other they are said to be *isomorphic*.[18] Since isomorphic ordinary feature structures are essentially indistinguishable, we will deal not with *OF* but rather with the set *AOF* of *abstract ordinary feature structures*, whose members are the isomorphism classes of *OF*. Under the partial ordering $\sqsubseteq_{AOF}$ (also called subsumption) induced by $\sqsubseteq_{OF}$ in the obvious fashion *AOF* becomes a poset, with bottom element $\bot_{AOF}$ the isomorphism class of the one-point ordinary feature structure with no arcs and no sort. As Moshier (1988) has shown, *AOF* is order-isomorphic with the set of *Nerode-Myhill triples* $< P, N, V >$ where P (the *pathset*) is a prefix-closed set of paths, N (the *Nerode relation*) is a right-invariant equivalence on P with finite index, and V is (the graph of) a partial function from P to *Sorts* that respects the Nerode relation (i.e., if $V(u) = a$ and $< u, v > \in N$ then $V(v) = a$). Under this identification, $< P, N, V >$ subsumes $< P', N', V' >$ just in case $P \subseteq P'$, $N \subseteq N'$, and $V \subseteq V'$.[19]

In fact, *AOF* is a meet-semilattice under subsumption, i.e., $\sqcap_{AOF} X$ exists for any non-empty finite subset X; the meet in *AOF* is usually called (feature structure) *generalization*. (As Moshier 1988 shows, finite meets correspond to componentwise intersection on Nerode-Myhill triples.) However, infinite generalizations may fail to exist in *AOF*. For example, the antichain $< A_i >_{i<\omega}$ whose first three elements are shown in (7) has no meet in *AOF*:

(7)
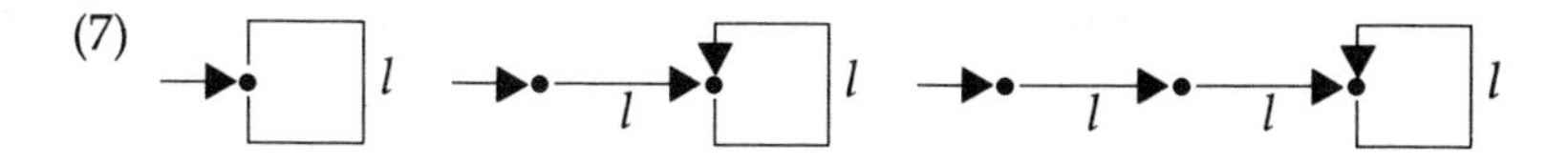

This difficulty can be overcome by embedding *AOF* into the complete meet-semilattice **OF** of (isomorphism classes of) ordinary feature structures with possibly infinite node sets. In terms of Myhill-Nerode triples, this corresponds to allowing Nerode relations with infinite index; it is easy to see that now arbitrary meets of nonempty sets can be obtained as before (via componentwise intersection). For example, the meet of the antichain (7) is the infinite structure (8):

(8) 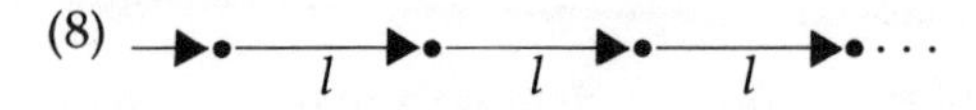

Alternatively, we can obtain **OF** as the *ideal completion* of *AOF*, i.e., the set *Ideals*(*AOF*) of ideals (lower-closed nonempty directed subsets) of *AOF* ordered by set inclusion, so that **OF** is an ω-algebraic cpo with *AOF* as its basis of compact ("finite") elements (see Birkhoff 1967: 187).[20] The embedding, which takes each member of *AOF* to the principal ideal which it generates, preserves (existing) arbitrary meets and finite joins (Gierz et al. 1980: 13); moreover, since in a poset with bottom, a nonempty intersection of ideals is still an ideal, arbitrary nonempty meets exist (and so do arbitrary joins of upper-bounded sets: just take the meet of all the upper bounds). Consequently **OF** is a countably based Scott domain. In this connection, it should be observed that an infinite subset of *AOF* may fail to have a join i.e., a unique minimal upper bound) in *AOF*, even if it is upper-bounded. For example, the chain $< B_i >_{i<\omega}$ whose first three elements are shown in (9) is bounded above by any of the members of the antichain (7), but lacks a join in *AOF:*

(9) 

Of course its join in OF is just (8). The join operation $\sqcup_{OF}$ is called (feature structure) *unification*. As shown by Moshier (1988), the domain **OF** is essentially the same as the *domain of descriptions* introduced by Pereira and Shieber (1984).

Disjunction and the Smyth powerdomain

For all its virtues, the domain **OF** is inadequate for modelling disjunctive information such as often arises in linguistics; e.g., the set of grammar rules and lexical entries for a given language can be regarded as a disjunctive specification of a possible linguistic object. To put it another way, generalization is too crude a meet operation; what is usually needed instead is true disjunction. For example, German *das Kind* "the child" may be nominative or accusative, but the generalization of

CASE *nom*

and

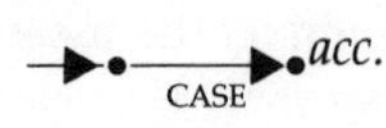

is the crude lower bound

$\longrightarrow\!\!\blacktriangleright\bullet\xrightarrow[\text{CASE}]{}\!\!\blacktriangleright\bullet$

which loses the case information contributed by *das Kind* that genitive and dative are not possibilities. Symptomatic of this problem is the fact that generalization and unification in **OF** do not distribute over each other. For example,

$$([\text{PER } 2nd] \sqcap_{\text{OF}} [\text{PER } 1st]) \sqcup_{\text{OF}} ([\text{PER } 3rd] \sqcap_{\text{OF}} [\text{PER } 1st])$$

turns out to be

$\longrightarrow\bullet\xrightarrow[\text{PER}]{}\bullet$

while

$$([\text{PER } 2nd] \sqcup_{\text{OF}} [\text{PER } 3rd]) \sqcap_{\text{OF}} [\text{PER } 1st]$$

does not exist. Intuitively these should coincide and equal [PER 1*st*].

Our solution to these difficulties, suggested by the "ϵ-types" of Ait-Kaci (1984), is to embed **OF** into a distributive algebraic lattice. We use the Smyth powerdomain construction (Smyth 1978), which is defined for any cpo D as follows. First, we form the poset P of compact elements of D.[21] Second, we denote by *Crowns*(P) the set whose elements are (possibly empty) finite sets of pairwise-incomparable elements ("crowns") of P, and by $\sqsubseteq_{Sm}$ the *Smyth* ordering on *Crowns*(P), i.e., $A \sqsubseteq_{Sm} B$ just in case each member of B is bounded below by a member of A relative to the ordering on P. It is easy to see that $<$ *Crowns*(P), $\sqsubseteq_{Sm}>$ is a meet-semilattice where the meet of two elements is given by taking their set union and then casting out non-minimal elements. This semilattice is clearly isomorphic in a natural way to the set *FUPF*(P) of (possibly empty) finite unions of principal filters ("FUPF's") of P ordered by *reverse* inclusion, with meet being set union. Finally, *Smyth*(D) is defined to be the ideal completion of $<$ *Crowns*(P), $\sqsubseteq_{Sm}>$ (or, equivalently, of *FUPF*(P)).[22]

Now suppose D is such that P has joins of all nonempty finite upper-bounded subsets (this is the case, for example, if D is a Scott domain). Then it is easy to see that *FUPF*(P) is a distributive lattice, with the join being set intersection, and that the isomorphism from $<$ *Crowns*(P), $\sqsubseteq_{Sm}>$ to *FUPF*(P) is a lattice isomorphism.[23] In this case, *Smyth*(D) itself is a distributive lattice;[24] what is more — and this will be of key importance later — it is a complete Heyting algebra.[25]

Returning now to the case where D is any cpo, we note that the natural embedding of P into $Crowns_{Sm}(P)$ (respectively, *FUPF*(P)), which takes each element to the singleton containing it (respectively, to the principal filter which it generates) is easily seen to preserve existing joins but, crucially, not meets. In general the image in $< \mathit{Crowns}(P), \sqsubseteq_{Sm} >$ of the meet of some set in P is bounded below by the meet in $< \mathit{Crowns}(P), \sqsubseteq_{Sm} >$ of the image of that set. That is the whole point of the construction: it gives better — more informative — meets.

In our present application, we take D to be **OF** and P to be *AOF*, the (countable) basis of compact elements of **OF**. Then the distributive lattice $< \mathit{Crowns}(AOF), \sqsubseteq_{Sm} >$ ($\cong$ *FUPF*(*AOF*)) and its ideal completion *Smyth*(**OF**) are denoted by *DAOF* (*disjunctive* abstract ordinary feature structures) and **DOF** respectively. The modelling convention here is simply that an element *A* of *DAOF* appropriately describes an object X just in case *at least one* of its members appropriately describes X according to the original modelling convention on p. 290; thus elements of *DAOF* that are smaller (as sets) are more informative in the sense that they eliminate more possibilities. The extreme case is represented by the empty set $\top_{DAOF}$, which cannot appropriately describe anything. The elements of **DOF** can then be thought of as (possibly infinite) disjunctions of (possibly infinite) ordinary feature structures; they are approximable by elements of *DAOF* (finite disjunctions of finite feature structures) in the sense that any element of **DOF** is the join of a possibly infinite chain of (images under the ideal-completion embedding) elements of *DAOF*. Since the embedding of *AOF* into *DAOF* (and **OF**) preserves existing joins, the join in *AOF* (and **OF**) is usually also called unification. Unlike the situation with respect to *AOF*, of course, finite joins always exist in *DAOF*. Finite meets also exist in *DAOF*, but of course the embedding of *AOF* into *DAOF* is not meet-preserving; in general, $\sqcap_{DAOF}$ yields a better lower bound than $\sqcap_{OAF}$. Correspondingly, $\sqcap_{DAOF}$ is called *disjunction* rather than generalization.

Semantics of grammars

By a *grammar*, we are going to mean a certain kind of formal specification that defines all the sorts in terms of each other by simultaneous recursion. We will then define the *denotation* of the grammar to be a mapping that associates with each sort a certain element of **DOF**. To put it another way, if the sorts are indexed from 1 to n, then the denotation of the grammar will be a certain element of $\mathbf{DOF}^n$, which is obtained as the least fixed point of a certain continuous mapping of $\mathbf{DOF}^n$ into itself. For each sort, we think of the element of **DOF** assigned to that sort as being our mathematical model of the type in the empirical domain that the sort names.[26] It is important to note that the use of powerdomain elements to denote grammars (or sorts) is independent of whether or not "set values" are utilized, though as we will show below (pp. 315-16), our Smyth-style modelling of types can still be carried out when set values are added. For example, in the approaches of Pereira and Shieber 1984 and Rounds and Manaster-Ramer 1987, the denotation of the grammar is defined as a (least fixed) point in the *Hoare* powerdomain (explicitly in the former, implicitly in the latter), although neither approach utilizes set values.[27]

To be more precise, we define a *grammar* over $< Sorts, Labels >$ to be an n-tuple of *rules* $G = << a_i, A_i >>_{1 \leq i \leq n}$ where, for each i, a_i is a distinct sort and A_i is an element of *DAOF*. In such a grammar, we think of the right-hand side of a rule $< a_i, A_i >$ as exhaustively specifying, in disjunctive form, what possibilities exist for something to be a well-formed object of type a_i. The recursion arises from the fact that the a_i can occur on the right-hand sides of rules. As we will see shortly, A_i (or, rather, its image in **DOF** under the ideal completion embedding) is a first approximation to the denotation of a_i. We will give the definition of this notion shortly, but first we introduce some notational conventions.

First, if u is a path and X is an ordinary feature structure, then $u \cdot X$, the *translation of* X along u, is the feature structure obtained by attaching X to the end of u; this notion is extended to elements of *AOF* and *DAOF* in the obvious fashion, and to *OF* and *DOF* by continuity (i.e., by requiring it to preserve joins of chains). Second, we use the vector notation $\mathbf{X} = < X_i >_{i=1,\ldots,n}$ for elements of $\mathbf{DOF}^n$. Third, if $X = < Q, q_0, \delta, \alpha > \in AOF$, we denote by $ap(X)$ the (finite) set of acyclic paths in X, i.e., those paths that never cross the same node twice. And fourth, we denote by $j_G(u, X)$ the index j such that a_j is the sort assigned to u in X; more generally, if the sorts are ordered, then we denote by $J_G(u, X)$ the set of all indices $j \in \{1, \ldots, n\}$ such that a_j is a supersort of $a_{j_G(u, X)}$.

***Definition* 3.** Let $G = << a_i, A_i >>_{1 \leq i \leq n}$ be a grammar, possibly one where the sorts are ordered, with $A_i = \{A_i^1, \ldots, A_i^{N_i}\}$, where each $A_i^l \in AOF$. Then the *denotation* of G, $\|G\|$, is the least fixed point of the mapping $\mathbf{T} : \mathbf{DOF}^n \rightarrow \mathbf{DOF}^n$ whose i-th component T_i is given by

$$T_i(\mathbf{X}) = \sqcap_{l=1}^{N_i} (\bigsqcup_{u \in ap(A_i^l)} (A_i^l \sqcup (u \cdot \bigsqcup_{j \in J_G(u, A_i^l)} X_j))) =$$

$$\sqcap_{l=1}^{N_i} (A_i^l \sqcup (\bigsqcup_{u \in ap(A_i^l)} (u \cdot \bigsqcup_{j \in J_G(u, A_i^l)} X_j)))$$

(In case the sort ordering is trivial, the last join on either side of the equation is replaced by $X_{j_a(u, A_i^l)}$.)

In the simplest case (where the sort ordering is trivial), the basic idea here is that $T_i(\mathbf{X})$ is the right-hand side of the grammar G after it has been "improved" by unifying every occurrence of each sort symbol a_j with X_j. By standard considerations about continuous self-mappings of cpo's,[28] the fixed point is the element of $\mathbf{DOF}^n$ obtained by taking the join of the chain $< \mathbf{X}^k >_{k \leq \omega}$, where $\mathbf{X}_0$ is the right-hand side of G, and for each $k \geq 1$, $\mathbf{X}_k$ is the result of improving G by $\mathbf{X}_{k-1}$. (In the case of a nontrivial sort, $\mathbf{T}$ unifies into each occurrence of a_j not just the j-th component of $\mathbf{X}$ but every component whose index is that of a sort that a_j inherits from.)

For this definition to make sense, we have to prove:

Theorem 1. $\mathbf{T}$ is monotone and preserves joins of chains in $\mathbf{DOF}^n$.

Proof. Monotonicity is obvious. Now let $<\mathbf{X}^k>_{k<\omega}$ be an ascending sequence. We have

$$\bigsqcup_{k<\omega} T_i(\mathbf{X}^k) = \bigsqcup_{k<\omega}(\sqcap_{l=1}^{N_i}(\bigsqcup_{u \in ap(A_i^l)}(A_i^l \sqcup (u \cdot \bigsqcup_{j \in J_G(u,A_i^l)} X_j^k)))).$$

But as we noted earlier, $\mathbf{DOF}$ is a complete Heyting algebra, so that finite meets distribute over arbitrary joins (Gierz et al. 1980, pp. 30-31). Hence, by the generalized associativity of joins in any complete lattice (Birkhoff 1967, p. 117), this is equal to

$$\sqcap_{l=1}^{N_i}(\bigsqcup_{k<\omega}(\bigsqcup_{u \in ap(A_i^l)}(A_i^l \sqcup (u \cdot \bigsqcup_{j \in J_G(u,A_i^l)} X_j^k)))) = \sqcap_{l=1}^{N_i}(\bigsqcup_{u \in ap(A_i^l)}(\bigsqcup_{k<\omega}(A_i^l \sqcup (u \cdot \bigsqcup_{j \in J_G(u,A_i^l)} X_j^k)))) =$$

$$\sqcap_{l=1}^{N_i}(\bigsqcup_{u \in ap(A_i^l)}(A_i^l \sqcup (\bigsqcup_{k<\omega} u \cdot \bigsqcup_{j \in J_G(u,A_i^l)} X_j^k))).$$

By the continuity of translation, and a final application of generalized associativity, this in turn is equal to

$$\sqcap_{l=1}^{N_i}(\bigsqcup_{u \in ap(A_i^l)}(A_i^l \sqcup (u \cdot \bigsqcup_{k<\omega}(\bigsqcup_{j \in J_G(u,A_i^l)} X_j^k))) = \sqcap_{l=1}^{N_i}(\bigsqcup_{u \in ap(A_i^l)}(A_i^l \sqcup (u \cdot \bigsqcup_{j \in J_G(u,A_i^l)}(\bigsqcup_{k<\omega} X_j^k))),$$

which is just $T_i(\bigsqcup_{k<\omega}\mathbf{X}^k)$.

By way of illustration, we consider a simple example of a context-free grammar in Chomsky normal form, given in (10):

(10) S → NP VP Vt → kissed
VP → Vt NP Vs → denied
VP → Vs S NP → John
  NP → Mary

To encode this as a grammar in our sense, we assume the sorts are s, vp, np, v_t, v_s, *kissed*, *denied*, *John*, and *Mary*; and that the attributes are LEFT, RIGHT, and MORPH. Our grammar G then takes the form:

$$\langle \ < s, \ \{ \ {}_{s}\begin{bmatrix} \text{LEFT} & np \\ \text{RIGHT} & vp \end{bmatrix} \} >,$$

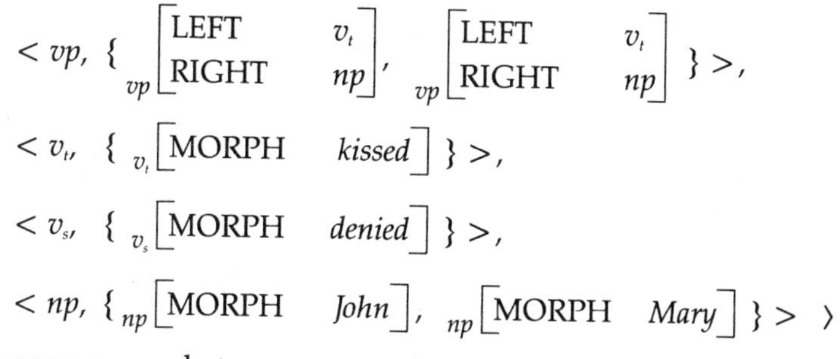

or, to use a somewhat more suggestive notation:

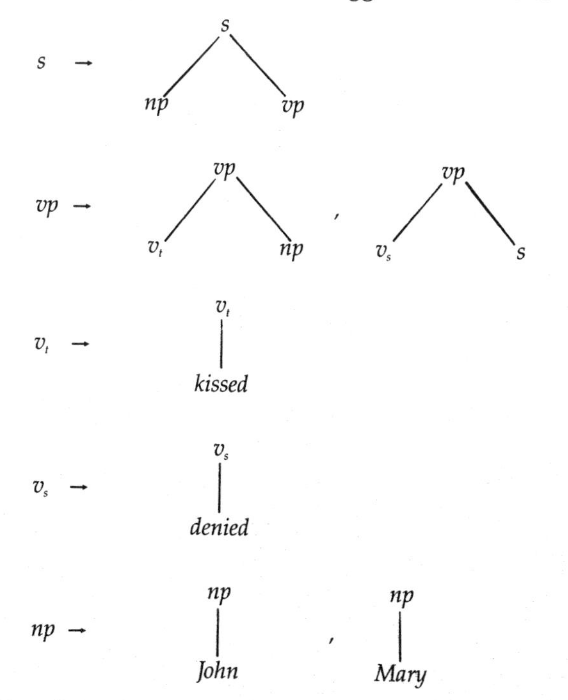

Recall that the grammar itself is a finite sequence of pairs, one for each sort. In each pair, the first element is the sort being defined and the second element is an element of *DAOF*, a set of pairwise subsumption-incomparable feature structures (elements of *AOF*) to be thought of as a partial but exhaustive disjunctive specification of all the possibilities that exist for something to be a (model of) a well-formed object of the type corresponding to that sort. In the present case each element of *AOF* is a (possibly partial) model of a phrase-structure-tree; the modelling is partial in the sense that some of the non-terminals may be unexpanded.

Then what is the denotation of this grammar? It will be an element of $\mathbf{DOF}^5$, in which, for example, the first component (corresponding to the sort *s*) is a certain element of **DOF**, in essence an infinite disjunction of elements of **OF**. In that infinite disjunction, there will be both compact elements (images of elements of *AOF* under the standard embedding of *AOF* into its set of crowns), which model finite terminated phrase structure trees with root labelled *s*; and there will also be noncompact elements, which model trees some of whose branches have infinite length.[29]

Let us now consider how the denotation of G, $\|G\|$, is obtained, by looking at its first few approximations by (images under the ideal-completion embedding of) elements of $DAOF^5$. Evidently, we need only consider the first two components (corresponding to the sorts *s* and *vp*) at each stage; the other three components obviously remain constant at all stages of the approximation. For $k = 1$, these will simply be the right-hand sides of the corresponding grammar rules, as shown in Figure 3.

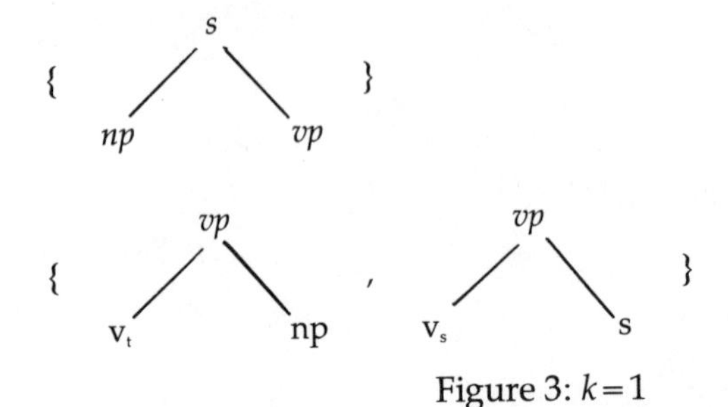

Figure 3: $k=1$

The next two steps are as shown in Figures 4 and 5.

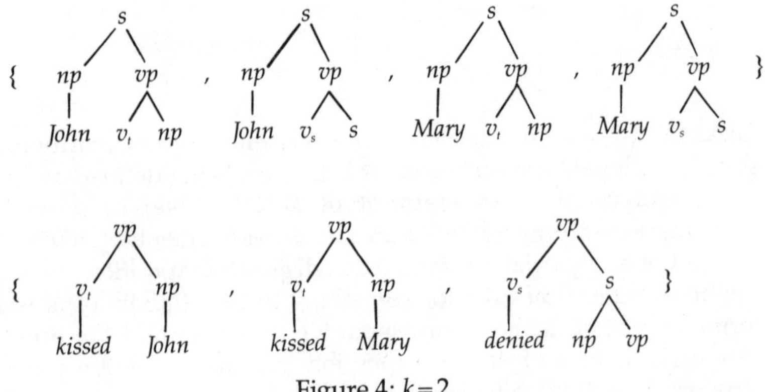

Figure 4: $k=2$

In general, at each step and for each component, some elements corresponding to terminated trees appear, and these will remain at all further steps. But — and this is characteristic of Smyth-style approximation — any element produced at a given step is subsumed by an element (of the same component) at every prior step. We approach the final result by starting with a space that contains the set of all possible final outcomes and successively narrow it down. (This is in contrast with Hoare-style approximation, where at any stage we may acquire elements approximating a member of the final result that were not approximated at any earlier stage.)

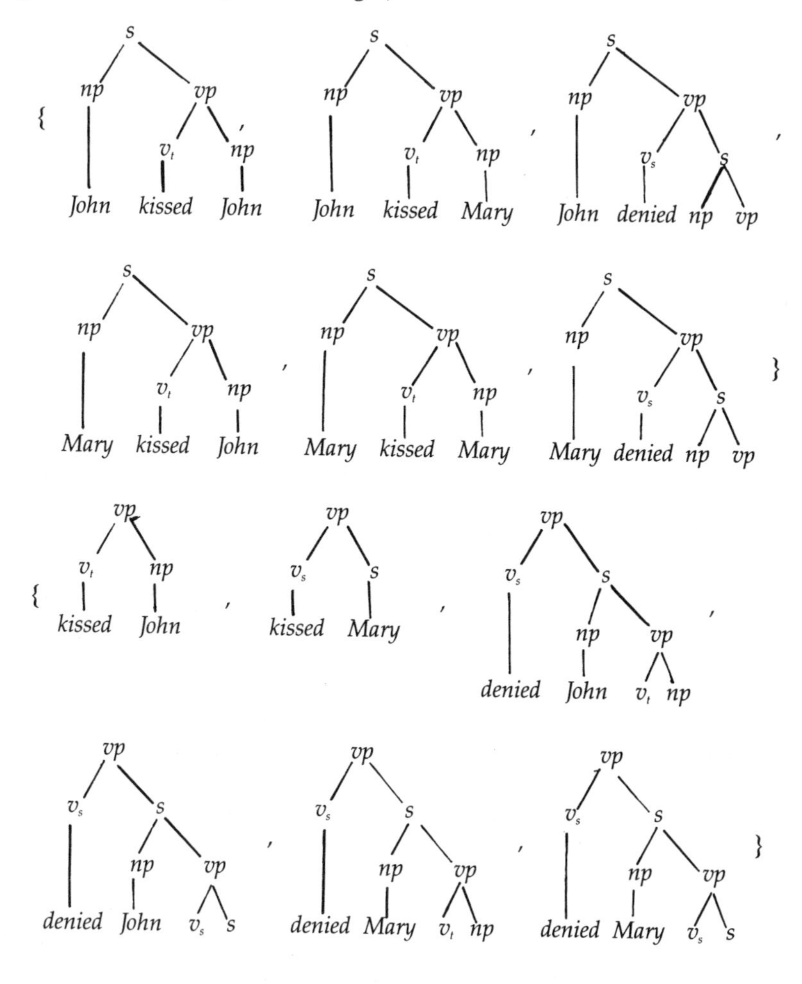

Figure 5: $k=3$

We note in passing that special-purpose formal languages, accompanied by equational deductive calculi, have been defined for expressing and computing with feature structures (see, e.g., Rounds and Kasper 1986). In Pollard (in press), it is shown how such a logic, augmented with sorts, can be used to implement natural-language parsers and generators based on Smyth approximation.

With this review behind us, we now generalize to feature structures with "set-valued" nodes.

FEATURE STRUCTURES WITH SET-VALUED NODES

We begin by extending the definition of feature structure to allow for set nodes as follows:[30]

Definition 4. An *feature structure* over $< Atoms, Labels >$ is a tuple $< Q_{ind}, Q_{set}, q_0, \delta, \alpha, \eta >$ such that:

(1) $Q = Q_{ind} \cup Q_{set}$ is a nonempty finite set of *nodes*, where Q_{ind} and Q_{set} are disjoint sets of *individual* nodes and *set* nodes respectively;
(2) q_0 is a distinguished element of Q called the *root* node;
(3) δ is a partial function from $Q_{ind} \times Labels$ to Q called the *transition* function;
(4) α is a partial function from Q_{ind} to *Sorts* called the *sort assignment* function;
(5) $\eta \subseteq Q_{set} \times Q$ is a *has-a-member* relation;
(6) (*connectivity*) for every node q, there exists a sequence of nodes $< q_0, \ldots, q_n >$ with $q_n = q$ such that for each $i = 1, \ldots, n$, either there is a label l such that $\delta(q_{i-1}, l) = q_i$ or else $q_{i-1} \eta q_i$.

The (countable) set of all feature structures is denoted by F; δ is extended to a partial function from $Q_{ind} \times Labels^*$ to Q as before. By an η-arc, we understand a pair $< q,p > \in \eta$. If $< q, p >$ is an η-arc, then we call p an *η-child* of q; the set of η-children of a set node q is denoted by *children*(q). Clearly a feature structure with no set nodes is just an ordinary feature structure. A feature structure is called *pure* if the only individual nodes are ones without outgoing (ordinary or η-) arcs; it is called *superpure* if it has no individual nodes. A *decoration* of a superpure feature structure A is a mapping d from the nodes of A to sets such that, for each each node q, $d(q) = \{d(p) \mid p \in children(q)\}$; a superpure feature structure A is a *picture* of a set S if it has a decoration that assigns S to the root. It is not hard to show (Aczel 1988: 5) that every set has a picture. Notice that the augmented version of our Modelling Convention requires that a superpure feature structure be a picture of a set that it models.

As with ordinary feature structures, cyclicity is allowed, but the existence of set nodes gives rise to the intriguing possibility of cyclic sequences of η-arcs, i.e., set nodes q such that $< q, q >$ is in the transitive closure of η. Intuitively, such nodes model sets which belong to their own transitive closures. As we do not wish to exclude such structures, and there are numerous applications which require them, as a matter of convenience we assume as our background set theory $ZFC^- + AFA$ (Aczel 1988), i.e., *ZFC* with the Axiom of Foundation replaced by Aczel's *Anti-Foundation Axiom AFA*. In our current terminology, this axiom says that every superpure feature structure has a unique decoration. Note that the existence part of the axiom says that *every* superpure feature structure is a picture of some set, which guarantees the existence of non-well-founded sets; the uniqueness part says that if two sets can be pictured by the same superpure feature structure then they are the same set.

We now ask: what is a natural notion of subsumption for all feature structures? Let us consider a couple of special cases. First, of course, subsumption should reduce to our old notion of subsumption in the case of ordinary feature structures. Second, by AFA a superpure feature structure models a unique set, i.e., it conveys complete (rather than partial) information about a set that it models. Therefore one superpure feature structure should subsume another just in case both depict the same set (in which case we must have subsumption in both directions). Now — assuming AFA — it can be shown that two superpure feature structures A and A' are pictures of the same set just in case there is an *Aczel bisimulation* between them, i.e., a relation R between their sets of nodes such that (i) $q_0Rq'_0$ and (ii) qRq' implies that every η-child of each q and q' is R-related to an η-child of the other.

In view of these considerations, we now define a relation $\sqsubseteq_F$, called *generalized subsumption*, as follows:[31]

***Definition** 5.* For A and A' in F, $A \sqsubseteq_F A'$ iff there is a relation $R \subseteq (Q_{set} \times Q'_{set}) \cup (Q_{ind} \times Q'_{ind})$ such that:

(1) $q_0Rq'_0$;
(2) if $\delta(q, l)$ is defined and qRq' then $\delta'(q', l)$ is defined and $\delta(q, l)\, R\delta'(q', l)$;
(3) if $\alpha(q)$ is defined and qRq', then $\alpha'(q')$ is defined and equals $\alpha(q)$;
(4) if $q \in Q_{set}$, $q' \in Q'_{set}$ and qRq', then every η-child of each of q, q' is R-related to an η-child of the other;
(5) R restricted to $(Q_{ind} \times Q'_{ind})$ is a total function; and
(6) for each $q \in Q$, qRq' for some $q' \in Q'$.

This definition can be generalized to nontrivial sort orderings by replacing equality in clause (3) with sort subsumption. The relations R

are called *partial simulations*, or *F-morphisms*. It is easy to see that clauses (1-5) guarantee that partial simulation reduces to ordinary subsumption for ordinary feature structures, and to Aczel bisimulation for superpure feature structures. Clause (6) is needed to ensure that each node of *A* that lies at the end of an ordinary arc gets related to something; otherwise, e.g., (12a) would wrongly subsume (12b) by $R = \langle q_0, q'_0 \rangle$:

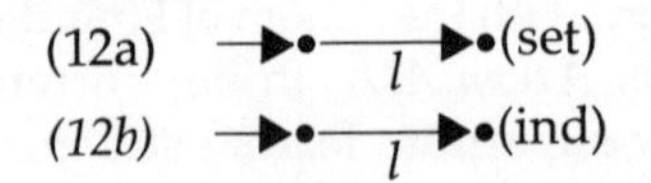

We will consider some other cases shortly; but first, we pause for a moment to consider the connection between partial simulation and the three standard powerdomain constructions.

Let *D* be a cpo with basis of compact elements *P*. Then as we saw, the Smyth powerdomain of *D*, *Smyth*(*D*), is just the cpo whose basis is $\langle Crowns(P), \sqsubseteq_{Sm} \rangle$, the crowns of *P* with the Smyth ordering. That is, we have

$$X \sqsubseteq_{Sm} Y \textit{ iff } \forall y \in Y \exists x \in X \textit{ such that } x \sqsubseteq_P y.$$

By contrast, the *Hoare powerdomain Hoare*(*D*) is the cpo with whose basis is $\langle Crowns(P), \sqsubseteq_H \rangle$, the crowns of *P* with the *Hoare ordering*, which is defined as follows:

$$X \sqsubseteq_H Y \textit{ iff } \forall x \in X \exists y \in Y \textit{ such that } x \sqsubseteq_P y.$$

(Equivalently, the basis of *Hoare*(*D*) can be identified with the set of finite unions of principal ideals in P ordered by set inclusion.) Finally, the *Plotkin powerdomain Plotkin*(*D*) (Plotkin 1976) is the cpo whose basis is $\langle Crowns(P), \sqsubseteq_{Pl} \rangle$, the crowns of *P* with the *Plotkin* (or *Egli-Milner*) *ordering*, given by

$$X \sqsubseteq_{Pl} Y \textit{ iff } X \sqsubseteq_{Sm} Y \textit{ and } X \sqsubseteq_H Y.$$

These three orderings can be extended from crowns of *P* to arbitrary finite subsets, leading to three different types of *simulations*. Thus, if X and *Y* are subsets of *P*, a relation *S* on $X \times Y$ is called a *Smyth simulation* provided for each element $y \in Y$, there exists $x \in X$ with *x S*-related to *y* such that $x \sqsubseteq_P y$, and analogously for Hoare and Plotkin simulations.

Now let us restrict our attention to feature structures where the root is a set node and all of its η-children are individual nodes with assigned sorts but no outgoing arcs; assume also that the sorts may have a non-trivial ordering. If we add the restriction that each sort symbol appear at most once in a given feature structure, then it is easy to see that partial simulations can be identified in a natural way with

Plotkin simulations on the signature. But this will not be the case if the restriction is lifted. To take a very simple example, consider the two pure feature structures (13a) and (13b):

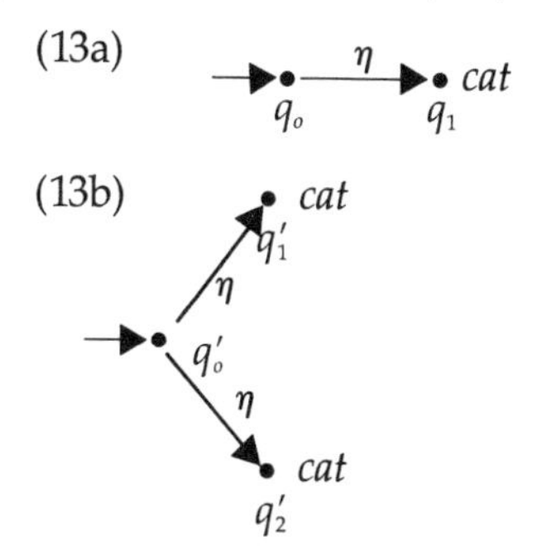

Now consider the relation $R = \{<q_0, q'_0>, <q_1, q'_1>, <q_1, q'_2>\}$. Both R and its inverse are Plotkin simulations. Indeed mutual Plotkin simulation would constitute a natural notion of feature structure equivalence if our modelling convention were such that (13a) and (13b) were both intended to model precisely the nonempty sets of cats.

But on our approach, (13a) models precisely the singleton sets of cats while (13b) models precisely those sets that contain either one or two cats. In particular (13b) should subsume (13a) but not conversely. And that is precisely the effect of clause (5) in our definition. Indeed, the inverse of R is a partial simulation from (13a) to (13b). But R itself is not a partial simulation, and indeed there can be no partial simulation from (13a) to (13b), for in this case clauses (4) and (5) make conflicting demands. Evidently we have a new kind of powerdomain construction, characterized by the property that a description places an upper bound on the cardinality of the set that it describes.[32]

We now continue with our program of generalizing the framework given above (pp. 296-306) to all feature structures. It is easy to see that the identity relation on the nodes of a feature structure is a partial simulation, and that the relational composition of two partial simulations is a partial simulation. Thus F forms a preorder under generalized subsumption, and consequently we obtain a poset AF of *abstract feature structures* if we identify two feature structures that partially simulate each other. Of course in this case mutual generalized subsumption is no longer just structural identity, since, e.g., (14a) and (14b) partially simulate each other:

(14a) $\rightarrow \bullet_{q_0} \xrightarrow{\eta} \bullet_{q_1}\ \emptyset$

(14b)

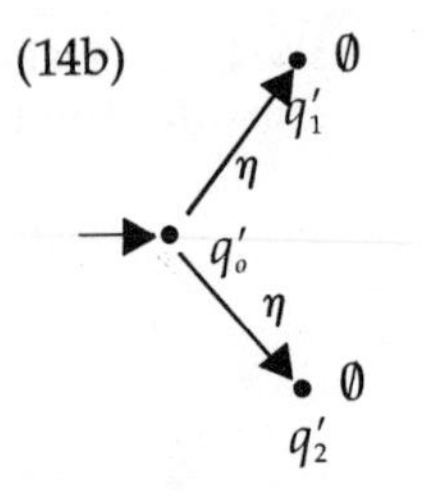

Here the ∅ notation indicates a set node with no η-children. Observe that the relation $R = \{< q_0, q'_0 >, < q_1, q'_1 >, < q_1, q'_2 >\}$ is a partial simulation from (14a) to (14b) and its inverse is a partial simulation from (14b) to (14a). It should be noted with care how this example differs from the superficially similar one described above where the empty set nodes are replaced by individual nodes labelled by the same sort. This difference reflects a fundamental way in which our use of graphs with nodes labelled by sort symbols to partially model sets differs from the way that graphs with nodes labelled by atoms are used by Aczel (1988) to model sets with urelements. An atom in an Aczel-style graph uniquely refers (perhaps autonymously) to an urelement. But a sort symbol does not pick out a definite individual; in a sense, it makes indefinite reference to an individual which is known only up to its type.

Continuing our strategy of generalization, we now relativize generalized subsumption to *AF* as we relativized subsumptions to *AOF*. A partial simulation in *AF* (or *AF-morphism*) is just an equivalence class of *F*-morphisms, where two *F*-morphisms $R : A \to B$ and $R' : A' \to B'$ are considered equivalent provided there are *F*-morphisms $S : A \to A'$, $T : A' \to A$, $U : B \to B'$, $V : B' \to B$, such that the following diagram commutes:

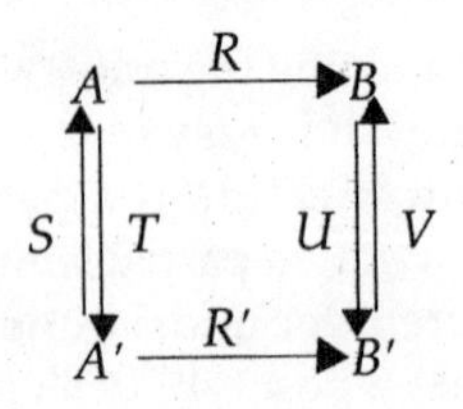

We then define the (relativized) generalized subsumption ordering $\sqsubseteq_{AF}$ as follows: for A and $B \in AF$, $A \sqsubseteq_{AF} B$ just in case there is an *AF*-morphism from A to B. The situation is a little more complex here than the corresponding situation in *AOF*, however, since for A and B in *AF* there can be more than one *AF*-morphism from A to B. That is, one feature structure can partially simulate another in essentially different ways, even after equivalence in the foregoing sense is factored out.[33]

Consider, for example, the situation where A and B are the mutual-generalized-subsumption equivalence classes of (15a) and (15b) respectively:

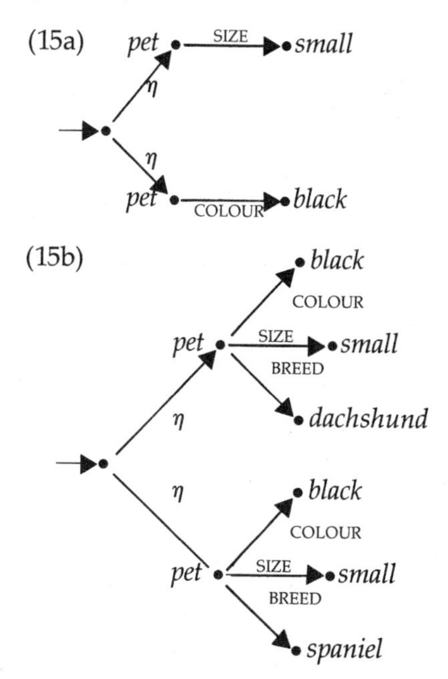

One partial simulation is the one that relates the small node to the small black dachshund node and the black node to the small black spaniel node; but there is another that relates the black node to the small black dachshund node and the small node to the small black spaniel node. In the two cases, the growth of information takes place in two essentially different ways.

A closely related complication is that, in general, finite joins do not exist in *AF*, even for nonempty upper-bounded sets (the empty join clearly does not exist, i.e., there is no bottom). For both of these reasons *AF* cannot be the basis of a Scott domain. To put it another way, a set of feature structures can have more than one minimal upper bound (a join is just a *unique* minimal upper bound). Intuitively, this is because the nonuniqueness of η-arcs gives rise to finitely branching nondeterminism when structures are unified. To take a simple example, consider the feature structures (16a) and (16b):

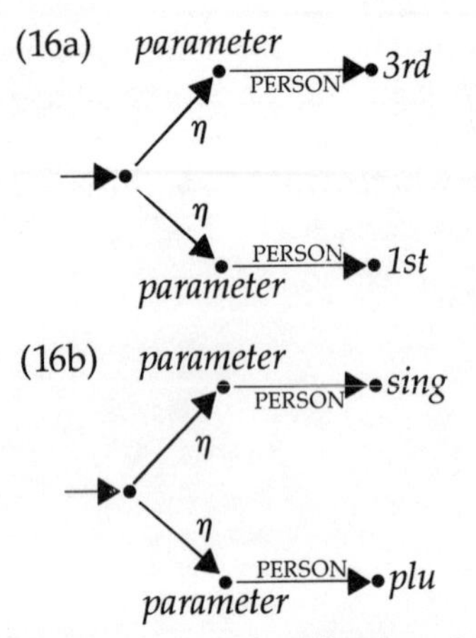

On a moment's reflection, it becomes clear that any set of parameters that could be described by *both* (16a) and (16b) can either be described by (17a) or (17b):

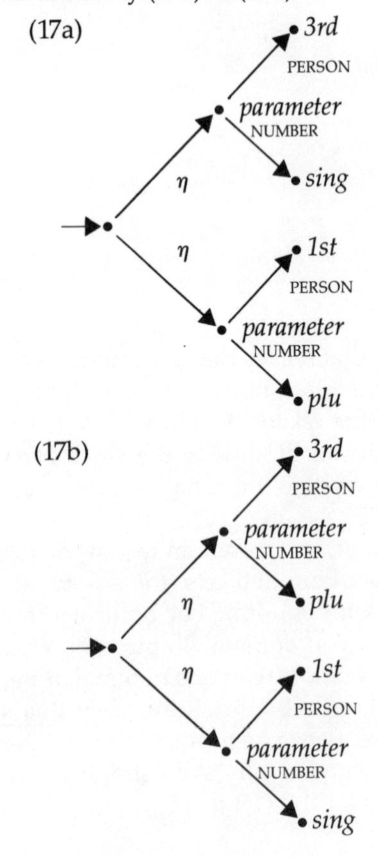

That is, although (16a) and (16b) lack a *least* upper bound, (17a) and (17b) constitute a complete set of *minimal* upper bounds, in the sense that (i) they are upper bounds, and (ii) any upper bound is partially simulated by one of them. This is a simple example, but it is not hard to see that the same general situation obtains for any nonempty finite set in AF. Unification is perfectly feasible for set-valued structures; but it does not produce a unique result.

Informally, we can describe a nondeterministic unification procedure for n feature structures in AF as follows: First, pick representatives $A_1, \ldots A_n$ in F. Think of the nodes as beads (different colours for different sorts, and all set nodes are black beads), and the arcs are threads (different colours for different labels, and all η-arcs are black threads). To produce a candidate upper bound, start by gluing the root beads together (but beads can never be glued together unless they are the same colour; if you ever wind up with no choice but to glue together unlike beads, then that execution of the procedure halts without producing a candidate). If they are individual nodes join all like-coloured threads together and glue together the beads so drawn together (again, only if they are the same colour!). If it is a set node, nondeterministically partition the set of black threads emanating from each root bead into a set of sets of threads that lead to same-coloured beads (the same number of sets in each partition); and then nondeterministically join each set of threads in one partition with exactly one set of threads from each of the the other partitions, gluing together the corresponding beads. Next, apply this procedure to each of the groups of glued-together beads (the beads connected directly to the root that didn't get glued to any other bead are done with altogether). Continue in this fashion until no further gluing is possible. The result is a candidate upper bound; obviously it is partially simulated by each of the A_i. It is easy to see that the set X of all candidate upper bounds (all possible results of successful executions) is finite. Finally, throw out of that set any member which is partially simulated by another (it is clear that partial simulation is decidable). The remaining set Y is the complete set of minimal upper bounds.

To see that the set of minimal upper bounds really is complete, let B be any upper bound. Then there are partial simulations $R_i : A_i \rightarrow A$ for all $i = 1, \ldots, n$. But the family of R_i provides us with the means for producing an $A' \in X$ (and therefore an $A \in Y$) that partially simulates B. Simply execute the unification procedure above, using the R_i as an oracle at each choice point (whenever the current set of nodes to be unified are set nodes) to decide which partitions to take and how to associate members of partitions across all the nodes in question. More precisely, η-daughters are identified ("glued together") whenever they are commonly related by the R_i to some node of B. By construction, this guided execution produces an element of X, and at least one $A \in Y$ must partially simulate it.

A cautionary note: there is no guarantee of uniqueness here![34] For example, let A_1 and A_2 be as follows:

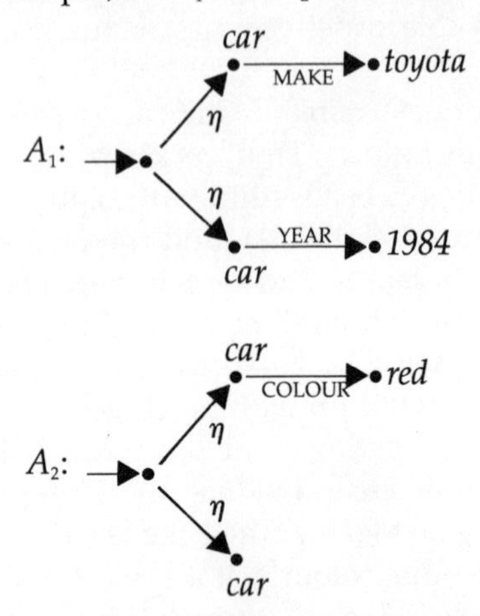

Now the minimal upper bounds of A_1 and A_2 are A and A':

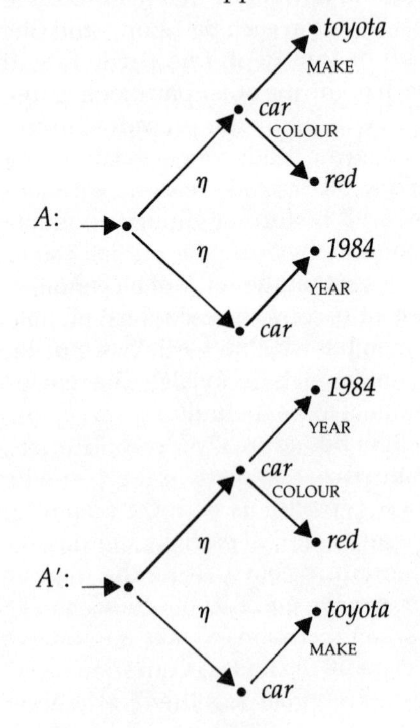

As it happens, A and A' in turn have two minimal lower bounds. One of them, call it B, is this:

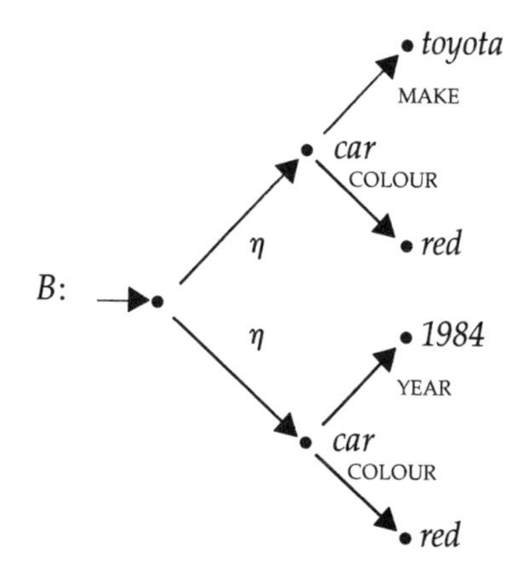

Thus we have partial simulations from A_1 and A_2 to B, which factor through both of the minimal lower bounds A and A'.[35]

Still, the situation is not altogether bleak. For although two elements A_1 and A_2 of AF with a common upper bound may fail to unify (have a unique minimal upper bound) in AF, the finite set of all their minimal lower bounds constitutes a perfectly respectable member of the poset of *disjunctive* abstract feature structures $DAF = <Crowns\ (AF), \sqsubseteq_{Sm}>$, where now $\sqsubseteq_{Sm}$ denotes the Smyth ordering induced on the crowns by the (relatived) generalized subsumption relation $\sqsubseteq_{AF}$. To put it another way, if we denote by **F** the ideal completion of AF, then the "unification" of A_1 and A_2 which is just a finite set of $\sqsubseteq_{AF}$-incomparable elements of AF, is just (the pre-image, under the ideal-completion embedding, of) a compact element of $\mathbf{DF} = Smyth(\mathbf{F})$. This should scarcely come as a surprise, for the minimal lower bounds were obtained as the result of a finitely branching nondeterministic process (in fact a complete one) of just the sort that the Smyth powerdomain is intended to model. In other words: to unify a set of feature structures in AF (or, more generally, in **F**), first take the standard embedding into the Smyth powerdomain and then do the unification over there. For example, the unification of A_1 and A_2 above, thought of as one-point crowns (or equivalently, as principal filters over AF), is just the crown whose two points are A and B (or equivalently, the union of the principal filters generated by those two points.)

There is still a little left to say, however, for we have not yet shown that DAF and **DF** are sufficiently well behaved to support the fixed-point construction that gives the denotation of a unification grammar. That is, we have to show the the proof of Theorem 1 still goes through when the notion of a grammar is generalized to allow elements of *DAF*, not just *DAOF*, on the right-hand sides of its definitions. If we can demonstrate this, then we will be guaranteed that the denotation of the grammar is well-defined as an element of $\mathbf{DF}^n$.

The required demonstration is easily seen to hinge upon showing that **DF**, like **DOF**, is a complete Heyting algebra. That in turn is guaranteed (by the theorem of Stone cited above) as long as $DAOF = \langle Crowns(AF), \sqsubseteq_{Sm} \rangle$ is distributive. So it is sufficient to prove the following:

Proposition. Let P be a poset such that every nonempty finite subset has a finite complete set of minimal upper bounds. Then $\langle Crowns(P), \sqsubseteq_{Sm} \rangle$ is a distributive lattice.

Proof. Given the identification between crowns and FUPF's, it is enough to show *FUPF(P)* is a lattice with set union and intersection as meet and join respectively; distributivity then follows since *Powerset(P)* is Boolean. Thus, let X and Y be two FUPF's of P. Their union is obviously a FUPF. As for the join (intersection), suppose $X = \bigcup_{i=1}^{n} Filt(x_i)$, $Y = \bigcup_{j=1}^{m} Filt(y_j)$, and let $mub(\{x_i, y_j\})$ be the set of minimal upper bounds of $\{x_i, y_j\}$. Then we have

$$X \cup Y = (\bigcup_{i=1}^{n} Filt(x_i)) \cap (\bigcup_{j=1}^{m} Filt(y_j)) = \bigcup_{i=1}^{n} \bigcup_{j=1}^{m} (Filt(x_i) \cap Filt(y_j)).$$

But the hypothesis on P implies that

$$Filt(x) \cap Filt(y) = \bigcup_{z \in mub(\{x,y\})} Filt(z),$$

so we conclude that

$$X \cup Y = \bigcup_{i=1}^{n} \bigcup_{j=1}^{m} \bigcup_{z \in mub(\{x_i,y_j\})} Filt(z),$$

which is a FUPF as required.

For ease of reference, we show in Figure 6 the relationships among all the spaces of feature structures constructed in this and the preceding section.

In this figure, the upper plane contains spaces constructed from ordinary feature structures, while the spaces on the lower plane are constructed from (generalized) feature structures that possibly have set-valued nodes. The dotted arrows indicate the projections of preorders of "concrete" feature structures onto the posets of "abstract" feature structures obtained by identifying two feature structures if the (generalized) subsumption relation holds between them in both directions. All other arrows indicate embeddings. The vertical arrows embed each space of ordinary feature structures into an analogous space of generalized feature structures. Among the horizontal arrows, those in the back plane are both natural embeddings of a poset into its Smyth-ordered set of crowns, whereby an element is mapped to the singleton set containing it; those in the front plane are both canonical injections of cpo's into their Smyth powerdomains. And all four of the arrows "facing out of the page" are embeddings of posets into their ideal completions, whereby an element is mapped to the principal ideal that it generates.

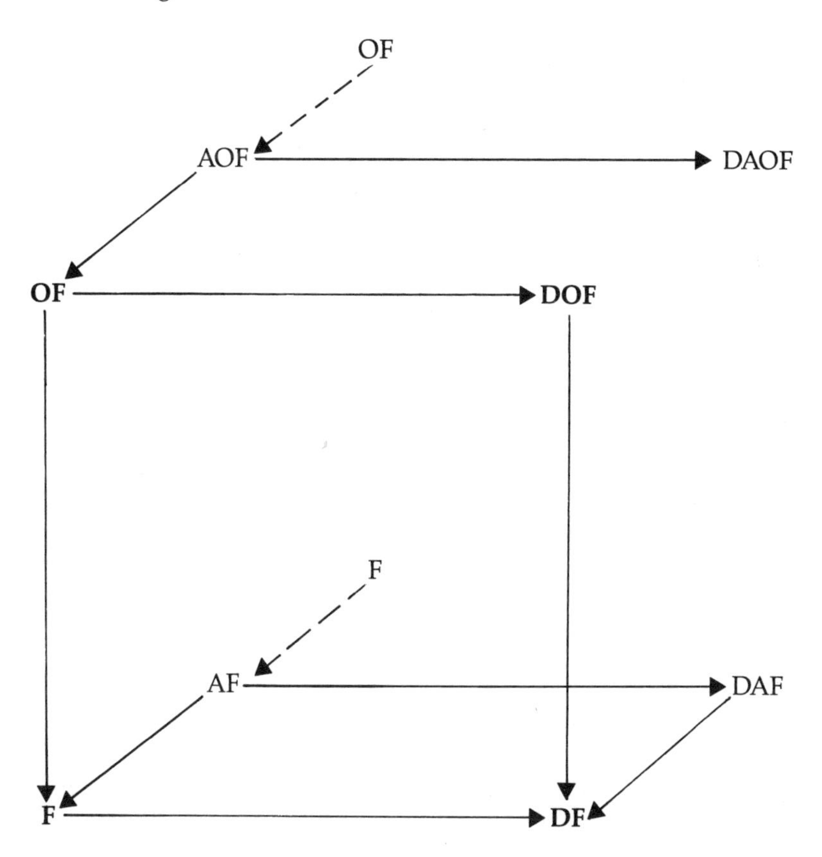

Figure 6: Relationships among spaces of feature structures

Conclusion

We have shown that grammars based upon feature structure unification can be given a mathematically rigorous and computationally feasible denotational semantics in terms of the Smyth powerdomain, in such a way that *sorts* of feature structures can be recursively defined in terms of each other, and we have sketched how this semantics can be extended to support notions of hierarchy and inheritance. In addition, we have introduced a linguistically motivated notion of set value, and shown that the denotational semantics can be straightforwardly generalized to accommodate them. Thus, we have provided a formal justification for a number of intuitively appealing augmentations of the standard feature structure apparatus which have been highly useful for the representation of linguistic knowledge but which as yet have only had the status of uninterpreted notation.

Much work remains to be done. Although we have generalized the fixed-point semantics to include set values, and thus shown grammars that employ them to be computable in principle, there is a considerable gap between principle and practice. The most obvious strategy for bridging it is to develop an extension to the Rounds-Kasper logic (and corresponding equational calculi) for talking about feature structures with set values, which is monotonic with respect to partial simulation. A similar program has already been initiated by Rounds (1988) with respect to Hoare and Smyth simulation; perhaps that line of work can be adapted to this purpose. There are also other feature structure augmentations that have been employed informally in linguistic analysis but which have not as yet been placed upon firm mathematical and computational foundations, including inequality (e.g., in stating noncoreference constraints on noun phrases), general path relations such as *append* (e.g., in "building-in" the theory of lists for handling word order), and "functional uncertainty" (i.e., disjunctions of path equations, as in the LFG account of long-distance dependencies). We intend to treat these and other extensions in detail in the future.

NOTES

1 Forerunners of this paper were presented to the U.S.-Japan Workshop on Natural Language Processing and Computer Software and to the Foundations of Grammar group at the Center for the Study of Language and Information. The authors wish to express their thanks to Carl Gunter, Gordon Plotkin, and Bill Rounds for illuminating discussions. The exposition was also much improved by the comments of Bill Rounds and an anonymous referee on an earlier draft. Pollard's research was supported by a grant from the National Science Foundation (BNS-87-18156).

2 For useful introductory surveys, see Shieber 1986 and Pereira 1987.

3 Linguists usually call the components of signs "levels of representation." We eschew this terminology, as it seems to reflect a mentalistic bias on the question of the ontological status of linguistic objects. At least for the purposes of this paper, we remain neutral on this question.

4 *Caveat*: For expository simplicity, we hereafter engage in a systematic abuse of language whereby a feature structure is not distinguished from the linguistic object that it models. This is *not* to be interpreted as indicating that we believe linguistic objects actually are feature structures, any more than we believe (say) that the motions of the solar system are Hamiltonian vectorfields on a symplectic manifold.

5 The head features employed here are essentially those of generalized phrase structure grammar (GPSG, Gazdar et al. 1985).

6 Here we use an encoding of lists that is standard in programming languages like LISP, wherein nonempty lists are represented by feature structures with the attributes FIRST and REST. The value of the FIRST attribute represents the first item on the list, while the rest of the list is represented by the REST value; the empty list is denoted by the special value *elist*, similar to LISP's NIL.

7 The signs that appear on the SUBCAT list of a sign correspond to the "subcategorized grammatical functions" of lexical-functional grammar (LFG, Bresnan, ed., 1982) with the subject appearing last, preceded by the object, second object, etc.; this list conception of subcategorization is derived from categorial grammar (e.g., Dowlty 1982).

8 In a more complete rendering, the CONTENT value here would contain an additional LOCATION attribute with a value of sort *parameter* corresponding to the space-time location of the walking, and the value of the CONTEXT attribute would contain a feature structure encoding the condition that this parameter must temporally precede the utterance location.

9 For extensive discussion of these and the various other kinds of situation-theoretic objects mentioned below, see Devlin (ms).

10 The value of the CONTEXT attribute will contain a feature structure encoding the condition that the object to which this parameter is anchored must be named "Bill."

11 For detailed arguments supporting the thesis that agreement features are attributes of the parameters in NPs, not of their syntactic categories, see Pollard and Sag 1988.

12 For detailed discussion of sorted feature structures and the associated mathematics, see Pollard (in press).

13 Well-known linguistic phenomena which exemplify long-distance syntactic dependencies include wh-questions, topicalization, relative clauses, and cleft and pseudocleft constructions.

14 This attribute generalizes the SLASH feature of GPSG.

15 Here the notation 't_1' indicates a trace ("gap") that is in a long-distance dependency relation with the NP "filler" bearing the same subscript (in the present case, *this bus*). In HPSG theory, certain attributes of both syntax and semantic content, including the NP parameter, are shared by a gap and its filler.

16 Thanks are due to Bill Rounds for pointing out the need for discussion of this point.

17 In a more detailed analysis, the context must also include the presuppositions that the feminine and masculine parameters can only be anchored to females and males respectively. This is because English is a natural gender language. In so-called syntactic gender languages (such as French and German), noun gender need not trigger such presuppositions.

18 Categorially, the terminology is justified since this can arise only if there are *OF*-morphisms $h : Q \to Q'$, $h' : Q' \to Q$, and in this case necessarily $h' \circ h$ and $h \circ h'$ are identities.

19 In the general case where there is a non-trivial ordering on the sorts, this last condition is replaced by the condition that $V(p) \sqsubseteq V'(p)$ for all $p \in P$.

20 Indeed, we need not have formed *AOF* at all, but could have obtained **OF** directly as the ideal completion of the preorder *OF*.

21 An element $x \in P$ is *compact* provided the join of a chain can never be equal to or

above x unless some element of the chain is. Intuitively, the compact elements are the ones that can be computed in finite time.

22 Thus *Smyth(D)* is an algebraic cpo, and it is ω-algebraic if P is countable. Incidentally, the Hoare and Plotkin powerdomains of D can also be defined as ideal completions of *Crowns(P)* under different orderings, as can the Scott topology of D.

23 In fact, these remain true under the weaker hypothesis that nonempty finite upper bounded subsets of P have a finite complete set of minimal upper bounds. We will take advantage of this fact in due course.

24 This follows from the fact (Birkhoff 1967: 114) that the ideal completion of a distributive lattice is itself a distributive lattice.

25 The theorem that the ideal completion of a distributive lattice is a complete Heyting algebra is due to Stone. For a proof, see Birkhoff 1967: 129.

26 For a justification of Smyth-style modelling of types in a somewhat different setting, see Rounds 1988.

27 The distinction between the Smyth and Hoare powerdomains will be explained on p. 308.

28 The least fixed point of a continuous mapping F of a cpo with bottom element $\bot$ into itself is given by $\bigsqcup_{k<\omega} F^k(\bot)$.

29 The latter will in general be of little interest to linguists. Langendoen and Postal 1984, however, argue for a version of linguistic Platonism in which expressions of infinite length are considered just as real as ones of finite length. Such noncompact elements can be viewed as models of the syntactic structures of infinite expressions.

30 This definition is similar to the definition of a *graph* in Rounds 1988, except that we disallow nodes with both ordinary outgoing arcs and η-arcs. This restriction does not involve a significant loss in generality.

31 The first four clauses of this definition coincide with the definition of feature structure bisimulation in Rounds 1988 (restricted to feature structures where no node has both ordinary arcs and η-children).

32 For a cpo D with basis P, the basis of the new powerdomain *Newpow(D)* is the set of finite multisets of P, with $X \sqsubseteq Y$ just in case there is a partial simulation between them, i.e., an onto mapping $v : X \to Y$ such that

$$\forall x \in X,\ type(x) \sqsubseteq type(\mu(x)).$$

In the special case where the multisets are sets, the partial simulations are precisely the Plotkin simulations that are (graphs of) functions. The mathematical details of this construction are discussed in Pollard (in preparation).

33 In categorial terms, *AOF* is a poset category. *AF* can also be viewed that way, but only by ignoring the possible multiplicity of morphisms between two given objects. When this is taken into consideration, *AOF* is a *generalized poset category* in the sense that, if $R : X \to Y$ and $S : Y \to X$ are morphisms, then they are identity morphisms. In generalized poset categories (as in posets), the only isomorphisms are the identities and all morphisms between two given objects go in the same direction. In Pollard (in preparation), the theory of partial simulations is developed in this setting.

34 Categorially, this means that *AF* has only weak finite colimits.

35 Thus *AF* cannot even be the basis of a locally algebraic domain (Gunter & Jung 1988): the principal ideal generated by B is not a lattice, for A_1 and A_2 are both in the ideal but lack a join.

REFERENCES

Aczel, P. (1988). *Non-Well-Founded Sets*, Lecture Note Series No. 14. Stanford, CA: Center for the Study of Language and Information

Ait-Kaci, H. (1984). *A Lattice-Theoretic Approach to Computation Based on a Calculus of Partially Ordered Types*, Ph.D. dissertation, Department of Computer Science, Moore School, University of Pennsylvania

Barwise, J. and J. Perry (1983). *Situations and Attitudes.* Cambridge, MA: Bradford/MIT Press

Birkhoff, G. (1967). *Lattice Theory*, 3rd ed. AMS Colloquium Publications, Vol. 25, Providence, RI: American Mathematical Society

Bresnan, J. (ed.) (1982). *The Mental Representation of Grammatical Relations.* Cambridge, MA: MIT Press

Chomsky, N. (1981). *Lectures on Government and Binding*. Dordrecht: Foris

Devlin, K. (ms.). *Logic and Information. Vol. I: Situation Theory*. Draft in progress

Dowty, D. (1982). Grammatical Relations and Montague Grammar. In P. Jacobson and G.K. Pullum (eds.), *The Nature of Syntactic Representation*. Dordrecht: Reidel, 79–130

Fenstad, J.-E., P.-K. Halvorsen, T. Langholm, and J. van Benthem (1987). *Situations, Language, and Logic*, Studies in Linguistics and Philosophy Series. Dordrecht: Reidel

Fitting, M. (1969). *Intuitionistic Logic, Model Theory, and Forcing*. Amsterdam: North-Holland

Gazdar, G., K. Klein, G.K. Pullum, and I.A. Sag (1985). *Generalized Phrase Structure Grammar*. Cambridge, Eng.: Blackwell, and Cambridge, MA: Harvard University Press

Gierz, G., I.K. Hofmann, K. Keimel, J. Lawson, M. Mislove, and D. Scott (1980). *A Compendium of Continuous Lattices*. Berlin: Springer-Verlag

Gunter, C. and A. Jung (1988). Coherence and consistency in domains (extended outline). In Y. Gurevich (ed.), *Logic in Computer Science*. IEEE Computer Society Press

— and D. Scott (1988) Semantic domains. Draft version of 22 April, 1988. To appear in *Handbook of Theoretical Computer Science*, North-Holland

Kamp, J.A.W. (1981). A theory of truth and semantic representation. In J. Groenendijk, T. Janssen, and M. Stokhof (eds.), *Formal Methods in the Study of Language*, Amsterdam: Mathematical Center Tracts, 277–322

Langendoen, T. and P. Postal (1984). *The Vastness of Natural Language*. Oxford: Basil Blackwell

Moshier, D. and W. Rounds (1987). A logic for partially specified data structures. In *Proceedings of the 14th Annual Conference on Principles of Programming Languages*

Moshier, D. (1988). Extensions to Unification Grammar for the Description of Programming Languages, Ph.D. dissertation, University of Michigan

Pereira, F. (1987). Grammars and logics of partial information. Technical Note 420. Menlo Park, CA: SRI International.

Pereira, F. and S. Shieber (1984). The semantics of grammar formalisms seen as computer languages. In *Proceedings of COLING 84*, 23–129

Plotkin, G. (1976). A powerdomain construction. *SIAM Journal of Computing*, 5, 452–87

Pollard, C. (in press). Sorts in unification based grammar and what they mean, in M. Pinkal and B. Gregor (eds.), *Unification in Natural Language Analysis*

— (in preparation). A linguistically motivated powerdomain construction

— and I.A. Sag (1987). *Information-Based Syntax and Semantics. Volume 1: Fundamentals*, Lecture Notes Series No. 13. Stanford, CA: CSLI

— (1988). An information-based theory of agreement, in D. Brentari, G. Larson, and L. Macleod (eds.), *Proceedings of the 24th Regional Meeting of the Chicago Linguistic Society*, Vol. 2: Parasession on Agreement in Grammatical Theory. Chicago: University of Chicago

Rounds, W. (1988). Set values for unification-based grammar formalisms and logic programming. Manuscript, CSLI and Xerox PARC

— and R. Kasper (1986). A complete logical calculus for record structures representing linguistic information. In *Proceedings of the IEEE Symposium on Logic in Computer Science*, June 1986

— and A. Manaster-Ramer (1987). A logical version of functional unification grammar. In *Proceedings of the 25th Annual Meeting of the Association for Computational Linguistics*

Scott, D. (1982). Domains for denotional semantics. In *ICALP 82*, Heidelberg: Springer-Verlag

Shieber, S. (1986). *An Introduction to Unification-Based Approaches to Grammar*. CSLI Lecture Notes 4, Stanford, CA: CSLI

— H. Uszkoreit, J. Robinson, and M. Tyson (1983). The formalism and implementation of PATR-II. In *Research on Interactive Acquisition and Use of Knowledge*, Menlo Park, CA: Artificial Intelligence Center, SRI International

Smyth, M. (1978). Power domains. *Journal of Computer Systems Science*, 16, 23–36

— and G. Plotkin (1982). The category-theoretic solution of recursive domain equations. *SIAM Journal of Computing, 11*, 4

Zeevat, H., E. Klein, and J. Calder (1987). Unification categorial grammar. In N. Haddock, E. Klein, and G. Morrill (eds.), *Categorial Grammar, Unification Grammar, and Parsing*, Edinburgh Working Papers in Cognitive Science, Vol. 1. Edinburgh: Centre for Cognitive Science, University of Edinburgh, 195–222

CHAPTER ELEVEN

Animals and Individualism

Kim Sterelny

INTRODUCTION

In the last few years, a significant number of philosophers have been sceptical about the explanatory value of an appeal to content in psychological theories.[1] Stephen Stich's *From Folk Psychology to Cognitive Science* is the best known, and perhaps most explicit, statement of scepticism but not the only one.[2] The central, though not the only, motivation for their scepticism[3] is an acceptance of the view that the natural kinds of psychology, though usually not the type identical to neurophysical kinds, ought to supervene on those kinds. The natural kinds of psychology should supervene on events within the skin. In the jargon, psychology should be *individualistic* or *narrow*. The central argument can be put in a relatively a priori (Stich 1978: Fodor 1981, Ch. 9, 1987, Ch. 2) or a relatively empirical way (Fodor 1987, Ch. 2). First, a statement of the a priori form: Psychology is the explanation of behaviour. But it seems that truth-conditional semantic properties are *irrelevant* to the causation of behaviour. For only states of the nervous system cause behaviour, but, notoriously, the semantic properties of our psychological states do not supervene on properties of our nervous systems. This is the message of the various Twin-Earth style thought experiments of Putnam and Burge. They are cases in which we hold the facts about the nervous system constant, while the semantics of the subjects' thoughts vary. Two physically identical people can have thoughts with different semantic properties, if they are embedded in different environments. A perfect copy of you living on a Twin-Earth doesn't have thoughts about New Zealand, as you do; his thoughts are about a different place, and have different truth conditions. But this fact about your thoughts has no effect on your behaviour, and if the semantic properties of thought are causally impotent, then it seems they are none of psychology's business. The taxonomies of psychological theory should be what have come to be called *narrow* taxonomies, taxonomies whose defining conditions depend solely on states of the nervous system, rather than 'wide' taxonomies, which also look to these states' environmental correlates.

Let's give a more empirical argument to the same effect. The best contemporary theory of mind is cognitive psychology. Cognitive psychology's central thesis is that cognition is computation. Computation, in turn, is the manipulation and transformation of symbols according to formal rules. To say these rules are 'formal' is to say they apply to symbols in virtue of their *syntactic* or *orthographic* properties. These symbols may have semantic properties, but they are manipulated in computational processes independently of such properties. But if cognition is just an ensemble of computational processes, then semantic properties are irrelevant to cognition. This argument does have to be qualified slightly; no one holds that *all* the processes relevant to cognition are computational. For transduction is not computational; no one expects to give a formal, algorithmic account of how, for instance, stimulation by ambient light on the retina results in a discrete input to the brain. But there is no comfort in 'slightly wide' psychology, for friends of content. Theories of transduction may concern themselves with some environmental facts, but nothing like the sorts of environmental facts needed to determine referential semantic properties.

To the extent, then, that cognition is computation, the skin is an appropriate boundary. For surely the syntactic properties of mental symbols are determined by the patterns of functional interrelations between those symbols. If so, a computational psychology is an individualistic psychology.

Burge's Standoff

In neither form is the argument decisive. The empirical argument depends on a parallel between cognition and computation. At most, it establishes that the *computational* subdomains of psychology traffic only in individualist kinds. But there is an entirely open question of how much of psychology is computational psychology. Fodor, for example, one of the high priests of the computational enterprise, attempted to demonstrate in *The Language of Thought* that concept acquisition cannot be a computational process.

In this section, I shall concentrate on the a priori argument, for it is intended to show the entire irrelevance of non-individualistic cognitive kinds.[4] I shall argue that Burge is right in thinking that the arguments for individualism are question-begging. Causation is local; hence, causation in the mind is local too. But individualists add to a harmless claim about causation controversial and undefended premises about individuation and explanation. But though Burge is right in rejecting the individualist's argument, his own defence of non-individualist psychological kinds miscarries.

The appeal to Twin Earth suggests that individualists have in mind this argument.

- P1 The behaviour of individuals on Earth is identical to that of their functional and physiological duplicates.
- P2 Psychology is the science of behaviour.
- C1 Since the behaviour of Earthians and their doppelganger is the same, the explanation of their behaviour, equally, is the same.
- C2 So "there is no room in the discipline for explaining their behaviour in terms of different mental states" (Burge 1986: 10).

But the first premise is question-begging, and the second is false. Consider first P1. What are the behavioural kinds? Suppose you are thinking of your mother, and engage in a chunk of mother-directed behaviour: you send a letter reminding her to make her will. Does your Twin behave the same way? He certainly doesn't send a note to *your* mother. If behaviour is described in intentional language, the behaviours of an individual and their Twin are *not* identical. If we give their behaviours individualistic descriptions, these indeed would be the same. But (a) it's not obvious that an individualistic language of behavioural description is *possible*. At the very best, it's very difficult to give a description of writing a letter that makes no commitments about the nature of the subject's environment. Second, even if we *could* construct individualistic descriptions of behaviour, it needs to be shown that that is the appropriate language for theoretical psychology. The construction and justification of an appropriate taxonomy is *not* trivial. On the contrary, it is one of the most difficult problems of ethology. Consider, for example, a cat scratching new furniture. Is the cat (1) sharpening its claws, (2) marking its territory, (3) stretching its tendons, or (4) none of the above. Human behaviour is far more complex. Its appropriate taxonomy cannot be settled by a few thought experiments or offhand remarks.

The second premise is false. It is not true that the sole, or main, purpose of psychology is the explanation of behaviour. Rather, it is the explanation of our psychological capacities. Nor are those capacities in general individualistically specified. Consider, for example, Fodor's modularity hypothesis. Fodor's modules are (among) our innately given capacities, and they are identified by appeal to their function.

That is, that aspects of our environment they respond to. We (probably) have face recognizers, phoneme-detectors, language parsers, and so on. These descriptions identify psychological capacities environmentally.

There is thus no reason to accept the view that psychology *must* be individualistic. But, of course, it *may* be individualistic. It has certainly seemed to some that the skin forms a natural boundary: the processes internal to that boundary ought to be explained by appeal to kinds internally definable. Michael Devitt, for example, has reasoned this way (Devitt, 1989). Burge attempts to show that psychology *must* be non-individualistic. I shall try to show that his argument fails.

Burge's stalking horse is the theory of vision developed basically by David Marr. Burge regards this both as an impressive theory, and one that is representative of theoretical psychology. Marr is concerned, first, to precisely specify the problems the human visual system solves, second, to outline computational algorithms that solve those problems, and third, to suggest neural mechanisms that implement those algorithms.

Burge claims that Marr's theory of vision is up to its neck in assumptions about our physical environment. I will briefly sketch Burge's case and then outline my central charge. In his discussion of Marr, Burge, in my opinion, misidentifies the individualist's claim. The individualist does not deny the importance of environmental considerations in giving a theory of vision. Rather, the individualist argues that the theory of vision can be factored into two components: an environmental component, and an internalist account of the workings of the perceptual system.

Let me be a little more specific. Burge claims of Marr "that the theory makes essential reference to the subject's distal stimuli and makes essential assumptions about contingent facts regarding the subject's physical environment" (p.29) and that "the theory is set up to explain the reliability of a great variety of processes and subprocesses for acquiring information, at least to the extent that they are reliable" (ibid.).

The individualist need deny neither claim. In explaining the interaction of mind and world, we must make essential reference to the world. The issue is whether our theory can be sorted into two elements: an account of the world, together with an individualistic theory of psychological processing. Let me illustrate with an example of Paul Churchland's. In *Stalking the Wild Epistemic Engine* he describes a certain neural specialization in rattlesnakes as a mouse-detector. For (1) this specialization is activated only by simultaneous inputs from infrared sensitive organs detecting warmth, and visual organs detecting movement, and (2) these joint inputs are typically caused by mice, the rattlesnake's standard prey.

Churchland thus gives a *non-individualistic* specification of a psychological mechanism. An individualist might offer a re-description of the same phenomenon, namely, an overview of the neural architecture together with the observation that a brain so set up is only adaptive, and could only evolve, in an environment in which mouseish objects were the typical small warm things. The issue is not one about the *importance* of the environment, it is about whether our *total* theory can be segregated into two components.

Let me take just one of the elements of Marr's theory, and illustrate my point about segregation. Marr thinks visual representation is computed in *stages*. The first stage after the retinal image Marr calls the "primary sketch." Burge gives an argument for a non-individualist individuation of one of the mechanisms involved in the production of the primary sketch. He thinks we should see one of these mechanisms as edge or contour-representers.

The story goes this way. The channels through which the primary sketch are constructed take their input from photosensitive retinal cells tuned to different frequencies. So there is a problem about the combination of their distinctive outputs. These channels are filtered through so-called "zero-crossings": the registration of an intensity discontinuity, in a given patch in the image, at the channel's frequency. The filter — the "zero-crossing" *enhances* the contrast at this point. Burge points out that there is no a priori reason why a zero-crossing at one frequency band in a retinal image is matched by a different band. The point is that if they were not, in general, matched, the outputs of the different channels wouldn't combine. And there would be no point in engineering a visual system this way: "zero-crossing" wouldn't name a mechanism of the visual system.

Though there is no a priori reason, these discontinuities have a standard distal cause: edges or contours. This physical fact does enable us to combine the output of the different frequency bands, and mark a boundary in the "primal sketch" of the retinal image.

Individualists can go along with all of this. They, of course, will not say that zero-crossings form representations of object boundaries. Instead they say that a visual system is engineered so that a processed image enhances contrasts on the pre-processed image. Zero-crossings are contrast enhancers. Such a visual system is adaptive and reliable only in certain environments, namely those — like ours — in which the lines of enhancement correspond to edges or contours in the creature's environment. Marr's theory is the beginnings of a systematic theory of how such correspondence is possible. The individualist, that is, distinguishes between (1) the physical world having certain properties P that permit certain processing to be adaptive, and (2) elements of the processing being representative of P, or individuated with respect to P.

I claim that insisting on this distinction undercuts Burge's positive case for non-individualism. I could only demonstrate that by going through *all* Burge's examples. That I will spare us.

So I shall construct a different case for non-individualistic, or broad, psychology. I don't, however, argue for the explanatory significance of full truth conditional content. I make a case for the theoretical utility of representational kinds; kinds identified in part by appeal to the environment. But since that case depends on attributing those representations to non-human minds — minds unlikely to be sophisticated enough to manifest the full complexity of PA psychology — I don't traffic in the traditional notion of content. My brief is for non-individualistic psychology: as it stands it won't justify the claim that intentionality is essential to human psychology.

I proceed as follows. In the next section I outline my claim. In the next, I consider a central objection. Finally, I distinguish this proposal from a similar one offered by Dan Dennett (1983), and conclude.

Biological and Psychological Semantics

In current work on psychological theorizing it is common to distinguish, in many different terminologies, three levels of psychological theorizing: "ecological level" theories, computational theories, and theories of the physical implementation of cognitive processes.[5] The ecological level is a specification of the cognitive capacities of the system in question that is neutral about the computational or algorithmic implementation of those capacities. This level is not individualistic: the capacities — perceptual, cognitive, behavioural — are specified in terms of the environment of the cognitive system. Identifying some mechanism as (say) a face-recognizer is to identify it not in virtue of its narrowly or individualistically specifiable workings but in terms of its biological function. That function is to detect and recognize certain features of the cognizer's social environment.

In this section I argue for the importance of these taxonomic kinds; they are important in enabling us to state cross-species commonalities, and in other ecological and environmental models. It is characteristic of many such models that they yield explanations that depend on the attribution of wide psychological states to animals. But though these explanations depend on the animals having these widely characterized psychological capacities, it does not matter *how* those capacities are computationally implemented.

Let's first consider the idea of cross-species psychological commonalities. If there are such commonalities across widely differing species, those commonalities can hardly be similarities in formal, computational structure. Suppose we want to say, for instance, that certain bats and birds can both perceive a certain kind of spider, and both know that it's venomous, (or dangerous, or to be avoided). But the bat and bird — and the host of other spider perceivers — share no internally definable psychological states. They share no language of thought and their internal organization, sense organs, and behavioural capacities will be distinct. No narrow psychology will be able to group all the different states that count as perceiving a spider approaching.

The real difficulty is to show that there is a genuine explanatory point to cross-species generalizations. I will try to make this idea plausible.

The cross-species psychological descriptions I have in mind are part of evolutionary and ecological explanations. For instance, returning to arachnaphobia, the fact that both the bat and a bird perceive that certain spiders are poisonous may be part of an account of why those species are dominant in their common environment. If they had not developed this psychological capacity, however realized, they wouldn't have survived. The bat and bird are *co-perceivers*, and through this they overcome a common danger. We may even want to say that *any* possible inhabitant of this environment must somehow realize these perceptual capacities (or an immunity to spider venom) or it won't succeed. The fact that these generalizations may be counterfactual-supporting highlights the abstractness of wide psychological characterization. In the case of these merely possible organisms we have no idea at all of how they manage to perceive the spider, we just know that they must *somehow* realize this perceptual state, or they won't survive.

This point can be generalized. Perception is a concept that is both important and species-general. Roughly, perception is the extraction of information by a creature from its environment. There is no narrow definition of perception: perceptual systems differ too much for that. Nonetheless, the concept is explanatorily important. I think we will need it to state ecologically important generalizations. For instance:

> The more mobile an animal is, the better developed are its perceptual capacities.

For phylogenetically comparable species, carnivores have better developed perceptual apparatus than herbivores.

These are only amateur hour examples, but they illustrate what I have in mind. Further, I expect there to be ecological generalizations covering other cognitive capacities: capacities whose interspecies definitions must be wide. Unless the evolution of our own cognitive capacities is an accident, there must be generalizations relating those capacities to survival strategies; generalizations that hold whatever the internal implementation of those cognitive capacities.[6]

Let's pass from illustration to theory; in particular, to evolutionary ethology.

One of the standing problems of evolutionary theory is to explain the evolution of altruistic behaviour. Prima facie, we would expect altruistic individuals to be out-competed; to contribute fewer descendants than their selfish, and hence more frequently surviving, rivals. The gene allowing altruistic behaviour would gradually disappear. In the last decade or two, a couple of theoretical models have shown how to tame this problem. The central notions are kin selection and reciprocal altruism.

One unsurprising form of altruism is child care. Child care is costly to the parent, and may reduce the number of its own offspring, but in some conditions at least, it's easy to see how a gene for child care could survive. Even though a child-caring rival may die earlier and with fewer children than its rival, the reduced mortality of the few may leave it with as many or more grandchildren — each tending to receive and pass on the child-care gene. Kin-selection generalizes this idea to other kin. A gene for helping those with similar genes to yourself can tend to increase the number of individuals with this genetic material — including the kin-altruism gene. The more distant the kin, the more of them must benefit, or the smaller the cost to the altruist must be. Now this model requires organisms to act differentially towards their kin. How do they manage that? Sometimes the mechanisms are not psychological; proximity does the job, for example, for sterile casts of social insects. But in many kin selection explanations, psychological mechanisms of recognition are clearly implicated: for example, in the co-operative behaviour of lionesses towards relatives, or in sociobiological explanations of the incest taboo. These explanations require that you can tell *who your relatives are*. They require a sufficiently reliable relative-detector. The existence of cuckoos shows that for some organisms this isn't easy. But these explanations are indifferent to the way the relative-detector works.[7]

The second strand of the explanation of altruism is the much more homely idea of exchanging favours: reciprocal altruism. Not surprisingly, this kind of explanation is hardly to be invoked for the dramatic examples of altruistic or apparently altruistic behaviour: female birds trying to draw predators away from nests etc. But it seems to find a home in the explanation — e.g., of the development of mutual grooming by primates, food-sharing, and the like,[8] that is, in cases where the likely costs aren't too great. Again, it is clear that these models have psychological preconditions: mutual altruism can develop only when animals have suitable recognitional and memory capacities.

A third model involving psychological assumptions is Maynard Smith's notion of an Evolutionarily Stable Strategy.[9] An ESS is a distribution of behaviour patterns in a community that is immune to invasion from within. Let me illustrate by the standard, simple example. Consider a population and a resource: say, nesting sites.

Suppose when two members compete for a resource, they adopt a dove strategy. They threaten and posture for a little while, then one gives up. On plausible assumptions about the costs of fighting and the payoffs of sites, a population of dove strategists isn't evolutionarily stable. A hawk mutation — an animal that actually fights — will win all encounters at no cost, since doves flee. Doves win only half their encounters, so hawk strategists are twice as successful as doves, and the hawk gene would rapidly spread. But unless the costs of actually fighting are low, or the benefits of the resource are very high, a population of hawk strategists isn't stable either. In such a population, a dove gene could invade, for it would not have to bear the high costs of fighting. It turns out that, in most cases, a mixed population is stable, the exact hawk/dove ratios depending on the details of the payoff assumptions. That population in uninvadable.

The point here is that the construction of many of these models make strong psychological assumptions. For stability depends on the *range of behaviours* within the repertoire of the population. Thus, on standard payoff assumptions, a hawk/dove population can be invaded by more sophisticated strategists; e.g., acting like a hawk towards doves and like a dove towards hawks. The bully strategy invades. So, strategies are uninvadable only relative to a range of alternative strategies; strategies aren't *absolutely* uninvadable. The range of relevant alternatives is determined, among other things, by the psychological capacities of the animals in the population. The smarter the animal, the wider the range of alternatives.

All these models attempt to explain the evolution of certain *specific* behaviours when applied to specific organisms (burrowing vs occupying wasp behaviour), or certain general patterns of behaviour when couched more abstractly. Qua theories of specific behaviour, they made psychological attributions to animals that are implementation neutral. They are *broad* psychological attributions: capacities to recognize, remember, and react to features of their (typically social) environment. Qua general models, they specify that any population of animals with certain needs, certain psychological repertoires, and certain resources will tend to evolve in specified ways. Again, the *way* the repertoire is implemented (so long as it's reliable enough and takes place in "real time") is irrelevant to its role in explanation.

MOTHER NATURE: BEHAVIOURIST?

Problem. In this paper I have been arguing that wide psychology finds its place in providing the conceptual equipment for generalizations — generalizations worth having — in ecology and evolutionary theory. But this story faces a serious problem. I am in the business of attributing non-individualistic or wide functional properties to organisms. These wide functional kinds are semantic kinds, but probably not the traditional ones of reference and truth. And the internal states possessing these properties are unlikely to be full-blown propositional attitudes. For it is crucial to my argument to be able to attribute such properties to creatures lacking the full splendour of human cognitive capacity.

An ugly worry about this line has been pressed in conversation by Steve Stich, and implied in papers by Elliot Sober and Dan Lloyd. How can appeals to kin selection, ESS, and the evolution of reciprocal altruism warrant the attribution of *any* property to internal states? Take, for example, the various *evolutionary* appeals we have made. Natural selection, you might say, is a Behaviourist. Only reproductive success matters for natural selection. And only behaviour matters for reproductive success. Questions about the causation of behaviour are "screened off" from the behavioural patterns themselves.

In our role of theorists of evolution and ecology it *doesn't matter* how behaviour is caused; it doesn't matter *what* mechanisms underlie advantageous or disadvantageous behavioural traits. So long as the trait is heritable, it wouldn't matter if baboons playing tit-for-tat kept track of renegers by magic. So the program as advertised looks to be in serious trouble. I say we attribute non-individualistic properties to internal states to serve the explanatory interests of evolutionary and ecological theorizing, especially but not solely by making possible inter-specific comparison. But, it is alleged, internal states are of no interest to these disciplines, ergo not their supposed non-individualistic properties either. To go back to my first example: it's avoiding spiders that our bat and bird must manage. Who cares how? That is, qua evolutionary theorist, who cares how?

This objection is important, but its scope ought not to be exaggerated. It doesn't touch the claim that Maynard Smith's notion of an ESS is up to its neck in broad psychology. Remember that strategies are stable only relative to a range of options, and that one central constraint on the option pool is the broadly characterized representational capacities of the organism: what they can perceive, represent, and remember. Still, the objection, if sound, does bear on much that I have said. It implies, for instance, that theorists of kin selection need consider only behavioural patterns, not the psychological capacities, however specified, that give rise to them. So I want to meet the objection if I can.

I need, therefore, some reason from within the theories of evolution and ecology for concentrating on mechanisms rather than the behaviour that those mechanisms produce. I think that can be done. One of the problems afflicting sociobiology is precisely its implicit behaviourism.[10] Standard sociobiological stories typically don't sharply distinguish between selective explanations of behaviour and selective explanations of capacities for behaviour. For special purpose behaviour rigidly produced by special purpose mechanisms this conflation is probably harmless. It doesn't matter whether we talk about the adaptive significance of bee dance or of the neural mechanisms underlying that dance. It is likely to be equally harmless to move between warning calls of birds and beavers and the mechanisms underlying those behaviours. In cases like these, capacities stand to behaviour 1 to 1.

But this conflation is not always harmless. Much behaviour of complex creatures is the result of interactions between mechanisms. And many mechanisms produce more than one type of behaviour. The behaviour a mechanism produces often depends on the internal psychological environment. This is just to make the hardly radical point that functionalism is the right theory of mind for complex and plastic creatures. When behaviour is the result of interaction between mechanisms, it won't have a Darwinian history: it won't be an adaptive behavioural trait with a history of natural selection to call its own. So, whenever we have interaction and plasticity — i.e., when the 1:1 correlation between behavioural pattern and underlying mechanism fails — qua theorists of natural selection we *are* interested in mechanism. It's *not* true that natural selection has no use for internal states.

Philip Kitcher, for example, has argued that while Richard Alexander is probably right in thinking that kin selective effects are crucial in the evolution of basic human psychological mechanisms — the mechanisms responsible for human sociality and social institutions — it is hopeless to look for *particular behaviours* that have kin selective explanations. The basic mechanisms will have such explanations, but only via a multitude of interactively caused behaviours.[11]

But this reply leads to a second, and more intractable problem. I think we can warrant appeals to capacity, not just behaviour in evolutionary theory. But the capacities in question are general mechanisms, not particular states. The capacity — say — to remember who your relatives are. Not the results of the exercise of such a capacity, such as remembering that that's Uncle Fred. Yet surely it's the *particular states* we want to attribute semantic properties to. Not even the stubbornist Stich simulation will object too vigorously to (say) specifying the visual system in wide functional terms, to saying, for example, that some aspects of *our* visual system are designed for the detection and recognition of rigid three dimensional objects moving in space; that another is to suppress illusions of apparent motion when our head moves, and so on.

What the semantaphobe denies is that there is explanatory mileage in claiming, for example, that an internal state of mine has had the property of *being about Stich*. And none of my examples gives any direct lie to this.

I am not sure how damaging a thrust this is. In part, I can meet it by distinguishing teleological theories of representation from teleological theories of the *point* of representation. It is one thing to give a teleological theory of what it is to be a visual system; much harder to give a teleological account of what it is to see Uncle Bill (though Ruth Millikan has tried it). But that's not what I am doing. I haven't said that it's constitutive of perceiving spiders that you have within you a normally functioning mechanism designed for spider detection. Rather, I claim that the *point* of saying critters have internal states with the content "that's a spider" is found in, for example, interspecific comparisons. These in turn are worth having in assorted biological enterprises.

Now it's true that particular contents aren't going to play much role in evolutionary discussion. Though they may play some: there's a fair chance that we and other animals share some innate concepts. But an appeal to specific and non-innate concepts could well play a serious role in ecology. "Garbage bin" isn't an innate concept of a possum, I am prepared to bet. But garbage bin detection, opening and emptying, plays a considerable role in the life-style of some suburban-adapted Australian possum species.

What I *don't* have to hand is a genuine inter-specific case of sharing non-innate concepts. For our purposes, it would be nice if kookaburras competed with possums for access to garbage bins, with systematic consequences for population structure, breeding habits, and the like. Sadly, they don't. Still, there is no general argument for thinking that there will not be cases of this kind.

It's time to give the score. I have been considering the objection that we can *at best* motivate wide functional characterization of *general* capacities. I don't have a convincing on-the-hoof counter-example to this claim. But nor is the objection decisive.

DENNETT'S COGNITIVE ETHOLOGY

There is a substantial convergence between my suggestion and a program urged by Dennett (Dennett 1983). Dennett argues that the intentional stance could play a useful predictive role in cognitive ethology; it enables us to frame hypotheses about animal behaviour, and test these, without commitment to any theory of the computational states responsible for this. This is close to my own program, as the following quotation illustrates:

> The intentional stance profile or characterization of an animal . . . can be viewed as what engineers would call a set of specs — specifications for a device with an overall information-processing *competence*. An intentional system profile says, roughly, *what information* must be receivable, usable, rememberable, transmittable by the sytem. It alludes to the ways in which things in the surrounding world must be represented — but only in terms of distinctions drawn or drawable, discriminations makeable — and not at all in terms of the actual machinery doing this work. (1983: 349)

There are, however, substantial differences between my position and Dennett's. First, Dennett defends intentional, or belief-desire, psychology whereas I defend non-individualistic psychology. Belief-desire psychology is a hybrid: two beliefs, for instance, are identical only if they have the same truth conditions (this is a non-individualistic condition), and if that condition is represented similarly in the two cognitive systems. So what I defend is not what Dennett defends.

It is implausible to attribute to animals — even animals as sophisticated as vervet monkeys — cognitive states as sophisticated as common sense beliefs; it is still less plausible to attribute second-order beliefs about beliefs. How, for instance, can we resolve the many indeterminacies: do vervets have beliefs about leopards, big dangerous things, or what? These problems do not really matter for Dennett's purposes, for he does not take these intentional states to be real psychological natural kinds. Rather, the attribution of belief-desire complexes to monkeys, beavers, or bees is a heuristic technique: a useful device for suggesting experiment and a staging post to a genuine theory of the creatures in question. It cannot be otherwise, in his opinion, for the intentional stance is committed to a rationality assumption, and nothing is genuinely, fully rational. Dennett is admirably explicit on the merely heuristic role of intentional attribution. Thus, for instance: "Beaty cites the adaptionists Oster and Wilson (1978) 'the prudent course is to regard optimality models as provisional guides to future empirical research and not as the key to deeper laws of nature.' Exactly the same can be said about the strategy of adopting the intentional stance in cognitive ethology" (1983: 353)

My approach contrasts with Dennett's. I claim that non-individualistic psychological properties are essential to ecological and evolutionary explanations. If I am right, non-individualistic psychology is not a halfway house through which ethology should move. Detecting a face, recognizing a relative, warning against a predator are genuine natural kinds.

This contrast leads to another. Since I think representational states of animals' minds play a role in explanation, I must answer the hard question an instrumentalist like Dennett avoids. I am committed to the existence of a reductive account of representation. I am equally committed to grappling with such hard issues as how far down the ladder of complexity it remains plausible to attribute content. The mere capacity to act differentially to something in one's environment does not require that one *represent* that thing in any sense, unless the virus which selectively attacks certain bacteria represents those bacteria to itself — "Egad! E. Coli!" Yet somewhere between the virus and ourselves talk of information and representation becomes appropriate, and somewhere closer to us talk of beliefs and desires does. I need a general theory of representation to elucidate these distinctions and to develop attributions of content less sophisticated and anthropomorphic than the folk-theoretic intentional apparatus.

This is not the place to grapple with this issue; I and others have struggled elsewhere with the question of whether a naturalistic account of (non-individualistic) representation is *possible*.[12] In this paper, I have made a case that we need a naturalistic account of non-individualistic psychological properties, for those properties have a genuine explanatory role in science.[13]

NOTES

1 This paper started life as a joint paper called "Semantic Psychology," written with Peter Godfrey-Smith of Sydney University. It was he who impressed upon me the strength of the objections raised in pp. 00–00 and remains unconvinced by my reply. His contribution to this version remains very considerable; I am in his debt.
2 More equivocal approaches to scepticism are to be found in Paul Churchland (1981), Patricia Churchland (1986), Field (1986), and Schiffer (1987). Michael Devitt is threatening to join these ranks in material in preparation.
3 Stich, for example, thinks as well that the notion of content is inevitably idiosyncratic and hence not an appropriate foundation for psychological theorizing; the Churchlands hold that content keeps company with a sententialist view of mental processing that they deplore.
4 See Jackson and Pettit (1988) for further arguments to this effect.
5 See, e.g., Marr (1982) Ch. 1, Newell (1982), and Dennett (1983) for differing versions of this distinction.
6 For a first pass at some of these problems, see Humphrey (1976) and Sober (1981).
7 For clear accounts of kin selection see Grafen (1984), Krebs and Davies (1981) Ch. 1, or Dawkins (1976) Chs. 6 and 7.
8 The central figure here is Trivers; see for example his 1971 paper. A well-taken example illustrating just the point I want, is reported in Parker (1977). Male baboons enter into temporary coalition against another, usually to obtain a mate. Preparedness to help another by entering a coalition is highly correlated with receiving help when soliciting for it.
9 There is already a huge amount of literature on this. The standard exposition is Maynard Smith (1982). Shorter accounts, exemplifying just the points I need are Maynard Smith (1983) and (1984).
10 See Sterelny (1987,1988).
11 See the last part of Chapter 8 and Chapter 9 of his 1986 book, *Vaulting Ambition*.
12 There are a number of theories on the market. One strand of Kripkian causal theories are naturalistic. See Devitt (1981) and Devitt and Sterelny (1987). Fodor has developed a correlationist causal theory in his *Psychosemantics*. Important alternative programs are Dretske's informational semantics (1981) and (1986) and Millikan's teleological theory of content (1984).
13 Thanks to Kathleen Akins, Michael Devitt, Jerry Fodor, Frank Jackson, David Lewis, Philip Pettit, Elliot Sober and, especially, Stephen Stich for comments on earlier drafts of this paper.

REFERENCES

Burge, T. (1986). Individualism and Psychology. *Philosophical Review* 95: 3–45

Churchland, Patricia (1986) *Neurophilosophy*. Cambridge, MA: Bradford/MIT

Churchland, Paul (1981). Eliminative materialism and the propositional attitudes. *Journal of Philosophy* 64: 67–89

Dawkins, R. (1976). *The Selfish Gene*. New York: Oxford University Press

Dennett, D.C. (1982) Beyond belief. In A. Woodfield (ed.), *Thought and Object*. Oxford: Clarendon Press

— (1983). Intentional systems in cognitive ethology: The "Panglossian Paradigm" defended. *Behavioural and Brain Sciences* 6: 343–90

Devitt, M. (1981). *Designation*. New York: Columbia University Press

— (1989). A narrow representational theory of the mind. Forthcoming in S. Silvers (ed.), *Representation: Readings in the Philosophy of Mental Representation*. Dordrecht: Reidel

— and Sterelny, K. (1987). *Language and Reality*. Oxford, Eng. and Cambridge, MA: Basil Blackwell/MIT Press

Dretske, F. (1981). *Knowledge and the Flow of Information*. Oxford: Blackwell

— (1986). Misrepresentation, in R. Bogdan (ed.), *Belief*. Oxford: Oxford University Press

Field, H. (1986). The deflationary conception of truth. In G. MacDonald and C. Wright (eds.), *Fact, Science and Morality*. Oxford: Basil Blackwell

Fodor, J.A. (1981). *Representations*. Cambridge, MA: Bradford/MIT Press

— (1987). *Psychosemantics*. Cambridge, MA: Bradford/MIT Press

Grafen, A. (1984). Natural selection, kin selection and group selection. In N.B. Davies and J.R. Krebs (eds.), *Behavioural Ecology*. Oxford: Blackwell

Humphrey, N.K. (1976). The social function of intellect. In P.G. Bateson and R.A. Hinde (eds.), *Growing Points in Ethology*. Cambridge, Eng.: Cambridge University Press

Jackson, F. and Pettit, P. (1988). *Functionalism and Broad Content*. *Mind* 107:381–400

Kitcher, P. (1986). *Vaulting Ambition*. Cambridge, MA: Bradford/MIT Press

Lloyd, D. (1986). The limits of cognitive liberalism. *Behaviourism*, 14, 1–14

Marr, D. (1987). *Vision*. W.H. Freeman

Maynard Smith, J. (1982). *Evolution and the Theory of Games*. Cambridge, Eng.: Cambridge University Press

— (1983). Game theory and the evolution of co-operation, in D.S. Bendell (ed.), *Evolution from Molecules to Men*. Cambridge Eng.: Cambridge University Press

— (1984). The evolution of animal intelligence, in C. Hookway (ed.), *Minds, Machines and Evolution*. Cambridge, Eng.: Cambridge University Press

Millikan, Ruth (1984). *Language Thought and Other Biological Categories*. Cambridge, MA: Bradford/MIT Press

Newell, A. (1982). The knowledge level. *Artificial Intelligence* 19: 87–127

Parker, C. (1977). Reciprocal altruism in *Papio Aniiubis*. *Nature* 265: 441–43

Schiffer, S. (1987). *Fragments of Meaning*. Cambridge, MA: Bradford/MIT Press

Sober, E. (1981). The evolution of rationality. *Syntheses* 46: 95–120

— (1988). Methodological behaviourism, evolution and game theory, in J.H. Fetzer (ed.), *Sociobiology and Epistemology*. Dordrecht: Reidel

Sterelny, K. (1988). Critical notice of Kitcher's "vaulting ambition." *Australasian Journal of Philosophy* 66:538–55

— (1987). Why's it all so hard. *Proceedings of the Russelian Society*, Sydney University, pp. 99–110

Stich, S. (1978). Autonomous psychology and the belief-desire thesis. *The Monist* 61

— (1983). *From Folk Psychology to Cognitive Science*. Cambridge, MA: Bradford/ MIT

Trivers, R.L. (1971). The evolution of reciprocal altruism, *Quarterly Review of Biology* 46: 35–57

CHAPTER TWELVE

When Is Information Explicitly Represented?

David Kirsh*

INTRODUCTION

Computation is a process of making *ex*plicit, information that was *im*plicit. In computing 5 as the solution to $\sqrt[3]{125}$, for example, we move from a description that is not explicitly about 5 to one that is. We are drawing out numerical consequences of the description $\sqrt[3]{125}$. We are extracting information implicit in the problem statement. Can we precisely state the difference between information that is implicit in a state, structure or process and information that is explicit?

Most discussions of implicit/explicit belief, knowledge, and representation confidently assume that we know what it is for information to be explicitly encoded; the problematic notion — if any notion is problematic at all — is implicit information.[1] What inspires this confidence is a particular vision of computation.

Let us suppose that to understand a computation it is necessary to track the trajectory of informational states the computing system follows as it winds its way to an explicit answer. If a computer is seen as a mechanism which applies rules to syntactically structured representations, it is natural to view explicit information as an encoding of information in syntactic structures that are interpretable according to a well-behaved theory of content, such as a truth theory. We can then point to a syntactic structure in the system and say 'that form encodes this content.' This, I believe, is our underlying idea behind our intuitions about explicit encodings.

As different kinds of computational mechanisms are discovered and explored — PDP systems, massive cellular automata, analogue relaxation systems — it is becoming increasingly difficult, however, to track the trajectory of informational states these mechanisms generate. There is no doubt that we must find some method of tracking them; otherwise there is no reason to think of them as more than complex causal systems.[2] But it is at present an open question whether the model of rules operating on explicit representations is a perspicuous model of their style of computation.

Once the lid has been raised on new styles of computation we are forced to re-examine our uncritical intuitions about basic notions. We already know that there are many ways information can be implicit in a state, structure or process, and that we are largely ignorant of the full variety of ways that information can be built into architecture, internal dynamics, and environment-system interaction. It seems to follow that one reason it is hard for us to track informational trajectories is that we don't yet know how to determine what information is in a system.

The same problem arises for computational systems that are familiar: we do not know how to determine unambiguously exactly what information is encoded in a system, even explicitly.

For instance, what information is contained in a system that has lost its pointers to one of its data sets? The data is still recorded in the system, in the sense that if the system could regain its obliterated pointers, the full data set could be retrieved. But *ex hypothesis* those pointers are unrecoverable. The states are unusable. Does the system still explicitly encode the data even though they are absolutely inaccessible?

Suppose the pointers are not simple address locations, themselves stored in a look-up table, but rather are calculated by solving a complex function. Or suppose the data is not found in a single location but spills over to many cell locations connected by pointers of the most complex sort. Is the data explicitly encoded?

Again, suppose a set of axioms is represented in a language as expressive as first order predicate calculus. Is the whole deductive closure of the set implicit? Even if that set is infinite, or would require exponential computation to derive its members? What about the $2^{100^{100}}$ digit of π? Is that implicit in the state of an arithmetical engine?

Or suppose that a system has highly ambiguous encodings and must deliberate in order to choose the right interpretation for a given word. What information is encoded explicitly? Is any? Must explicit encodings be non-ambiguous?

That such questions arise is proof that we have unsettled intuitions about the meaning of explicit and implicit information even in familiar programmable symbol manipulating systems. Computer and cognitive scientists talk as if they have a precise idea of these concepts. But they do not.

My intent in what follows is to articulate a particular conception of explicit information that at least may serve as a stable basis for subsequent inquiries into the meaning of implicit information. When I began this inquiry I too assumed that our notion of explicit was unproblematic. No longer. The paper is divided into three sections.

In the first, I show, in greater detail, why the notions of explicit and implicit need elucidation. Our intuitions are not consistent; nor is there any settled view how to resolve them. Yet the concepts are important for both computer and cognitive scientists.

In the second section, I explore efforts to identify explicit information with syntactically and semantically well-defined representations. It is hard to imagine any more natural image of explicit encodings of information than sentences in a natural language. But as we shall see it is not enough that information be present in an encoding; it must be usably present, it must be 'instantly' accessible. This condition of use places a heavy constraint on what sorts of representations can encode information explicitly.

In a brief third section, I mention some of the implications of my view. My approach throughout will be informal.

Our Intuitions About Explicitness Are Inconsistent

Perhaps the simplest and most persistent intuition we have about what explicit means is that information is explicit if it is there, for all to see, much like an unambiguous word in a book. This image of words in a text has four properties which it is tempting to ascribe to explicit representations generally:

(1) locality: they are visible structures with a definite location;
(2) movability: no matter where in a book a word is to be found, or where in a library the book is stored, that word retains its meaning and retains its explicitness;
(3) meaning: words have a definite informational content;
(4) availability: the informational content of a word is directly available to the system reading it; no elaborate translation or interpretation process is necessary to extract the information it represents.

On the surface, these four properties seem to explain some obvious facts. For instance, we believe that the number 3 is explicitly represented as being *in* the set A: {1,5,3,7,4,4} because the information that 3 is a member of the set is *on the surface* of the data structure. The meaning of the numeral '3' is readily understood by any numerically literate agent, so its informational content is directly available. It is not so context sensitive that the agent must read the entire data structure or the entire contents of current memory to determine that meaning. And it has a specific location in the data structure. We can point to what in the data structure explicitly carries meaning.

By contrast, if we say that an element is a member of a set iff it satisfies a given list of properties, say $\{x \mid x$ is an even integer and $0 \leq x \leq 9\}$ we designate a unique set but in a manner which requires computation on the part of the user. The elements cannot be just read off. It is true that we are stating the property list explicitly as opposed to using the elliptical ... notation to denote the elements, as in $\{0,2,4,\ldots\}$, which specifies the properties of the elements implicitly. But both specifications fail to present the elements in a manner that can be directly read off. They do not explicitly encode the elements of the set.

The same difference holds between a table of trigonometric functions where there is a separate entry for each value of sin $n°$, and Euler's equation $e^{ix} = cos x + i \sin x$ for generating trigonometric values. Euler's equation is a *compact* way of describing a trigonometric table. But it does not explicitly represent the table. Taken to a first approximation, and restricted within certain bounds, Euler's equation offers the same information as the table. Yet it is in a different form. The informational content of the table is not directly available in the equation; an elaborate reasoning process is necessary to extract the information explicitly contained in the table.

Another way a representation can carry information inexplicitly is by display. For instance, the number of elements in A: $\{1,5,3,7,4,4\}$ – A's cardinality — is not explicitly represented by A, even though each digit in A is explicit and ready to be counted. A *displays* its cardinality; it does not explicitly represent or encode it.

Let us say that information is displayed if there is a process in a system which can, in short order, *extract* that information, while it is explicitly represented if there is a process in the system which can *immediately grasp* the information. Information that is displayed lies just beneath the surface. Information that is explicitly represented lies on the surface.

The trouble with using immediate graspability, or better *immediate readability* as the mark of explicitness is that we run into problems as soon as we ask whether to count accessing time as part of the reading process. Are the elements in large sets immediately readable? Suppose we have a matrix 10,000 by 10,000 and we want to know the identity of the element in position (6754,9629). Even if each position in the matrix is marked by two numbers representing row and column, we shall have to expend some computational energy in locating the right position and identifying the symbol found there. The task of finding a position seems no different in principle from determining cardinality, both involve counting, or some other numerical operation. Both require computation. But then if we deny that cardinality is explicitly encoded because it can be determined only through computation, shall we not also have to deny that the value at location (i,j) is explicit?

The point at issue here is whether symbols which are *on the surface* in a structural sense may be below the surface in a process sense. I believe they can, and that this difference between structural immediacy and process immediacy lies at the heart of confusions about explicitness.

From a process perspective information is explicit only when it is *ready to be used*. No computation is necessary to bring the content into a usable form. From a structural perspective information is explicit when it has a definite location and a definite meaning. The confusion arises when a representation that seems to be in a usable form when viewed structurally turns out to be in a non-immediately usable form procedurally.

For instance, imagine a reader who wishes to use the suggestions in a book to help him solve a particular engineering problem. The suggestions are there, in a phrase or a line somewhere, but if the book has no index or no obvious ordering, the reader will have to scan an arbitrary amount of the book to find the information he needs. Should we say that the sought-after information is explicit for that reader? Relative to his goal of problem-solving, an indexless book is an inefficient representation of the information he needs. It fails to record the information in an easy to use form.

It will no doubt be objected that there is an important difference between finding information and using it once found. Most everyone will agree that if information is encrypted in a baroque code requiring lengthy decryption before being comprehensible then that information is not explicit. Encrypted information requires preprocessing. The information is present, in some sense, but not present in a usable enough form to be deemed explicit.

But in the case of our imagined engineer, there is no question that once the representation is actually retrieved it is easy to read. The question is whether the accessing process should be viewed as part of the representation's readableness. Is there a relevant difference between spending time and effort to locate information, and spending time and effort to decrypt? Both retrieval and decryption are algorithmic processes; both involve some form of pattern matching or network following.

My own view is that there is not a relevant difference. Explicitness is tied to usability. And usability implies a match up between the procedures available to the agent and the forms the content is encoded in. Granted, these are matters over which we have no fixed intuitions. But we have biases. Are words that are hidden in a tangle of other words any different than encryptions? A standard method of passing secret information is to send a book to one's ally with the unwritten understanding that message words are found in certain spots. Was the secret message encrypted? The question is open to dispute. Suppose the reader must *deliberate* to determine which passages in the book are the relevant ones. Does the book explicitly encode the information he needs?

From a purely computational standpoint there is no fundamental difference between spending time and cycles in finding a datum in space (memory) and spending a similar measure of time and cycles computing that datum in time. It may seem that there is a principled difference here, just as it seems that there is a principled difference between space and time. But we have learned otherwise. Accordingly, just taking computational effort as the measure of explicitness, there is no way of choosing whether to represent a given block of information by a powerful procedure plus limited data or by a weak procedure plus exhaustive data. Either form may be able to provide the information we want when we want it and in the form we want it. Accordingly, out structural notion of explicitness may run at odds with our procedural notion. Despite out intuitions about what is on the surface we cannot decide what is explicit without knowing in detail how a system works.

Our Intuitions About Implicitness Are Inconsistent

Our intuitions are even more unsettled concerning *implicit* information. It is natural to suppose that information is implicit if it is *mediately* readable; the information is structurally hidden and/or procedurally distant but nonetheless *recoverable* by additional processing. It can be made explicit. Thus, it is often thought that the hallmark of implicit information is that it is not explicit but could be made explicit.

To take the canonical example, in formal logic it is generally assumed that formulas which are not part of a given axiom set are implicit if they lie anywhere in the set's deductive closure. Structurally they are absent but procedurally they are recoverable. Given enough processing they can be brought to the surface and represented explicitly.

This definition, however, runs into problems as soon as we try to say what 'in principle, recoverable' means. Returning to deduction, shall we say that certain formulas are implicit regardless of how much effort is required to recover those formulas? Our intuitions are not decisive here. For any set of premisses, there are going to be certain theorems that are easy to prove — nearby in lemma space — and certain others that are computationally distant.[3] Are all these theorems equally implicit? Perhaps implicitness is a matter of degree? But in that case what shall we say about theorems that are infinitely distant, or infinitely hard to reach? And what shall we say about as yet unproven theorems? Is Fermat's last theorem implicit in Peano's axioms? We know that most interesting representational systems are incomplete. It is possible, then, that Fermat's last theorem is neither provably true nor provably false. Assuming that it is true but not provable is it implicit? On one account it is not, for it does not lie in the deductive closure of the axioms. Yet if the theorem is true (and constructivists are right), then there must exist some non-deductive processes that can extract that information from the axiom set. Shall we say that a given datum of information is implicit in a representation only relative to a set of operations?

To press the point, consider a system able to discover generalizations. For such a system Euler's equation might be discovered by reasoning about a trigonometric table. Euler's equation is potentially implicit in a trigonometric table for that system. Yet whether we think the equation is actually implicit depends on how much other knowledge we believe is necessary to make the discovery. A system which can draw generalizations has a *chance* at discovering Euler's equation. But such a discovery would be remarkable. For one thing, the equation contains more information than the table itself — it applies to any real value of x — so it represents a powerful abstraction. The generalization is more than a simple interpolation of the data. It generalizes to new entities. No one knows what is required for such abstractions. Often they are inspired guesses. This is particularly true where the generalization refers to a fundamental concept, such as e, which is not in the descriptive language of the data. e is like a theoretical entity: supported by observations but not reducible to them. How these new concepts are discovered depends on so many factors that it is impossible, in general, to determine whether a given agent will ever stumble on the correct generalization. But then should we allow that generalizations are implicit in data?

Given these difficulties it is hard to make precise a notion of recoverability which can serve to demarcate the realm of the implicit accurately. I still think it is hard to imagine a more natural criterion of implicitnes that has any chance of working than *that which is not explicit but which could be made so*. But not everyone would agree.

For example, it has become popular in some circles[4] to call information implicit if it is latent in a system even if it is unrecoverable. It is well known that a computation can often be made more efficient by exploiting regularities rather than by explicitly representing them. These regularities are really assumptions about the environment, or about the interaction of the agent with the environment. For instance, a vision module designed to extract a 3-D shape from two stereoscopic images works rapidly if it is equipped with an algorithm which differentiates. Such an algorithm will work if the assumption about the world — in this case, that objects change in shape smoothly and continuously — is true.

Smoothness is a condition of the world that justifies differentiation. It is a *success condition* determining whether the algorithm will work. If the condition is false the algorithm will fail to compute a correct answer. A designer who wishes to determine the algorithm's reliability will need to know how often the assumption is correct. For the truth of the assumption is the theoretical justification of the algorithm. But shall we say that the algorithm implicitly represents the assumption? Or that information about the visual world is implicitly encoded in certain of the states, structures or processes of the visual system? Clearly, this assumption is not recoverable by the system itself because the vocabulary of early vision does not include terms such as smoothness. We find expressions of values for wavelength, physical intensity, zero-crossings, surface contours, depth measures, and so on. But nowhere in this vocabularly does a term for smoothness appear. Nonetheless, it has become fashionable to speak of the system as having an implicit theory of the world. Some would argue that the information is causally effective. Is this just sloppy language?

To take another case, some robots currently under research navigate without maps. Such systems are equipped with a compass, with knowledge of their orientation with respect to an origin, and suitable instructions to find their way from any point in a maze to any other. These robots explicitly represent information of the form *if at position A then to get to B orient* 90° *go 10 steps, turn* 120° *then go 15 steps*. It is easy to prove that the total information contained in such instruction sets is sufficient to define a structural map giving the position of all points and identifying all open corridors. A structural map is, in principle, *recoverable* from the instruction set, though not recoverable by the system itself unless it has certain analytic skills.

Should we say that information about structural relations is implicit in the instruction set? By our condition of explicit recovery we must not. According to the condition of recoverability, a system does not encode information implicitly unless it can recover that information and explicitly encode it. Because the robot lacks the ability to translate from its instructions to structural representations of its environment we are obliged to say the instructions do not contain structural information implicitly. Yet what shall we make of the intuition that it is because the instructions *do* contain structural information implicitly that when they are interpreted correctly they generate the right behaviour? If the instruction set did not conform to the structure of the world what could explain its success? *Prima facie*, the instructions succeed because they implicitly encode information about the structure of their environment. They contain structural information.

What these examples show, I believe, is that we have not yet any settled view about our intuitions about explicit and implicit information. On the one hand, it is reasonable to require that information be *actually* recoverable to be implicit. Yet, on the other, it is reasonable to grant that information can be embedded in a system so that it is causally efficacious despite being unrecoverable. Recoverability is contingent on a host of other arguably incidental processes. This, at any rate, is a position some would like to defend.

Why It Matters Whether Our Intuitions Are Unsettled

Such inconsistencies would be unproblematic if terms like explicit and implicit never appear in psychological and philosophical theories. But they do.

Fodor,[5] for instance, maintains that mental states are functional relations to explicit representations. To know or believe a certain fact is to be in a certain computational relation to a representation which explicitly encodes information about that fact. Explicitness is important to Fodor. Yet he never states what explicitness amounts to short of saying it must satisfy some ill-defined formality condition. This leaves us groping for a workable criterion of formal. We do not know, for example, whether cardinality when displayed rather than represented directly is a formal property. The same applies to the relation of having a location in a matrix, or to the relation of being connected to, or being beside. Some of these properties are usually represented, others are displayed. Are they formal properties? How we answer these questions has deep consequences. For, according to Fodor, it determines what shall count as an episode in our mental life.

Gibson and his followers, too, make strong claims about implicit information. They maintain that information about shape and distance, for example, is implicit in the ambient flux of light energy striking our retinas. The visual system, we are told, does not explicitly represent edges etc., then construct further representations of shape. It picks up the invariants implicit in visual energy fields and puts that information to use in controlling behaviour "directly," without ever representing it. Because such information is put directly to use, the visual system has no facility for explicitly naming invariants; it never represents them explicitly. Visual invariants in Gibson's sense, fail to be implicit by our condition of recoverability. For by that condition a system can implicitly encode information only if it can, in principle, explicitly encode it as well, since to recover information it must be possible to represent that information explicitly. How shall we interpret Gibson's remarks? Can information about visual invariants be implicit in light energy?

More recently, Perry, Smith, and Rosenschein[6] have contended that information can be implicit in a system because that system is embedded in a particular environment. A system well-adapted to its environment contains information about that environment and its momentary relations to that environment, even though that information is built into the design of the system and so is in principle inaccessible. On their account, information need not be amenable to eventual explicit representation to be implicit. The information we decide is present in a sytem is not identical to the sum of information that is explicit plus the information that is recoverable. Once again information can be implicit but unrecoverable. Yet again what are we to make of these claims?

The upshot, it seems to me, is that we know somewhat less about information processing than we suppose. Information processing has always been understood as a process of transforming representations — transforming explicit representations. But owing to the variety of physical mechanisms that are often interpreted to be computing functions, this view is no longer universally held. This does not mean that information processing is *not* explicit representation processing. But until we have an adequate theory of the relation of implicit to explicit information we cannot decide the issue.

TOWARDS A THEORY OF EXPLICITNESS

Because implicit is defined largely negatively as information that is present but not explicitly encoded, any inquiry into implicit information must presuppose a theory of explicit information. If I am right, however, our structural and procedural intuitions about explicitness are so confused that there is no easy theory to offer. We may offer a stipulative theory but it will never satisfy all our intuitions.

I shall argue that the source of confusion lies in the bewitching image of a word printed on a page. Words on a printed page have four properties that make them an attractive model of explicitness: locality, movability, meaningfulness, and immediate readability. When we look closely at each attribute we find each is, in some manner, misleading.

In the remainder of this paper, I will discuss each attribute in an effort to separate truth from fantasy. My own positive account emerges along the way.

Four Conditions on Explicitness

Locality. It is a fact of English and all other natural languages that words are represented by written tokens that are spatially compact and readily separable from their spatial neighbours. It is tempting to suppose that all explicit encodings of information share this property. Explicit information, after all, is encoded in codings. These codings must present the information in a modular, readily surveyable manner; otherwise they could not present the information in a ready to use form.

If locality were true, however, it would eliminate, in one stroke, the possibility of distributed connectionist systems ever having explicit representations. This seems to me a good reason to deny it. The kernal of truth in the locality condition is that a word, however it be represented, must be a *determinate* something. It must have identity conditions. It is pointless talking of a state, structure or process encoding information explicitly unless we can be precise about which state, structure or process does the encoding. But it certainly does not follow that these identity conditions mandate spatial isolation. They don't. They require, of course, that the system using the representation must have operators which can recognize those representations. Humans find spatial boundaries especially easy to use for individuation. But there is no *a priori* reason why symbols must be spatially separate. To demand spatial separation places an unmotivated restriction on the range of recognition devices that non-humans might use in reading and communicating.

For example, we can readily imagine a system which encodes information in the wavelengths of the visible spectrum. Since many wavelengths and intensities can be superposed on the same spatial region but later filtered out, one colour, such as a shade of white or pink, can actually explicitly encode many different pieces of information. A system appropriately endowed with colour filters can read off the information immediately. It "sees" separate colour tokens and reads them directly. To anyone without the filters, however, the information is invisible.

The same principle applies to information encoded in scatter diagrams. We can imagine a fax machine, for instance, which distributes information like buckshot sprayed over a page. Several such pages could be superposed. To a system appropriately set up, each page is separable without loss. But again, only to a system with the appropriate operators to read the buckshot distribution. Humans would find such distributions unreadable because they lack edges and simple spatial forms. The problem would be compounded by superposition. Yet what humans are able to see is irrelevant. There are many codes we cannot read unaided.

The only constraint on explicitness that locality imposes, then, has little to do with spatial forms, spatial cohesiveness or spatial size. It is about identity conditions: a representation that explicitly encodes information must be made up of tokens that are readily separable from their surroundings. This separation process may vary arbitrarily from system to system. To be sure, there are limits. A system must be able to identify tokens quickly without engaging in substantial computation. Later I will state more precisely what this means. But for now the point is that locality does not mean spatial isolation. It means separability by the host system.

This then is the first condition on explicit encoding of information:

> **Condition (1)**: The states, structures, or processes — henceforth symbols — which explicitly encode information must be easily separable from each other.

Movability. The ideal of *operational* identity conditions also lies at the bottom of the movability condition. In its simplest form the movability condition states that a word can occur in more than one location and still carry its meaning. That is, the identity of a word is largely independent of context.

There is a profound justification for this idea. If we grant that there is no fundamental difference between transmitting information across space, and transmitting it across time, then transmission across time is storage, while transmission across space is spatial communication. Words serve as compact vehicles for meaning. They are the carriers of information. It is natural, then, to assume that we can use them both to store and to send information.

This view leads quite naturally to the idea that a word retains its meaning whether on page 1 or page 601. In principle, it could be sent from one page to the other. Either in token or in type. One consequence of this view, however, is that context is largely irrelevant to word content. This follows because if the information content of a word changed once it was transmitted we would have no way of reliably sending information. The very idea of a word is of a physical vehicle that holds its meaning across situations. But then the identity of a word must be largely independent of where and when the word appears in a system.

Now symbols which are totally mobile and which retain their identity whatever their context lie at one extreme of a continuum of symbol systems. Such symbols can never by polysemous; they can never have indexical elements; and they can never be read differently by different operators. In short, an information processing system using totally mobile symbols must be uniform throughout: there can be no regions where the symbol is interpreted differently, and no states which the system can enter into which cause exactly similar symbols to be read differently.

To see just how restrictive this constraint is I shall briefly examine several representational languages. We can then test the reasonableness of movability as a condition on explicit encoding. For if we believe that we *can* encode information explicitly in a language that violates the movability condition we have grounds for rejecting movability as just construed as a condition on explicit encoding.

Which languages satisfy movability? By a language I mean any set of individuatable states, structures or processes which can be paired with meanings. In the simplest cases, the theory which specifies a language consists of two components: a notational component and a meaning component.

The notational component tells us what to accept as allowable variation in the structure of a token. It is the theory of symbol separation. So, for example, although my writing changes from page to page as I change my posture, a good notational component would specify enough variability in tokens to cluster my written words into correct equivalence classes. It gives the identity conditions of atomic symbols.

The meaning component tells us what information is carried by each symbol. So, for example, a meaning theory for the symbols on a map will tell us that '•' means *cities with populations in excess of 50,000*.

Now just how restrictive is a language that allows no ambiguity whatsoever? Restrictive, very restrictive. Such systems can have no syntax, even a simple syntax. For instance, simple languages such as Arabic notation for numbers will have too complex a structure. Perfect notational mobility implies that the '5' in 105 and the '5' in 501 must carry the same information. It implies that identical notational elements encode identical information whatever their context. But of course the meaning of '5' changes with position. In 105, '5' means units, in 501, '5' means 500.

In order to capture this variability of meaning with position, we need to introduce a syntax. The point of syntax is to allow us to determine how to adapt the information content of a symbol to its position. For languages with syntax, the meaning component must factor in the contribution to meaning which the word's syntactic role makes. Since most languages do have syntax, a theory of language will usually contain, in addition to its notational and meaning components, a third component — a syntactic component — which describes the syntactic role symbols play when combined into compound structures.

In the case of Arabic notation the contribution to meaning made by position is trivial. Defining position as location in a string read from right to left, we then interpret '5' as follows:

'5' in position i means $5 \times 10^{i-1}$

The language for counting based on Arabic numerals has a very simple syntax. Yet even this syntax, we have seen, violates the movability condition. I think that few of us would deny that Arabic numerals represent numbers explicitly. So movability in its extreme form is too restrictive. But there are many other natural languages in which position is not all that can affect meaning. The symbols which come before or after may matter. Can these encode information explicitly?

For instance, a standard trick for extending an instruction set beyond the limits set by the individual keys on a keyboard is to use some characters as *switches*. A control character, for example, allows any character that follows it to be interpreted as a command rather than as a typed letter. *A* in the context of *Control-a* does not have its normal meaning. The control character switches *a*.

To accomodate switches we need to increase the complexity of the syntax of languages discussed so far. Up until now a syntactically primitive symbol has had no apparent structure. Our notational theory told us exactly what symbols were syntactically atomic. Now, however, we must treat certain *strings* of characters as syntactically simple. They are to be accorded the same syntactic role as atomic symbols despite their apparent molecular structure. This will have no real effect on our meaning theory. For once a language has syntax, its meaning rules are defined over syntactic elements. Thus, in our meaning theory we will find axioms such as:

'Control-a' means *move cursor to line's beginning*
'Control-e' means *move cursor to line's end*

These revisions extend the representational power of a language by adding more primitive symbols to its list of meaningful expressions. Because these syntactically primitive symbols are not notationally primitive, however, the host system has a slightly more complex recognition task, for it must first recognize the notation for 'control' and the notation for 'a' before it can recognize the presence of 'control-a' as a syntactic primitive.

Is the presence of a set of switches in a language sufficient to prevent it from encoding information explicitly? That depends on how local is the connection between a switch and the notational element it changes. For instance, if a switch may be set at an arbitrarily early point in a sentence, then the user of the language must keep track of which switches have been set. This never gets complicated in a computational sense because the user need only keep a stack of active switches and compare each current element against the stack. In fact, even if a switch once turned on may be switched off by a later switch, the job of tracking which switches are on is still simple. For again, the user need only check the current element against the stack. If that element is a switch, the user removes a member from the stack. If it is a non-switch, the user either changes its meaning (and removes a switch), or the user interprets the element with its standard meaning. Accordingly, the only taxing feature of this language is that one must remember which switches have been set. If there are few switches, and each switch is a distinct notational element such as Meta or Control which cannot itself be switched in meaning, this task should be trivial. If humans do not, in fact, find reading this language trivial, that they may tell us something about how long, and how large is the stack which humans can manage with ease.

Switches are a simple method for expanding the vocabulary of a language while keeping notation concise. But they do not allow us to compact our meaning theory: for every switched symbol there must be a meaning axiom.

In most economical languages, however, the same switch can effect different symbols in similar ways. For instance, the manual describing the commands available for my editor tells me that 'a' and 'e' when following 'Control' and 'Meta' mean *beginning* and *end* respectively, while 'Control' and 'Meta', when preceding 'a' and 'e,' indicate *line* and *sentence* respectively. Thus *'Control-a'* means *move cursor to line's beginning*, while *'Meta-a'* means *move cursor to sentence's beginning*.[7]

Such systematic variation in meaning allows us to compact our meaning theory. We can save memory. But it does not save computation. Once again, a stack will be needed to mark the switches that are on, and once again each notational element will be compared against it. Memory is saved because we can get by with fewer meaning axioms. But now instead of determining the meaning of *Control a* by looking up that entry in a memory intensive meaning theory, we must compute its meaning by looking up *Control* and *a* separately, and combining those meanings according to a general rule of combination. Computation increases.

Should we say such languages explicitly encode information? Again, it seems an open question. As codes go, such languages are no more compact than ordinary switched languages. They do get by with a more memory-compact decoder; but that decoder may take slightly longer to apply than one which just looks up the answer. Everything turns on how much time it takes to determine meaning. If the set of switches is small, or the host's decoder is highly parallel, then meaning may be determined almost instantly.

Let us grant that there is room for doubt whether languages with dedicated switches can be counted on always to encode information explicitly. There can be no doubt, however, that if we allow that *any* symbol may serve as a switch for any other, then information will sometimes be hidden. For now every symbol may, in principle, be ambiguous.

The net effect of unconstrained ambiguity is that syntactic rules may be arbitrarily complex because the disambiguations they must help to perform may be arbitrarily complex. The set of syntactic rules necessary, for example, for deciding which type of 'a' we find in a given expression, may enjoin us to examine many letters, or combinations of letters, before and after *a*. Because each of these letters in turn might be ambiguous we might eventually require a set of syntactic rules that computes a function of staggering complexity. For instance, to decide what *a* means in a certain context we might have to compute a function such as $a = f(b, c, f_1(d), f_2(f_3(f_4(e))), \ldots)$ to determine the particular *a* we are dealing with, and then look up its meaning.

To take an example from English, the sentence *Police police police police police*[8] is in principle grammatical. Read as: Police who are policed by policemen, are themselves policers of policemen, we see that each occurrence of 'police' has a unique syntactic and semantic identity. In some contexts 'Police' means *policeman*, in others it means *to enforce*. Complexity enters because there are so many combinations of meanings to consider. The amount of computation needed just to determine what a single expression means rises exponentially with the length of the sentence. Should we say that each occurrence of 'police' is explicit when it is so hard to identify the symbol?

If there remains any doubt here it is because we have not completely resolved whether to base our judgment of explicitness on the computational cost of recognizing and using a symbol or on the fact that the notational and syntactic elements are well-defined. On the other hand, it is tempting to go with our eyes and say that if we can see there in front of us a well defined symbol we know to be meaningful, it must be explicit. We can see the term *police*, so its meaning must be explicitly encoded. Its identity is well defined, though hard to determine. On the other hand, if efficiency is important then it is not enough that there exists *some* mechanism, however complex, for recognizing the symbol. For by that token, structure hidden arbitrarily deeply in a represention could also be called explicit. Thus, an edge might be said to be explicitly encoded on a retinal intensity matrix because there is an effective procedure for extracting it. This is absurd. Surely there is a substantive difference between information that is explicitly encoded and information that must recovered? But then efficiency does matter; it is the driving condition.

Accordingly, each step away from total movability — total context independence — is a step which increases the complexity of the processes which recognize a symbol and its meaning. At some point these recognition costs become too high and a language becomes unable to encode information explicitly. Such languages are too complex to be read and understood immediately. The truth in the movability condition then is this:

> **Condition (2)**: An ambiguous language may explicitly encode information only if it is trivial to identify the syntactic and semantic identity of the symbol.

Immediately readable. I have been arguing that for an expression in a language to encode information explicitly it must be trivial for a user of that language to recognize the notational components of the expression; trivial for that user to recognize the syntactic role of those components, and trivial as well for it to recognize the meaning of both the components and the expression as a whole. In short, I have been arguing that the expression must be immediately readable.

But what does it really mean for a recognitional process to be trivial? Can we say in more precise computational terms what immediately readable amounts to?

The definition I propose is that we call a recognitional process *trivial* if there is a mechanical process that identifies the relevant property in *constant time*. Constant time is a measure of the absolute computational complexity of a process. It means that the number of computational operations needed to solve a problem is a constant independently of the size of the problem instance. For example, the time needed to recognize whether a binary number is even is constant regardless of size because the test for evenness is local, it involves checking the last digit. Similarly, to decide if an encoding is all 1's, all we need to do is add 1 and check to see if the new number overflows, i.e., has a longer encoding length. This too can be done in constant time.

Constant time is the smallest complexity order known. Few computations fall within it. Accordingly, in saying that an expression explicitly encodes information only if it can be parsed and interpreted in constant time, we are placing a strong and precise condition on explicitness. Instead of vague intuitions about structures being *immediately usable* or permitting their information to be *directly read off*, we now have a precise principle for interpreting immediacy.

The criterion of constant time, like all complexity orders, is meaningful only if we know what can be done in a single step. For example, on some machines a piece of information can be retrieved in a few steps regardless of how big the memory storage unit is. On such machines, retrieval is a constant time operation. On other machines, as memory size increases the number of steps needed to find information rises logarithmically. On these machines, memory retrieval is a log time operation.

The one weakness in using constant time as a mark of immediacy is that sometimes we may recognize small inputs immediately, even though the recognitional process for arbitrary inputs takes non-constant time. Strictly speaking, complexity is not meaningful for finite inputs; for in principle the answers to any finite problem can be stored in a giant look-up table, where the minimal amount of computation required to find an answer would be approximately the same across all problems.

Yet sometimes this is exactly what we believe is the case: patterns are recognized immediately because they are matched in memory.

To accommodate this intuition let us think of operators as having a certain spatial *attention span*. We may think of the attention span of an operator as its input window, the number of basic notational elements that may be in focus at any time. In a sense, it is the measure of parallelism inherent in the operator.

For example, to determine whether a given encoding is symmetric — as in *0110*, where one half is a mirror image of the other — a system with an attention span of 1 will iteratively compare numerals at each end. In our own case, however, if the number of digits is small, we can tell at a glance, without aid of iteration, if the two are mirror images. We gestalt this *global* property of the numeral. And so we can decide symmetry in constant time for any numeral up to some length n. Once n is reached there are too many digits to fit inside our attentional field. At that point, we too must iteratively scan. And so the property is no longer explicit. We no longer just 'see' it.

Attention span sets an upper bound on what can be immediately gestalted. That means that the net affect of attention span for larger problems is, at best, to reduce by a constant factor the absolute number of steps required.[9] This has the effect that complexity orders are unchanged by attention span. If a problem was order log it remains order log, except for problem instances that fall within the attention span. The upshot is recognition processes that are normally non-constant time, such as recognizing switched symbols, remain so, unless the symbols are, for instance, close together in time or space.

Accordingly:

Condition (3): symbols explicitly encode information if they are either:
- readable in constant time; or
- sufficiently small to fall in the attention span of an operator.

Meaning. The final condition of explicitness is that every expression that contains explicit information must have a definite information content. It is tautological that a symbol can explicitly encode information only if there is some information that it encodes. Although I have been arguing that the question of explicitness really concerns how quickly information can be made available in an encoding, I have yet to explain what I mean by *information*.

Just what it is for a state, structure, or process to express meaning remains the premiere issue of twentieth-century philosophy. My objective here, though, is not to clarify what information means, but to show that whatever theory of meaning one holds, the same concept of immediate apprehension can apply. Accordingly, let us consider some contenders.

The first theory of meaning is the most widely held:

> A system immediately recognizes the meaning of a symbol if it grasps the contribution which the symbol makes to the meaning of the larger expression of which it is a part.

For instance, recognizing the meaning of 'cat' in *the cat in the hat*, on this view, consists in entering a state which contributes to the larger state of knowing the truth conditions of the whole sentence. To discharge the question-begging term 'knowing' we might reformulate the claim in more verificationist terms. Thus, recognizing the meaning of 'cat' in *the cat in the hat* consists in turning on a subset of the abilities involved in recognizing when a cat is in a hat. Associated with 'cat,' then, would be a set of abilities, or dispositional states, some of which are perceptual and motor, which are triggered (in constant time) by the appearance of the symbol 'cat' and which can combine with other abilities triggered by other symbols. The substance of the theory lies in first identifying the relevant abilities, and second explaining how they come to be triggered in just the right way and just the right order to produce appropriate composite abilities.

When understood in its referential version, this theory requires that the agent be able to 'know the referent' of an expression in constant time. If we identify this condition with the agent's being in a certain state, then we can say that a symbol explicitly encodes a certain datum of information for a system *S* only if *S* can enter a state of knowing the truth conditions of the expression in constant time. It is an empirical and conceptual problem to determine what this state is for any given system.

The second theory of meaning also attempts to explain in process terms what understanding consists in. Unlike the first, this theory places most emphasis on reasoning skills.

> A system immediately recognizes the meaning of a symbol if it directly enters a state that rationally constrains the system's future inferences and judgements.

For instance, recognizing the meaning of 'cat' in *the cat in the hat* may consist in entering a state which constrains the class of deductions and inductions that the system might make about cats, hats, being inside, and so on. These possible inferences and judgements are somehow *rationally regimented* by the semantic import of 'cat.' The substance of the theory lies in explaining first, the norms of reasoning and judgment, and second, how a given proposition fares in the cognitive economy of a rational agent. If we accept the second theory of meaning, then, we will say that a symbol explicitly encodes a certain datum of information for S only if S can, in constant time, enter a state which appropriately constrains S's possible trajectories of reasoning and judgment. It is an empirical and conceptual problem to determine what these constraints are for any given system.

The third theory of meaning I shall mention, unlike the other two, does not attempt to provide a full blooded account of meaning that grounds understanding in a set of abilities to recognize truth conditions or to reason rationally. This theory offers no explanation of the abilities an agent must have to understand a concept. It does, however, tie understanding to the activation of symbols. These symbols in turn might activate abilities.

> A system immediately recognizes the meaning of a symbol if it accesses (in constant time) any relevant associated symbols.[10]

For instance, recognizing the meaning of 'cat' in *the cat in the hat* may consist in retrieving certain other symbols. Eventually this process must ground out in basic abilities to act, perceive, and reason. Because these grounding abilities might be slow to activation they cannot be part of the explicit content of a symbol. They may be part of the symbol's total meaning. But they are not explicitly encoded.

Thus to take an example from English once again, the symbol "him" in the sentence 'Then John read *him* his rights.' explicitly encodes only that information that can be directly read from it. The referent is not part of this information, for a parser may have to look arbitrarily back among other sentences to locate it. Sometimes this process of locating the referent of a pronoun can be done by syntactic means. There are binding rules and in certain cases these will suffice to determine reference. In short sentences, though, the referent may be within the attention span of the agent. Hence explicitly encoded. For example, *John shot himself*. In other cases, the parser may have to rely on extra-syntactic knowledge which no agent could have within its attention span. For instance, to determine the referent of 'it' in 'Christine put the candle onto the wooden table, lit a match, and lit *it*,' the interpreter must exploit knowledge about tables and candles. Tables are rarely lit, and especially not simply by applying matches to them. The knowledge that is employed to decide that *it* refers to the candle, is located somewhere in the system, but there is no predetermined place it must be. It could be encoded in the state of the interpreter, in rules for meta-level control, or in a lexical data base. Determining where this state is could take arbitrarily long. Accordingly, its content is not explicitly present.

Sketchy as these accounts of meaning are they provide us with a hint at the fourth condition, for they share a common feature: namely that whatever meaning is, the states, structures, or processes that instantiate apprehension of that meaning must be able to be turned on in constant time. Thus:

> **Condition (4)**: The information which a symbol explicitly encodes is given by the set of associated states, structures, or processes it activates in constant time.

IMPLICATIONS

I have claimed throughout this paper that our intuitions about explicit information are confused. Explicitness really concerns how quickly information can be accessed, retrieved, or in some other manner put to use. It has more to do with what is present in a process sense, than with what is present in a structural sense. Three notable consequences follow.

The first consequence is that not everything which we can assign a meaning to is explicitly encoded. If a sentence takes longer than constant time to interpret, then its meaning is not on the surface. Again this holds whether the sentence is a declarative, such as *Buffalo buffalo buffalo buffalo buffalo*[11] or a procedure. For instance, the sentence (add(square 2) (5th-root 3125)) in one sense means (add 4 5). But not explicitly. On the surface, it means to add the square of 4 to the 5th root of 3125.

On the other hand, if (add 4 5) were a constant time procedure we would be obliged to say that its evaluation *was* explicitly represented. This has the consequence that information may be explicit even if it is not represented by its canonical symbol. Returning to our adding example, (add 1 1) explicitly encodes 2 for any system that can add in constant time or which has memorized the answer and can trivially retrieve it. Normally, we would expect that the difference between 2 and (add 1 1) is precisely that the first explicitly encodes 2 while the second explicitly encodes information about a procedure. Yet, while it is true that (1 + 1) does explicitly encode information about adding, it also explicitly contains information about the evaluation of the procedure. This ambiguity rarely occurs, however, because usually, procedures do not evaluate in constant time. So the information they contain is not explicit. It is only when the procedure is a constant time procedure, or when the performance of the procedure is in the attention span of the host, that the symbols become ambiguous. In such cases, there is no reason not to regard the procedure call as both a call to a certain procedure with certain values, and a special name for its evaluation.

If this seems unreasonable, consider a related case. In binary notation the last digit which ordinarily encodes explicitly the number of units, also encodes whether a number is even or odd. Should we say information about evenness is explicitly present? I think we must. We would not likely find an entry in a Tarski meaning theory stating that '. . . 0 means even number.' But if the host system has an understanding of what even or odd is, and it tests for evenness by checking the last digit, then what grounds have we for denying that the last digit carries two meanings explicitly? It is reasonable to suppose that '1' in the units location just is how we represent oddness. To be sure, a system may not always make use of information that is explicit. Being explicit is not the same as being *occurrent*. Occurrent states are occurrent because control has passed to a procedure which interprets the symbol. The control makes occurrent an interpretation. But what a symbol explicitly represents is independent of what is happening in the control structure. The only connection is that if we do not *know* all the constant time procedures that might act on the representation we cannot know all the bits of information the representation might encode explicitly.

The upshot is that an ordinary meaning theory for a language will not specify all the meanings of a symbol unless that theory was constructed with foreknowledge of the constant time procedures present in the system, and foreknowledge of what can be the contents of a span of attention.

This need for foreknowledge is one reason procedural semantics is hard. But by restricting the focus of procedural semantics to the constant time effects a representation can have rather than to arbitrarily long term effects, the project of procedural semantics may become more tractable. This, I take it, is one substantial virtue of the proposed theory of explicitness.

A second consequence of tying explicitness to rate of access is that there is no principled distinction between declarative and procedural representations. It is irrelevant whether 'place 10' means a location in some vector, or it means some compiled procedure which when invoked starts a set of processes for placing object 10 somewhere. Either meaning may be explicitly encoded. What counts is that the link to the vector location or to the compiled procedure be negotiable in constant time.

This focus on process is significant because it encourages us to think about information processing in terms of informational movement. Representations are inert unless coupled with processes which interpret them. This trivial point is often ignored in correlational theories of meaning. Thus we find correlationists looking for static structures to interpret as representations even in essentially dynamic systems such as relaxation networks. The truth, however, is that for such dynamical systems, information content is to be found in the coupling of structure with process. It is the union of structure and process which can explicitly encode information.

The final and most philosophically significant point is that the

Language of Thought as usually conceived is capable of generating representations that are not actually explicit in the proprietary sense I have been discussing. This has the consequence that, at a minimum, one of the following is *false*:

- the language of thought is the best level of analysis to represent perspicuously the episodes in our *mental life*;
- the events in our mental life are identical with operations on explicit representations;
- the language of thought perspicuously describes human information processing.

The primary motive for postulating a language of thought, it will be recalled, is that there are important regularities in thought, belief and action that would otherwise be missed were we to look for explanations at the neural level alone or at the competence level alone. Competence theories are above all structural theories: they tell us what is computed, what knowledge a system must have to be able to perform those computations, and, in the best competence theories, they tell us something about the decompositional structure of the computational process. But the details of the actual computational trajectories they leave undescribed. The opposite is true at the neural level: so many details are available about particular computational trajectories individual people follow that it is hard to find a level of characterization which perspicuously captures the important informational transitions that generalize well. The language of thought is meant to be the right level of abstraction to describe these informational transitions.

Does the language of thought successfully describe this middle level? In most discussions, the language of thought is assumed to be as complex as first order predicate calculus. Yet we know that parsing and assigning a semantic interpretation to first order predicate calculus is a non-constant time process. Something must give. We cannot believe the language of thought to be simultaneously a complete description of human information processing and a complete description of human mental life. If we opt for the language of thought as a complete description of the mental, then we must forsake identifying mental activity with computation on explicit representations.

If any of these three conclusions seems counterintuitive it is because we tend to think of explicitness as a local property of a data structure: something which can be ascertained without studying the system in which it is embedded. But that is a mistake.

ACKNOWLEDGEMENT

My thanks to Eric Saund, Greg Smith, and Bob Stalnaker for many helpful hours of discussion.

NOTES

* Support for this work has been provided in part by DARPA under Office of Naval Research contracts N00014-85-K0124, and by the Army Institute for Research in Management, Information and Communication Systems contract number DAKF11-88-C-0045.

1 See, for instance, Dennett, Three Kinds of Intentional Psychology, reprinted in *The Intentional Stance*, Cambridge, MA: Bradford/MIT Press 1987), 55–6; Hector Levesque, and Ron Brachman, "A Fundamental Tradeoff in Knowledge Representation and Reasoning," reprinted in *Readings of Knowledge Representation*, eds. Brachman and Levesque (1985), 41–70; Rob Cummins, "Inexplicit Information," in *The Representation of Knowledge and Belief*, eds. M. Brand and R.M. Harnish. (Tuscon: University of Arizona Press 1986).

2 Even if we choose to see them as computers merely because we can interpret them at an input output level as implementing functions, we certainly can never know that they are computing those functions correctly unless we can interpret intermediate states.

3 The issue here is not whether we can decide *after* a successful proof has been found just how many inferential steps are required to reach a given theorem. The issue is that there is no means of determining during the course of searching for a proof how far away that particular theorem remains. We cannot tell whether we are getting closer or moving farther away from the theorem.

4 Cf. Noam Chomsky, *Rules and Representations* (Oxford: Blackwell 1980); Stan Rosenschein, "Formal Theories of Knowledge and AI and Robotics." (*SRI* 1985; Hubert Dreyfus, *What Computers Can't Do*, (New York: Harper and Row 1979); and Terry Winograd and Fernando Flores, in *Understanding Computers and Cognition* (New York: Addison-Wesley 1987).

5 For instance, in *Psychosemantics: The Problem of Meaning in the Philosophy of Mind* (Cambridge, MA: MIT Press 1987), 25, Fodor states: "According to the Representational Theory of Mind, programs — corresponding to the 'laws of thought' — *may* be explicitly represented; but 'data structures' — corresponding to the contents of thoughts — *have to be*" (Fodor's emphasis).

6 Brian Smith, "On the Threshold of Belief," to appear in *Foundations of Artificial Intelligence*, special edition of *Artificial Intelligence*, ed. D. Kirsh; Rosenschein, "Formal Theories of Knowledge in AI and Robotics."

7 It is clear that the designers of this editor chose command names they thought reflected some systematicity, in order to make the commands easier to remember. The commands themselves, however, they implemented by a look-up table, thereby treating names as fusions. This design choice was a typical store versus compute choice. It is clearly more efficient to store all the combinations of letter sequences as molecular primitives, if the number of combinations is small, rather than parse each sequence each occasion of use. But as a vocabulary increases it pays to parse. For bigger applications it may be wiser to use compositional axioms, instead of a look-up table.

8 This example, and the proof that unbounded ambiguity leads to exponential computation is drawn from *Computational Complexity and Natural Language* by Barton, Berwick, and Ristad (Cambridge, MA: MIT Press 1987).

9 Of course, if attention span could be flexibly expanded, then the complexity

profile could be altered. A machine that is able to dedicate the ever-increasing hardware and memory required to enlarge its input window can potentially "see" the answer to ever larger problems. But the overhead in terms of the number of wires and connections needed to sustain an enlarged window of attention will grow at the same rate as the number of steps grows for systems with fixed attention span. So there will always be relatively narrow attention spans.

10 Cf. Allen Newell, "Physical Symbol Systems," *Cognitive Science* 4 (1980): 135–83; and Allen Newell, "The Knowledge Level," *Artificial Intelligence* 18: 1 (1982): 18–127.

11 Meaning: Buffalo from Buffalo outwit buffalo from Buffalo.

CHAPTER THIRTEEN

Psychological Inference, Constitutive Rationality and Logical Closure

Ian Pratt

SYNOPSIS

In this paper I wish to examine an argument that has come to command wide acceptance within the contemporary philosophical community. So plausible has this argument seemed that it has sometimes served as a foundation for quite amazingly counterintuitive claims. I hope to show that the argument is flawed, and thus save us from the counterintuitive claims.

Here is the argument. Whenever we describe someone as being in a certain state of mind — as doubting, perceiving, affirming, denying, willing, not willing various things — our description has this curious property: it implies, in the first instance, not how that state of mind *will* change, but, rather, how it *ought* to change. We can say of a student pondering a mathematics problem what solution he *ought* to arrive at, because we can say whether the proposition he is trying to establish really does follow from the axioms he is assuming. Or we can say of a detective engaged on a murder hunt which of his suspects he *should* regard as the most likely culprit, because we can say whether the detective's evidence confirms — according to our chosen inductive method — the proposition that this or that individual is guilty. But whether we think the student or the detective will *actually* arrive at the proper conclusions is another and less certain matter, one which depends on our crediting them with the propensity to reason as they ought. More generally, we can say of a person characterized in psychological terms how he ought to reason, by tracing the logical (or confirmation-theoretic) interrelations between the contents of the propositional attitudes — beliefs, desires, intentions, and so forth — attributed to him. Thus, propositional attitude attributions determine, in the first instance, not how their subjects will reason, but only how their subjects will reason if they reason aright.

It follows — so goes the argument — that if psychological descriptions are to have any explanatory or predictive power, they must be accompanied by an assumption of general rationality — an assumption which implies that the subjects being described will, probably at least, reason as they ought. For, without such an assumption, there will simply be no observable predictions to be made. Add to this conclusion the view that the content of psychological descriptions is exhausted by the observable predictions which they facilitate, and you arrive at the following result: to ascribe psychological states is to ascribe rationality. In other words, it is literally nonsense to say of people (or any other creatures or entities) that they have beliefs, desires, intentions, and the rest of it, unless you are also prepared to credit them with a propensity to revise those psychological attitudes in a generally rational way.

This basic argument, plus or minus various details, underlies many current philosophical accounts of the psychological attitudes: since psychological descriptions only determine, in the first instance, what their subjects will do under some assumption of optimal or rational functioning, some such assumption must be part and parcel of our willingness to engage in that kind of description at all.[1] Not that the argument lacks detractors. Cognitive psychologists have accumulated a large body of experimental evidence purporting to show that errors in human reasoning, far from being occasional blots on a basically commendable copy-book, are, on the contrary, many and systematic.[2] And these psychologists have been joined by a number of philosophers who are less than sanguine about an a priori argument for human rationality.[3]

The aim of this paper is to oppose the argument for general rationality given above, not, as many others have done, by citing evidence to the contrary, but rather, by challenging the premise on which it rests — that psychological descriptions without an accompanying assumption of general rationality are devoid of explanatory or predictive power. The challenge takes the form of drawing attention to a rival view of how psychological descriptions figure in reasoning, a view which can be found in Hobbes, and which has since surfaced periodically in the philosophical, psychological and, more recently, artificial intelligence literatures. Indeed, it is of the essence of this theory of psychological inference that such inference exploits the weakness of human reasoning rather than its strength. In the next section, I describe an aspect of human cognition which I believe underlies psychological inference, and which I call *virtual reasoning*. The account of psychological inference proper occurs in the following section, and in the remainder of the paper, I compare the approach developed here with that taken by the defenders of the argument for constitutive rationality.

One point of terminology. I call the concepts of belief, desire, hope, fear, intention, and so on *psychological concepts*. The psychological concepts are just the ones which we use distinctively to reason about other people's (or our own) minds. Reasoning involving such concepts I call *psychological reasoning* or *psychological inference*.

Virtual Reasoning

Let us begin with an experiment. Imagine that, upon unlocking the door of your office one morning, you are confronted by a scene of chilling gore: the bloodied corpse of a colleague lying face up on your desk, a dagger driven into his chest up to the hilt.

What is your reaction? Surprise. Horror. Evidently, the man has been murdered. The police must be summoned; nothing must be touched. Who could have done it? Only someone with a key to the office. The secretary? The department chairman? Speaking of which, the door was locked just now, which means that whoever did it locked the door after him. And why do that?...

Thus one's mind runs on. Or approximately thus. What people think of when asked to imagine a dead body greeting them as they arrive at work varies from person to person and from time to time, but something along the lines sketched in the previous paragraph is presumably what goes through most people's heads when asked to contemplate such a scenario. Imagine finding a body on your desk, and you find yourself imagining your reactions — inferring that the man was murdered, resolving to call the police and not to disturb any evidence, wondering who did it (and making a few immediate deductions), puzzling over the locked door. Of course, you do not really infer anything or resolve to do anything. It is simply that, when you imagine a situation such as the above, you complete and extend the situation by imagining that you infer and resolve to do certain things.

I say "imagine inferring," "imagine resolving," but that's not exactly right. The sort of thing I have in mind is not at all like, for example, imagining taking a drug which causes one to infer that taxis are made of newspaper and to wonder why the sky is the colour of marmalade. The kind of imagination I tried to elicit in asking you to imagine finding a corpse in your office is quite different: it is not a matter of imagining engaging in certain pieces of reasoning, but rather, of *reasoning within the scope of an imagined cognitive predicament*. As such, the reasoning purports to be, in some sense, rational. It would be rational, if one *really did* see a corpse, to wonder how it got there, to resolve to call the police, and — given one's other beliefs, perhaps — to make certain specific deductions as to who was responsible. Likewise, when one *imagines* the same thing, one's mind runs on in a correspondingly rational way. Hume tells us (though with more than a hint of irony) that nothing is more free than the imagination of man. Possibly so. But the cases I want to draw attention to here are those where imagination is a reasoned, rather than a rhapsodic affair.

We need a name for this kind of thinking. I propose to call it *virtual reasoning*. (I should have preferred the phrase *imaginative reasoning*, but it already has another well-established English meaning.) Virtual reasoning is reasoning within the scope of an imagined cognitive predicament. Like ordinary reasoning, it includes adopting new beliefs, rejecting old ones, forming intentions, asking questions, and so on; except that all these changes are confined within the scope of an imaginary scenario. In order to distinguish these activities from their counterparts in thinking for real, I shall dub the latter, *agentive reasoning*. Agentive reasoning standardly leads to action in a regular and straightforward way. If you really see a corpse, and thereupon form the intention of calling the police, then you will reach for the telephone. If, in addition, you suspect the secretary, you may inform the authorities of that suspicion. Not so with virtual reasoning, needless to say. Another difference: agentive thinking has regular and straightforward lasting effects on one's psychological state, particularly in respect of adopting or retracting beliefs. Virtual reasoning, by contrast, is a relatively transitory affair: all "inferences" and "changes of mind" are confined to the reverie, and vanish when it vanishes, leaving only a memory of the thinking's having occurred. If you really discover the body of a colleague, you permanently modify your beliefs about him, permanently cancel your plans to meet with him, and so on; but when you *imagine* discovering the same thing, your long-term beliefs and intentions will not be affected in the same way. It is as if, in virtual reasoning, one makes a temporary copy of one's beliefs and intentions, to which changes can be made without affecting the original.

Of course, the difference between what I have been calling agentive and virtual reasoning cannot be enshrined in any sentence-long definition. And it would be a simplification to say, for example, that agentive reasoning both leads to action and affects one's long-term beliefs, whereas virtual reasoning does neither of those things. For one thing, imagining seeing a corpse in one's office might lead to the formation of contingency plans capable of affecting one's future actions. For another, merely imagining seeing a corpse might colour one's future reasoning in case one really does see a corpse — for example, by making one more receptive to suggestions that conform to what one earlier imagined. Nevertheless, the way in which reasoning about imagined situations guides our actions and changes our beliefs has a subtlety and indirectness about it which sets it apart from the more regular and straightforward way in which ordinary (what I call agentive) reasoning operates. The difference between agentive and virtual reasoning is one which I can only hope to capture by appeal to the reader's experience: the rough-and-ready characterizations given above are intended as an aid to such an appeal.

Virtual and agentive thinking are not, however, totally dissimilar. On the contrary, there are phenomenological correspondences between the two. The surprise and horror one imagines upon imagining seeing a corpse are *somehow* like real surprise and horror — except that they are less intense and immediate. The puzzlement arising from the locked door is *somehow* like the puzzlement that arises from a real conundrum — except that it is less vexing and urgent. The make-believe inference that the secretary might have been responsible is *somehow* like a real inference to the same effect — except that it has less force and vivacity. As Hume put it:

> I say, then, that belief is nothing but a more vivid, lively, forcible, firm, steady conception of an object, than what the imagination alone is ever able to attain. This variety of terms, which may seem so unphilosophical, is intended only to express that act of the mind, which renders realities, or what is taken for such, more present to us than fictions, causes them to weigh more in the thought, and gives them a superior influence on the passions and imagination.[4]

The themes of virtual reasoning can be as mundane as they can be sensational. For example, one can imagine replying to a certain chess gambit with which one is experimenting, observing how things look from the opponent's viewpoint as the gambit unfolds. Or one can imagine being faced with difficult questions at a meeting, or an unruly student in class, and so on. Virtual reasoning, or reasoning within the scope of a make-believe situation, is a faculty which normal people employ naturally and frequently.

I wish to put forward two hypotheses concerning virtual reasoning. Both are empirically testable; both are (I claim) intuitively plausible; in neither case, however, can I support them with hard — that is to say, with non-introspective — evidence. They are:

> Accuracy:
> When one engages in virtual reasoning within the scope of some imagined cognitive predicament, one reasons very much as one *would*, or at least, *might well* reason if one really *were* in that cognitive predicament.
> Efficiency:
> Many of the cognitive processes responsible for virtual reasoning are also those responsible for agentive reasoning.

According to the accuracy hypothesis, the sorts of (virtual) thoughts that run through your mind when you imagine finding a corpse in your office are similar to the (agentive) thoughts which would (or at least might well) run through your mind if you really did see a corpse in your office. Thus, should such a situation befall you, you probably would be surprised, horrified, curious as to who did it, puzzled about the locked door; you probably would resolve to call the police, not to disturb any evidence, and so on. Similarly, *mutatis mutandis*, for the correspondence between, on the one hand, imagining replying to a chess gambit, or imagining answering a difficult question in a meeting, and, on the other, what you would have been thinking had you really been countering the gambit or fielding the question. In short, the accuracy hypothesis states that virtual reasoning can be a source — though not an infallible one, to be sure — of counterfactual knowledge about one's own agentive reasoning.

The efficiency hypothesis is harder to justify. But it would at least be unsurprising if many of the same cognitive processes that are engaged when one imagines seeing a corpse in one's office — those processes, that is, whereby one makes the various virtual inferences and resolutions as outlined above — are those very cognitive processes which would be responsible for the corresponding real (i.e., agentive) inferences and resolutions if one really were to see a corpse in one's office. Unsurprising also, if the cognitive faculties that are engaged in countering imaginary chess positions are those very cognitive processes that are engaged in countering real chess positions, i.e., the cognitive processes underlying one's ability to play chess. The efficiency hypothesis is thus an explanation for the accuracy hypothesis: virtual thinking is a reliable source of knowledge as to how one would think in the imagined situations, simply because the same mechanisms are involved in reasoning within the scope of an imagined situation as would have been involved if that situation had been for real. The efficiency hypothesis would also explain the phenomenological correspondences between virtual and agentive thinking, as chronicled by Hume.

But I have no desire to press these points: I am willing to let the efficiency hypothesis remain a hypothesis open to empirical confirmation or disconfirmation. The hypothesis will, however, figure crucially in the account of psychological reasoning given in the next section.

The following observations concerning virtual reasoning will stand us in good stead for the remainder of this paper. First, virtual reasoning, like agentive reasoning, has to do with more than deciding whether given propositions are true or false. For example, when one imagines seeing a corpse, any ensuing bafflement (virtual bafflement) about the locked door can only be as a result of having spontaneously inferred (virtual inference) first that there was a murder, and second that the door had been locked, probably by the murderer. One does not need to be asked these things; nor need one imagine being asked: the inferences can come unprompted. Likewise, the processes whereby one resolves (virtual resolution) to telephone the police and not to disturb any evidence are *reasoning processes*, in the wide sense in which I use the term, but they are not processes for deciding whether a given proposition is true or false. There is much more to reasoning — virtual and agentive — than trying to settle given questions: unprompted inference, the raising of questions, the resolving of contradictions, the formation and revision of intentions, are also important.

Second, as is well-illustrated by the case of the imagined chess gambit, one has a sense, in virtual reasoning (as in agentive reasoning), of which inferences are easy and obvious, and which are difficult and obscure. The pawn threat to the queen may be such that any fool could have seen it; but seeing the long-term strategic threat in the center may take some inspired thinking. All this one experiences for oneself in the process of imagining facing a given board position. Thereby, one can know which inferences, had the imagined situation been real, one would have made easily, and which, by contrast, would have been difficult. Thereby also, one can draw conclusions as to which inferences one would have made for sure (the easy ones), and which might well have eluded one (the hard ones).

Third, as the chess example also illustrates, one may, in imagining replying to the gambit, imagine doing so without knowing beforehand how it is intended to unfold. Of course, since it is a gambit of one's own devising (hence the need to test it in this way), one is aware from the start what it involves, what its main threats are designed to be, what defensive value it has, and so on. Nevertheless, one can, when imagining replying to it, suspend that knowledge temporarily

— exclude it from one's imagined cognitive predicament. Thus, in imagining replying to the gambit, one confronts the problem of figuring out (virtual figuring-out) the intentions of one's imaginary opponent as one 'sees' his position develop. And this exercise of imagining that one does not know the intended threats in advance can, I claim, be a source of information as to how one would have — or, at least, might well have — responded had one really been ignorant in the way imagined. For example, one might conclude that one would have been unable, in such circumstances, to divine the plan behind the gambit before it was too late. (More on the utility of such knowledge anon.)

Presumably, the more radical the ignorance one imagines labouring under, the less reliable, as a source of counterfactual knowledge, will be any virtual inference that ensues. Imagine that you barely know the rules of chess, have not a smattering of elemental strategy, but are nevertheless contemplating a given board position. What you imagine, if anything, is unlikely to be a reliable guide to how you would have reasoned in such circumstances. Nevertheless, virtual reasoning can proceed in some situations where one imagines oneself lacking beliefs one in fact has (as well as having beliefs one in fact lacks), provided that the adjustments one's imagination is called to make are sufficiently small.

Fourthly, one thing that virtual inference is not. It is sometimes important to engage in contingency planning. By considering what one should do should one see a corpse in one's office or discover a fire in a labyrinthine building, one can rehearse thinking which one might not have the time to produce in a real emergency — the most appropriate course of action, the nearest escape. The results of such rehearsed thinking can then be distilled and preserved for later use as and when required. In the same way, devising replies to standard chess openings or answers to anticipated questions can be of use should one ever encounter them in the future. Thus, contingency planning increases the efficiency of thinking by spreading the cognitive load more evenly over time.

None of this is quite what I have in mind when I speak of virtual reasoning. Consider again the corpse-discovery example. When I imagine being confronted with the body, I can perform the reasoning in which I might well engage if I were in that situation — and this is reasoning *on the spot*, reasoning which must be performed quickly in (make-believe) distressing and possibly dangerous circumstances. Performing such virtual reasoning is not a matter of reflecting at leisure on what would be the best thing to do in such circumstances. Of course, when reasoning in the imagined situation, one is *trying* to decide on the best course of action; but to try in difficult circumstances under pressure of time — even if those circumstances and that pressure are imaginary — is not necessarily to succeed. Indeed, it may be as a result of doing badly in such a piece of virtual reasoning that one decides to devise an optimum plan. Similarly for the case of imagining discovering a fire: if you want to work out the best course of action in such circumstances, I do not recommend simply imagining being in them and observing your reactions. *Supposing* that a situation obtains is one thing; *imagining* being in it, quite another.

The point emerges even more clearly in the chess gambit example. When I imagine encountering the gambit as one who has never seen it before, I can consider what I might well do if deprived of the knowledge of how the gambit is to develop. It is quite possible that I should fail to notice (virtual noticing) the threat which it is the whole point of the gambit to pose. It is quite possible, then, that I should imagine responding with an inappropriate move. Clearly, this is not contingency planning. Contingency planning makes sense only if one uses all the information at one's disposal, and only when one is interested in inappropriate actions or erroneous conclusions so as to avoid them. Thus, virtual reasoning incorporates limitations, especially of time and knowledge, which contingency planning should have every interest in assuaging. Of course, virtual reasoning *is* reasoning, and, as such, involves the same attempts at accuracy and correctness that infuses all reasoning; it is just that virtual reasoning can be carried out in imagined situations which recognise features likely to be detrimental to these goals.

Psychological Inference

Preliminaries having been dealt with, I now address the topic of psychological inference. The central insight I want to develop is this. If virtual reasoning can be a reliable guide to how one would have thought had one been in a certain cognitive predicament, then it can be a reliable guide to how *another person* would have thought had *he* been in such a cognitive predicament. In which case, it can be a reliable guide to how another person *will* think if he is in such a cognitive predicament.

Let me illustrate with an example. Suppose that, while playing chess, you see an opportunity to capture the white queen. The plan involves advancing pawns so as to wall in the queen, enabling a piece of lower value such as a knight to move in for the kill. As things are, however, it will take several moves to close the trap, during which time white could still move to safety. Accordingly, you decide to move the pawns furthest from the white queen first, so that your intentions will be less obvious; in addition, you leave some other pieces vulnerable to a development which your opponent has set in train, and for which he must keep his queen where it is. This kind of thinking should be familiar to any chess player: in chess, it is no good posing threats to which obvious responses exist, but only those which are difficult to see until it is too late. Thus, one way to get the better of an opponent is to be able to reason about which things will be easy for him to see, and which difficult. Of course, this is not reasoning about chess per se; it is reasoning about your opponent's likely thoughts. That is, it is psychological reasoning.

I claim the following. Sometimes, we can tell whether a threat is obvious or obscure for our opponent because we imagine facing the situation he is facing and, when we do so, experience either ease or difficulty in seeing (virtual seeing) that the threat exists. Thus, you can tell that your plan to capture your opponent's queen might prove difficult for your opponent to see, because you can imagine yourself in his situation — i.e., faced with a knowledge of the dispositions of the pieces on the board but not of the black player's intentions — and, reasoning within the scope of this imagined situation, you yourself find the threat unobvious. My suggestion, then, is that we can tell the obvious from the obscure by employing our faculty of virtual thinking.

Again, consider the position of a murderer contemplating fleeing the country for fear of being caught by the police. He may know what evidence the investigating detective has, and he may know that that evidence implicates him. But that does not answer his question: the important thing is whether the detective is likely to use the clues at his disposal to full effect. And the murderer may, on considering the detective's cognitive predicament (by imagining being in it), decide that the inference in question would never occur to him in such circumstances. On this basis, he may conclude that a suitable chain of reasoning is unlikely ever to occur to the detective either.

The same kind of reasoning occurs in everyday situations. Why might I worry that my friend may doubt that I like him? Because I have, unavoidably, missed several recent appointments with him, uttered comments which, on reflection, might have been misinterpreted, and so forth. I know that the comments might have been misinterpreted because I can imagine hearing them (and can imagine having the particular concerns and interests with which I know my friend to be currently preoccupied), whereupon the offence-giving interpretations, which had not occurred to me before, now become obvious. I know that these interpretations, together with the missed appointments and so forth, might cause my friend to doubt my loyalty because, when I imagine having experienced the things I believe him to have experienced, that thought occurs (virtual occurrence) to me. Notice that the fact that I can devise a clever argument to demonstrate my loyalty, or the fact that I am assured of it anyway, are nothing to the point. It is enough that, in contemplating the situation from my friend's point of view, I can well imagine coming to that conclusion.

These examples illustrate a faculty that I believe is universal among normal persons. One can know how other people will — or at least may — think, and hence how they will or may behave, because one can imagine oneself in their cognitive predicament and do the relevant reasoning for oneself. In other words, one can use one's own cognitive processes to simulate those of others. This, I suggested above, is how one can know what is or is not an obvious threat in chess, whether the police are likely to be onto one, whether one's friend doubts one's fidelity. I also think that this is how one sometimes decides whether a lecture will be comprehensible to one's students, a journal article to its intended audience, and a remark in a conversation to one's interlocutor. To quote Hobbes:

> But there is another saying not of late understood, by which they might learn truly to read one another, if they would take the pains; and that is, *Nosce teipsum, Read thy self* . . . [which is meant] to teach us, that for the similitude of the thoughts, and Passions of one man, to the thoughts, and Passions of another, whosoever looketh into himself, and considereth what he doth, when he does *think, opine, reason, hope, feare &c*, and upon what grounds; he shall thereby read and know, what are the thoughts, and Passions of all other men, upon the like occasion.[5]

Predicting the course of another's reasoning by simulating it with your own will succeed only if certain uniformities obtain between your cognitive processes and his. If you want to predict the likely course of your opponent's thinking in a game of chess, and thence his likely moves, it is no good your deciding whether *you* would have seen the threat implicit in a certain chess position, if that is unlikely to bear any relation to whether *your opponent* would have seen it. Indeed, anyone who has played a pocket computer at chess may have noticed how, especially in the end-game, it can be defeated by ploys which we know would be obvious to (and therefore useless against) good human players. And, as for chess, so for anything else: if you try to predict someone's likely thinking by means of virtual reasoning, you will succeed only insofar as his cognitive processes function as yours do.

Not that you have to be soul-mates, you and he. For virtual reasoning is a powerful and flexible cognitive tool. As we have seen, we can, in imagining ourselves in another's cognitive predicament, imagine believing things which we in fact doubt, imagine also lacking knowledge to which we are in fact privy. We can imagine that we, like he whose thinking we are out to simulate, have a distaste for long, involved calculations (or a passion for them), or that we, like he, are of an impetuous disposition (or of a cautious one), and so on. If we think our chess opponent stupid and dull, we discount, in imagining ourselves in his situation, our more inspired assessments of the state of play, and take into account only routine and immediate thoughts. If we think the detective on our tail is a Clusot, rather than a Poirot, we can allow for that too. Of course, the more we have to stretch our imagination to accommodate the differences between ourselves and our subject, the less reliable — or so I assume — will be the results of the process of psychological simulation I have just described. But total certainty is, needless to say, not what we need be after. A generally correct idea of the likely possibilities will do.

The view I am putting forward, then, is that psychological inference is sometimes psychological simulation. In order to know how someone will — or may well — reason, we imagine ourselves in his position and simulate that thinking by means of our faculty of virtual reasoning.

Psychological simulation can be an efficient method for predicting the behaviour of others. This is because, in simulating another's thought, we need no *theory* of how human reasoning — either ours or anyone else's — is done; we simply need to be able to do it. And if the efficiency hypothesis of the second section is true, then many of the same cognitive mechanisms that we use in simulating another's thought are the cognitive mechanisms used in ordinary (non-psychological) agentive thinking. For instance, when I predict my opponent's likely responses to a chess gambit, I use many of those cognitive mechanisms which enable me to reason non-psychologically about chess. Whatever cognitive mechanisms that allow me to scan a board for threats and opportunities, compute the values of exchanges of pieces and so forth, now find alternative employment: they enable me to make inferences about how my opponent will or may think in a given situation. And this is so because my ability to engage in virtual thinking means that I can use those cognitive mechanisms to simulate the likely workings of their counterparts in my opponent's head. And, in this way, I can experience his likely difficulties first hand.

This is the crux of the matter: when one predicts that another person might well fail to see a threat in chess, or fail to make an inference, or otherwise engage in some seductive but erroneous piece of thinking, one need not have any sort of *theory* of how that person thinks. Such knowledge is unnecessary because psychological inference can simply trade on the fact that one person's cognitive mechanisms are really very much like another's. An inference that is obvious to one person is likely to be obvious to all; an inference that will require one person to search through and evaluate a large set of complicated possibilities will impose the same requirement on all. Not that people all think exactly alike: it is just that people think in sufficiently similar ways to allow one to guess at another's reasoning by imagining oneself in that other's cognitive predicament, and by doing the reasoning for oneself.

I say again: none of the above implies that psychological inference requires us to *know how* anyone's cognitive mechanisms work; i.e. none of the above implies that psychological inference requires us to possess a theory of reasoning. On the contrary, it is of the essence of the view of psychological inference as psychological simulation that such knowledge is unnecessary. Whilst a theory of reasoning is not required, however, neither is one precluded. It is possible that sometimes, psychological simulation is overridden by explicit theorizing; possible also that our psychological simulations are somehow "infiltrated" by — and thus affected by — our explicit beliefs about the way people reason. Exactly when and how such overriding or infiltration takes place is a good, and partly empirical, question. But it is one I cannot delve into here.[6]

Let us review the chess example to get a fuller picture of what is going on. From your knowledge of your opponent — that he has an unobstructed view of the board, that he has been playing as if to win, that he has made no illegal moves — you are able partially to characterise his state of mind — that he *knows* the locations of the pieces, *desires* to win, *is acquainted with* the rules of chess. Perhaps you have also inferred from his previous moves that he is a good player, and that he is following a certain plan. The point of this psychological characterisation of your opponent, i.e., of the ascription of beliefs, desires, and intentions to him, is that it enables you to know, in the relevant respects, what it is like to be he. It enables you to put yourself in his cognitive predicament, whence you can predict, by means of virtual reasoning, the likely evolution of that predicament. Having thus reached a conclusion about the various things that he may or may not think, you can then predict his likely behavior. For instance, if you think he is likely to intend to move his queen so as to threaten your bishop, then you can conclude that he may well execute such a move.

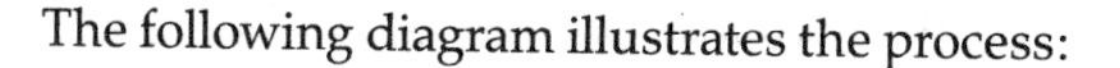

The following diagram illustrates the process:

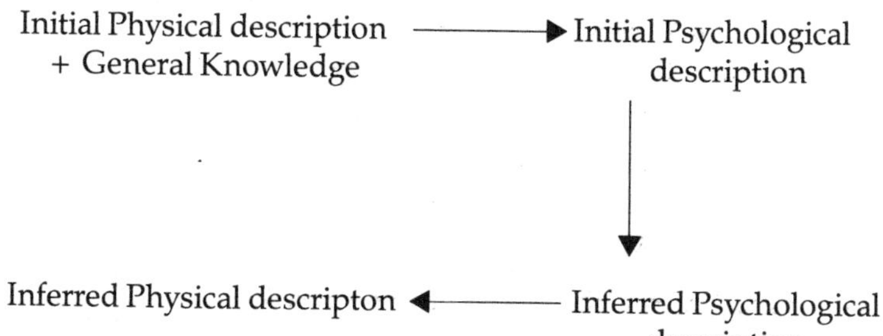

From behavioural description plus general knowledge, infer psychological characterization; from psychological characterization employ virtual thinking to predict likely future psychological characterisations; from likely future psychological characterisations, infer likely behaviour.

This raises the question of how we obtain the psychological characterisation of our subject in the first place, and of how, once having obtained it and having decided (by means of virtual inference) on some of the likely ways in which that psychological state might develop, we draw the appropriate conclusions about the subject's likely behaviour. I suggest that we can get by with fairly crude pieces of common knowledge. For example, it is common knowledge that, if someone obeys the rules of a board game, he almost certainly *believes* that the rules are thus-and-so; common knowledge that persons in our culture who play board games desire seriously — though not maniacally — to win; common knowledge that people staring at chess-piece-sized objects twenty-five centimetres away tend to know where those objects are (under suitable descriptions); and so on. And of course, if somebody says something, that can also be an indication that they believe it. Common knowledge about what beliefs people have is, on the present view, no different in principle from common knowledge about anything else.

One can also ascribe beliefs and other psychological attitudes to a person on the basis that such attitudes would explain his behaviour. Thus, if I know that the investigating detective has been asking lots of questions about me, and has sent his sergeant to tail me, I can infer that he probably suspects me. For when he asked the questions about me and had me tailed, he presumably did so *intentionally* (it being common knowledge that these are among the sorts of things that are done intentionally). Having attributed these intentions to the detective, I can then use my faculty of virtual thinking to decide which possible beliefs (compatible with what I already know about the detective) might, if *I* had had these beliefs, have led *me* to form such intentions. The same beliefs, I can then conclude, might be responsible for the detective's actions. Thus virtual reasoning can be important in constructing the initial psychological characterisation, as well as in predicting its evolution.

As for the prediction of behaviour from psychological descriptions, I suggest that similarly mundane bits of common knowledge come into play. Thus there are certain things which we know — be it from the cradle, or be it by bitter experience — that people are liable to do if they fully intend to: moving chessmen according to a plan is one of them; meeting submission deadlines for papers is not. But again, these are just pieces of common knowledge like any other.

The essence of psychological concepts, I claim, lies in the primary function these concepts perform in human thought. And that primary function is to facilitate the prediction of behaviour by means of virtual inference. A psychological description just is a description which tells us, in some sense, what it is like to be in the position of the subject.[7] It is a description which enables us immediately and unproblematically to imagine being in the described state and which therefore enables us, so I have been arguing, to use our own cognitive faculties to predict its likely development, whence we can predict the behaviour it is likely finally to induce. Of course, a physicalistic (or information-theoretic) description of our subject lacks this property: to describe someone's physical constitution and dispositions (or to detail the flow of information contained in his internal states) is not to be able to imagine being in his situation, unless we first infer a psychological characterization from that description.

Constitutive Rationality and Logical Closure

Virtual inference lets us use the limitations of our own cognitive abilities to probe the likely limitations of those of others. Paradigmatically, our difficulty in seeing the long-term threat implicit in a chess position tells us that others might experience the same difficulty. In this respect, our ability to perform psychological inference trades on the fact, if it is a fact, that people think very much alike; it does not invoke any assumption about how rational the common inferential dispositions are. In particular, according to the view of psychological inference as psychological simulation, there is no assumption of general rationality to mediate any inference from how the subject ought to reason to how he will reason;

on the contrary, psychological simulation is as good for predicting error as it is for predicting sound thinking.

In this respect, the view of psychological inference propounded above stands diametrically opposed to what now almost constitutes an orthodoxy in contemporary philosophy of mind, an orthodoxy maintained by such figures as Dennett, Davidson, Lewis, and Stalnaker. Writes Dennett:

> Suppose we travel to a distant planet and find it inhabited by things moving about its surface, multiplying, decaying, apparently reacting to events in the environment, but otherwise as unlike human beings as you please . . . . Once we have tentatively identified the perils and succors of the environment . . . we shall be able to estimate which goals and which weighting of goals will be optimal relative to the creatures' *needs*. . . . Having doped out these conditions (which will always be subject to revision) we can proceed at once to ascribe beliefs and desires to the creatures. Their behaviour will "manifest" their beliefs by being seen as the actions which, given the creatures' desires, would be appropriate to such beliefs as would be appropriate to the environmental stimulation. Desires, in turn, will be "manifested" in behaviour as those appropriate desires (given the needs of the creature) to which the actions of the creature would be appropriate, given the creature's beliefs.[8]

In other words, the attribution of beliefs and desires is constrained by the requirement that the beliefs and desires so attributed render the actions of the attributees rational — and therefore explicable — given those beliefs and desires.

To be sure, all who claim that assumptions of general rationality are constitutive of psychological concepts confess the possibility of occasional irrationality or non-rationality: of shoddy reasoning, of failing to believe logical consequences of what we believe, of forgetfulness, of destroying the things we love. Up to a point, that is. That such failings are possible means that the constitutive principles of rationality must be hedged with provisos that they only apply for the most part and in certain circumstances, and must be bolstered by principles (perhaps specific to an individual or culture) describing the more systematic ways in which subjects are likely to err. But there are limits to the extent to which constitutive rationality can be usefully etiolated in this way. Again, to quote Dennett: "Not surprisingly, as we discover more and more imperfections (as we banish more and more logical truths from the creature's beliefs) our efforts at intentional prediction become more and more cumbersome and undecidable, for we can no longer count on the beliefs, desires and actions going together that *ought* to go together."[9]

Or: what is the point of attributing beliefs and desires when we, the attributors, are not going to make significant use of the logical (or confirmation-theoretic) interrelations between their contents, and how can we make use of those interrelations without some assumptions of general (if hedged) rationality?

(To which I answer: the point of attributing beliefs and desires is that the resulting descriptions are ones we can imagine to be true of ourselves, thus enabling us to predict their likely evolution by simulation.)

This brings us to the further question of logical closure. Common sense has it that, if someone believes that p, and if p implies q, then that person does not necessarily believe q. Not so, the advocates of constitutive rationality sometimes say: belief, contrary to what many suppose, is *closed* under logical consequence. At one point, Dennett writes:

> one gets nowhere with the assumption that entity x has beliefs p, q, r . . . unless one also supposes that x believes what follows from p, q, r . . . otherwise there is no way of ruling out the prediction that x will, in the face of its beliefs p, q, r . . . do something utterly stupid, and, if we cannot rule out *that* prediction, we will have acquired no predictive power at all.[10]

This argument is similar to the one rehearsed in the first section of this paper: that since the contents of psychological attitudes determine in the first instance, via the logical interrelations between their contents, how their possessors *ought* to use them, the explanatory and predictive utility of psychological characterizations depends upon an assumption of general rationality. Similar, but not quite identical: an argument for constitutive rationality is not by itself an argument for the logical closure of belief. For it is not obvious — as Dennett has since conceded — that simply trying to believe lots of logical consequences of one's current beliefs is a particularly rational cognitive policy. In a more recent article, Dennett qualifies his earlier position on the relationship between deductive closure and constitutive rationality:

First, a few words on what rationality is *not*. It is not deductive closure.

> Nor is rationality perfect logical consistency, although the *discovery* of contradiction between propositions one is inclined to assent to is always, of course, an occasion for sounding the epistemic alarm

> alarm .... Inconsistency, when discovered, is of course to be eliminated one way or another, but making the rooting out of inconsistency the pre-eminent goal of a cognizer would lead to swamping the cognitive system in bookkeeping and search operations to the exclusion of all other modes of activity....[11]

Believing consequences of one's beliefs, and avoiding inconsistency among them are laudable activities, but only up to a point: if carried to excess they are likely to mar one's chances of doing anything useful.

Does that mean that Dennett thinks that we need not presume general rationality among believers after all? No, it just means that rationality cannot be equated with maintenance of consistency and striving for completeness. It may be rational, says Dennett, quoting Cherniak with approval,[12] to be sometimes inconsistent because memory must be compartmentalized in order to make retrieval processes tolerably efficient, and because a compartmentalized memory will give rise to inconsistencies. Rational also, not to spend one's every waking hour making pointless logical deductions; rational to satisfice rather than maximize, and so on. Thus Dennett is denying the necessity of deductive closure and consistency without retracting the necessity of general rationality.

This move, Dennett thinks, blunts much of the criticism (notably, that levelled by Stich) that his views have received. If the thesis of constitutive rationality seemed almost trivially false, that was because what had seemed to be paradigms of rationality — deductive consistency and logical closure — are in fact caricatures. If we adopt more realistic standards, Dennett assures us, the thesis of constitutive rationality is not so counterintuitive.

Not all constitutive rationalists take this view, however. Stalnaker, for example, is explicit in arguing that if beliefs are attributed in such a way as to render the actions of their attributees explicable and predictable, then belief is logically closed:

> To say that an agent believes that P is to say something like this: the actions that are appropriate for that agent — those he is disposed to perform — are those that will tend to serve his interests and desires in situations in which P is true. But this is not quite right for the following reason ... [s]uppose I believe, as I do, that someone will be elected President of the United States in 1988. One way in which that proposition could be realized is for *me* to be the one elected, but I know that that is not the way my belief will come true. For my actions to be appropriate, given that I have this belief, it is surely not

> required that I take account of that possibility, since it is excluded by other of my beliefs. The actions that are appropriate for an agent who believes that P depend not only on what he wants but also on what else he believes. So it is necessary to define appropriateness relative to a total set of beliefs, or a belief state. And all that matters about such a belief state, as far as the appropriateness of actions or the agent's dispositions to act are concerned, are the entailments of the belief state. So there is no basis, on the dispositional account, for excluding from the set of an agent's beliefs any propositions that are entailed by his beliefs.[13]

Stalnaker argues from a constitutive principle of rationality (one that guarantees conformity of actions to beliefs and desires) to the conclusion that the objects of belief have a certain structure — a structure that takes a belief state to be a logically closed set of propositions.[14] He is of course aware that this conclusion is apparently at odds with common usage, and he offers us an ingenious explanation of their compatibility by complicating the former and judiciously reinterpreting the latter.

But I do not wish to review these arguments here. Nor do I wish to adjudicate between Dennett's and Stalnaker's differing stances on the argument from constitutive rationality to logical closure, for my main argument is with the thesis of constitutive rationality itself. I mention Stalnaker's arguments on this question because his position contrasts so sharply with the view developed in this paper. On the psychological simulation view, it is of the essence of belief that one does not always believe everything one's beliefs entail. For it is, on this view, a paradigm of psychological inference to conclude that one's chess opponent might miss the point of one's new gambit because, putting oneself in his position, one does not find the threat it poses obvious; a paradigm also to conclude that the detective will not infer one's guilt, not because his evidence does not implicate one, but because one cannot imagine managing to make the required inferences oneself.

More generally, the contrast is this. For the constitutive rationality advocates, the happy hunting grounds of psychological inference are where we can assume optimal or rational functioning. Thus, we can ascribe beliefs, desires, and intentions to a person or an alien or a computer with whom (which) we are playing chess, just as long as he plays well enough for the logical relations between the contents of these attitudes to be a useful guide to his behaviour. If we want to predict his mistakes, we have to resort to a theory of how our opponent functions (what Dennett calls the "design stance"); that is, we must leave behind psychological reasoning proper.

On the psychological simulation approach, by contrast, psychological inference only really comes into its own when we have to predict sub-optimal functioning. Thus, ascribing beliefs, desires, and intentions to a person (though not perhaps to an alien or computer) with whom we are playing chess can be a useful guide to the threats and opportunities he is likely to miss, because such a psychological characterization enables us to put ourselves in his cognitive predicament and experience his likely difficulties first hand. Prediction under assumptions of optimal functioning, say I, is a different matter, one which need make no interesting use of psychological notions at all. Such prediction is just a matter of the kind of look-ahead which is best done with a large memory and not much imagination; it does not require virtual reasoning.

In fact, Dennett briefly considers the idea that psychological inference might proceed by simulation, an idea he finds "lurking in Stich's view." But, says Dennett: "How can [psychological simulation] work without being a kind of theorizing in the end? . . . If I make believe I am a suspension bridge and wonder what I will do when the wind blows, what "comes to me" in my make-believe state depends on how sophisticated my knowledge is of the physics and engineering of suspension bridges. Why should my making believe I have your beliefs be any different?"[15]

Is reasoning-by-simulation a kind of theorizing, as Dennett suggests? In particular, when I reason about another's beliefs by means of virtual reasoning, does what "comes to me" in my make-believe state depend on my knowledge of the laws of human psychology? Well, yes and no. One could say that, in a sense, if people can predict the reasoning of others (even by simulation), they must *know* the laws of psychology, though it would of course have to be admitted that this is knowledge so implicit that anyone who could articulate it as a theory (in the everyday sense) would have the profession of psychology at his feet. This way of talking, although not the one I have been using, I have no serious objection to: the interesting question then becomes that of how this "knowledge" is stored in people's brains. And my answer is to refer the reader to the conjectures offered above in the section on virtual reasoning (pp 00–00).

But whether information thus encoded deserves to be called *knowledge of the laws of psychology* is beside the point. For Dennett's original argument had it that psychological ascriptions would be almost totally unconstrained were it not for some general principles mediating psychological inference, and that such principles could only be provided by some sort of general rationality assumption. Ergo, constitutive rationality. The point about psychological inference by simulation is that it enables psychological inference to trade on an assumption, not of general rationality, but of general uniformity. Whether or not that counts as having a psychological theory, rationality never gets into the picture.[16]

The situation is this, then. The constitutive rationality advocates think that psychological inferences and explanations are mediated by assumptions of general rationality, whence they argue a priori for (some sort of) general rationality, and whence some of them argue for logical closure; I, on the other hand, think that psychological inference proceeds by psychological simulation, whence I reject the argument for general rationality, and whence I argue against closure. Perhaps there is some useful notion which the constitutive rationality advocates are eliciting. The notions of "belief," "desire," "intention," and so on which they characterize may indeed — though I have yet to be convinced — prove useful for predicting and explaining the behaviour of intelligent aliens, chess-playing computers, and even humans. Furthermore, I can agree that these notions are useful for predicting the behaviour of *optimally functioning* creatures or entities, as Dennett suggests, and are subject to a form of logical closure, as Stalnaker argues. But I demur from calling these notions "belief," "desire," "intention," and so on (except within scare-quotes), and I demur from calling episodes of their employment "psychological inference." On the question of logical closure in particular, I prefer to side with the vulgar: of course people do not believe all the consequences of their beliefs. That, I have been arguing, is the whole point of the psychological notions, the feature without which the faculty of virtual reasoning would find no employment in psychological inference. The reason some constitutive rationality advocates depart from common usage in this respect is that they believe it impossible that psychological characterizations could have any predictive power — and hence psychological concepts any content — except under assumptions of rationality which ultimately violate this common sense. The burden of this paper has been to argue that this belief is untrue.

CONCLUSION

I trust I have not given the impression that it is *obvious* how psychological inference can proceed by means of virtual reasoning along the lines described. For I have not given an account of how it is that we imagine we have beliefs, desires, perceptions, plans, and predilections that we in fact do not have, or of how it is that we imagine we are oblivious of that which we know full well, while still managing to reason within such make-believe situations. Nevertheless, I have tried to make plausible the idea that we can reason effectively about the thinking of another person precisely because we can compare it with our own reasoning, the fact that he does not share all of our attitudes notwithstanding.

If so, I argued, then we can perform psychological reasoning even though we have no theory of how it is that people reason (why bother with a *theory* of reasoning when we can just *do* the reasoning?), and even though we make no presumption of general rationality on the part of those we are reasoning about. So conceived, psychological reasoning represents an efficient use of our cognitive resources. The psychological processes involved in predicting our chess-opponent's likely responses are, in large measure, just those psychological processes we use to evaluate chess positions ourselves.

It is this kind of thinking that psychological notions are *for*. For it is of the essence of describing someone as having beliefs, desires, intentions, and what have you, that one can immediately and unproblematically know what it is like to have those attitudes oneself. In so doing, one can immediately and unproblematically put oneself in a position to simulate the thinking that those attitudes are likely to induce. The same is not true of a description in terms of behaviour and dispositions to behave: one must first infer what psychological attitudes such a behavioural description is likely to correspond to before one can imagine being in the described situation oneself.

If this account of psychological inference is correct, it follows that the argument proffered by the constitutive rationality advocates for general rationality — an argument which assumes that psychological descriptions can wield explanatory or predictive power only under an assumption of general rationality — is undercut. So is the argument for logical closure. And the argument for constitutive rationality is undercut whether or not it is taken to imply logical closure, and whatever, exactly, the relevant benchmarks of rationality are taken to be.[17]

NOTES

1 See, for example, Daniel C. Dennett, "Intentional Systems," *Journal of Philosophy* 68:4 (1971), reprinted in *Brainstorms* (Montgomery, VT: Bradford Books 1980); Donald Davidson, "Mental Events," esp. pp. 221–3, in *Essays on Actions and Events* (Oxford: Clarendon Press 1980); D.K. Lewis, "Radical Interpretation," in *Philosophical Papers* Vol. 1 (New York: Oxford University Press 1983), and *On the Plurality of Worlds* (Oxford: Blackwell 1986), section 1.4; Robert C. Stalnaker, *Inquiry* (Cambridge, MA:MIT Press 1984).

2 For example, P.N. Johnson-Laird and P.C. Wason, "A Theoretical Analysis of Insight into a Reasoning Task," in their collection *Thinking: Readings in Cognitive*

Science (Cambridge, Eng.: Cambridge University Press 1977); also Amos Tversky and Daniel Kahneman, "Judgment under Uncertainty: Heuristics and Biases," *Science* 185 (1974): 1124–31, reprinted in the same Johnson-Laird and Wason collection.

3 For example, Stephen P. Stich, "Could Man be an Irrational Animal?: Some Notes on the Epistemology of Rationality," *Synthese* (1984). See also Paul Thagard and Richard Nisbett, "Rationality and Charity," *Philosophy of Science* 50 (1983), where only a very limited form of constitutive rationality is endorsed.

4 David Hume, *Enquiry Concerning Human Understanding*, 3rd ed. (Oxford: Clarendon Press 1975), 49.

5 Thomas Hobbes, *Leviathan*, ed. C.B. Macpherson (Harmondsworth, Eng.: Penguin Books 1968), 82. For a discussion of Hobbes' idea and its relation to consciousness, see Nicholas Humphrey, *Consciousness Regained* (Oxford: Oxford University Press 1983), Chs. 1 and 3; see also William Lyons, *The Disappearance of Introspection* (Cambridge,MA: MIT Press 1986). The most influential modern champion of reasoning-as-simulation is Kenneth Craik, whose views on this subject are expounded in his book, *The Nature of Explanation* (Cambridge: Cambridge University Press 1943); see esp. pp 50–61. Following in Craik's footsteps, Kahneman and Tversky develop a similar idea in their essay, "The Simulation Heuristic," in Daniel Kahneman, Paul Slovic, and Amos Tversky, *Judgment Under Uncertainty: Heuristics and Biases* (Cambridge: Cambridge University Press 1982). Another writer who touches on the idea of modelling the thoughts of others by means of one's owns thoughts is Quine: "[I]n indirect quotation we project ourselves into what, from his remarks and other indications, we imagine the speaker's state of mind to have been .... Correspondingly for the other propositional attitudes, for all of them can be thought of as involving something like quotation of one's own imagined verbal response to an imagined situation" (W.V.O. Quine, *Word and Object* [Cambridge, MA: MIT Press 1960], 219). And this theme is developed (though in a different direction from that taken in this paper) by Stephen Stich in "Dennett on Intentional Systems," *Philosophical Topics* 12:1 (1981), and in *From Folk Psychology to Cognitive Science* (Cambridge,MA: MIT Press 1983), Ch. 5. See also David Lewis' "Prisoners' Dilemma is a Newcomb Problem," in *Philosophical Papers* Vol.2 (New York: Oxford University Press 1986), 299–304.

Well developed views on psychological inference by means of psychological simulation are to be found in the artificial intelligence literature: see Lewis G. Creary, "Propositional Attitudes: Fregean Representation and Simulative Reasoning," *Proceedings, Sixth International Joint Conference on Artificial Intelligence* (Tokyo 1979), and Andrew R. Haas, "A Syntactic Theory of Belief and Action," *Artifical Intelligence* 28:3 (1986). More generally, the study of simulation as a technique in artificial intelligence has recently been gathering momentum: the journal *Simulation* now has a regular AI feature, often devoted to some aspect of reasoning-as-simulation.

Some people have commented that there are points of contact between the view taken in this paper and the writings of the *Verstehen* theorists, or — less kindly — that I have "rediscovered hermeneutics." In fact, I think the connection superficial. However, I prefer to leave the question to the reader's judgment, rather than to engage in an amateurish exposition of material I cannot understand.

6 Paul Thagard has pointed out to me that examples of this phenomenon are provided by cases where our psychological predictions appear to be affected by false beliefs about the way people reason. An example would be the Milgram experiments on obedience to authority. What do you think most people would do if asked to administer lethal electric shocks to helpless vicitms? Reflecting momentarily on what you would do, you conclude (by virtual reasoning, maybe), that you, and hence presumably they, would at the very least vigorously decline. But that is not the way many people react (they throw the switch after only minimal persuasion), and it is very possibly not the way you would react. If so, that suggests (but does not imply) that your inference to the contrary is not the result of

pure virtual reasoning as described above, but is conditioned by a widespread and fallacious belief about people's moral sensibilities.

7 Thomas C. Nagel, "What Is It Like to Be a Bat?," *Philosophical Review* 83 (1974).

8 Dennett, "Intentional Systems," 8. For relevant publications of the other authors just mentioned, see note 1.

9 Ibid., 11.

10 Ibid.

11 Ibid., "Making Sense of Ourselves," in Daniel C. Dennett, *The Intentional Stance* (Cambridge, MA: MIT Press 1987) 94,95.

12 C. Cherniak, "Rationality and the Structure of Memory," *Synthese* 57 (1983).

13 Stalnaker, *Inquiry*, 82.

14 Such a conclusion does not make allowance for attitudes that are, to use David Lewis' phrase, irreducibly *de se*, though Stalnaker has made such allowances elsewhere. But this detail is unimportant for present purposes. David Lewis (see works cited in note 1) holds a very similar position to that of Stalnaker.

15 Dennett, ibid., 100.

16 In the same article, a few pages earlier, Dennett appears to endorse a line of argument according to which what people actually do think is evidence for how they ought to think, the reasoning being that our judgments about the latter rely on our "intuitions about what makes sense." Thus (p. 98): "When considering what we *ought to do*, our reflections lead us eventually to a consideration of what we *in fact do*. . . ." So perhaps rationality is connected with psychological inference thus: psychological inference requires a theory (in a suitably broad sense of the word) of reasoning, and that theory of reasoning — i.e., our view of how we actually do reason — is ultimately the only evidence for our view about how we ought to reason.

Two points. First, even on the large assumption that Dennett is correct about the ultimate source of evidence about what we ought to do, it certainly does not follow that the principles of rationality which we arrive at will generally endorse human reasoning. Second, Dennett must beware of a strategic difficulty. He originally argued that psychological ascriptions would be unconstrained if we did not assume general rationality; now he claims that the ultimate evidence for our beliefs about rationality comes from how we think we (and others) *actually* think. But our having a view about how we (or others) actually think just *is* a matter of our making lots of psychological ascriptions. And what is to constrain those?

17 I am indebted to Gilbert Harman, Paul Thagard, Dieter Münch, and an anonymous referee for valuable help and suggestions.

CHAPTER FOURTEEN

Intrinsic Information

John D. Collier

INTRODUCTION

In everyday usage, information is knowledge or facts acquired or derived from study, instruction, or observation. Information is presumed to be both meaningful and veridical, and to have some appropriate connection to its object. Information might be misleading, but it can never be false. Standard information theory, on the other hand, as developed for communications (Shannon & Weaver 1949), measurement (Brillouin 1962), and computation (Solomonoff 1964; Kolmogorov 1968; Chaitin 1975), entirely ignores the semantic aspects of information. Thus it might seem to have little relevance to our common notion of information. This is especially true considering the range of applications of information theory found in the literature of a variety of fields. Assuming, however, that the mind works computationally and can get information about things via physical channels, then technical accounts of information strongly restrict any plausible account of the vulgar notion. Some recent information-oriented approaches to epistemology and semantics go further.

The usual sense of information is intentional: information is a property of representations that are meaningful to some subject. Systematic studies in epistemology (Dretske 1981) and semantics (Barwise & Perry 1983), though, suggest that information can be conveyed from the world to our minds. If so, concrete situations in the world can bear information. Despite this, we don't normally think of concrete situations as either representational or intrinsically meaningful. Dretske tries to resolve this problem by basing information on causal laws, which he claims *are* intentional. His claim is dubious unless the concept of intentionality is stretched beyond the limits of cognitive penetrability. Barwise and Perry also connect information to causality, holding that meaning exists in the world of inanimate objects, as well as in our thoughts and ideas. Whether or not the move to place meaning out in the world is accepted, if we assume even a modest realism there must be some property of objects that allows us to have information about them. This property must be causally based and causally communicable to us, as well as being commensurate with information in the vulgar sense. This paper explores the nature of this property.

I argue below that even if Dretske, Barwise, and Perry are right in thinking that meaning is out in the world, the property of interest must be characterizable in non-intentional terms. In the following sections I largely ignore questions of meaning, focusing only on non-intentional aspects of information. I will use the resources of contemporary information theory to sketch an account of information intrinsic to external objects. My project is modest. I wish to specify the characteristics of things in the world that allow them to be objects of representations.

There are reasons to suppose the concept of information can be usefully extended to the non-intentional world. First, cognitive systems are physical (and also biological). This is not analytic, but it is a "deep" fact about our world. Whatever allows meaningful information at the cognitive level is constructed from physical resources. Physical reality constrains the way cognitive systems can work, including how they can process and interpret information. It is possible that there are no interesting correlates of cognitive information outside of representations, but I believe this is empirically false.

The second consideration is less direct. Any explanation of the veridicality of our representations must appeal to a connection between them and the world. This connection, if knowledge is not purely coincidental, must be fairly regular and predictable. The only plausible candidate is some sort of causal relation. The connection need not be direct, since indirect or common causes would serve as well for information transmission (Dretske 1981:38). The world must contain either information itself, or else something that when properly connected to our cognitive processes is converted into information. This something is transmitted by physical and biological means; received information at the cognitive level must interface in a law-like manner with the transmissions.

I am going to stipulate that what is transmitted *is* information, irrespective of whether there is a cognitive receiver. This notion of information is more general than the common one, containing it as a species. It might be less confusing to use a new term for this broader notion, but technical usage has already extended information to the non-intentional realm. This move has produced insights into the interpretation of randomness and probability theory (Kolmogorov 1965, 1968; Chaitin 1975), as well as to computational complexity and the thermodynamics of computation (previous references; and Bennet and Landauer 1985; Bennett 1982, 1987). The extension of the concept is both relatively harmless and highly productive. Whether this indicates some deep truth about the world, I leave to the reader.

Information Viewed From The Top And From The Bottom

There are two approaches to understanding information which might be loosely labelled "top-down" and "bottom-up." The top-down approach starts with statements or other representations for which intentionality is taken for granted, and tries to specify what determines their information content. This approach has both formal and "informal" versions. The bottom-up approach starts with an account of the information content of concrete objects and works up to intentionality and beliefs. I will argue that the purely formal top-down approach fails because there is no equivocal way to assign empirically relevant a priori probabilities to statements. The informal top-down approach is more promising, but needs to be supplemented by the bottom-up approach. A pure bottom-up approach is likely possible (evolution seems to have accomplished it), but impractical at present.

An early formal version of the top-down approach is the Carnap/Bar-Hillel account of semantic information (Bar-Hillel 1964). They use the resources of inductive logic to define the information content of a statement in a given language in terms of the possible states it rules out. For "technical reasons" they calculate the states ruled out as a number of *state descriptions*. A state description is a conjunction of atomic statements assigning each primitive monadic predicate or its negation (but never both) to each individual constant of the language. The information content of a statement is thus relative to a language. Evidence, in the form of *observation statements*, contains information in virtue of the class of state descriptions the evidence rules out. (They assumed that observation statements can be connected to experience unambiguously.) Information content, then, is inversely related to probability, as intuition would suggest.

It turns out, though, that our pre-systematic intuitions confuse two different measures of information content, both of which have plausible but incompatible properties. The first measure of the information content of statement i is called the *content measure*, cont(i). It is defined as the complement of the a priori probability that i is true:

$$\text{cont(i)} = 1-\text{p(i)} \quad [1]$$

This measure fails the *additivity condition,* according to which the combined information content of two inductively independent statements[1] should be the sum of their individual information contents (Bar-Hillel 1964:302). It also fails some natural assumptions about conditional information. These problems motivated the introduction of another measure, called the *information measure,* inf(i):

$$\text{inf(i)} = \log_2 (1/(1\text{-cont(i)})) = \log_2 \text{p(i)} \quad [2]$$

The value of this measure is in bits. Although inf satifies additivity and conditionalization requirements, it has a property that some people find counter-intuitive. If some evidence e is negatively relevant to a statement i, then the information measure of i conditional on e will be greater than the absolute information measure of i. This violates a common intuition that the information of i given e must be less than or equal to the absolute information of i. The content measure, cont(i), does satisfy this intuition (Bar-Hillel 1964:306-7). Personally, I do not share this widespread intuition since it requires effort to correct the inference based on e that i is less likely. The issue requires further study, but is not relevant in what follows.

Elegant though the Carnap/Bar-Hillel account may be, it has problems. Because they use methods restricted to applied first order languages with identity, Carnap and Bar-Hillel cannot deal with complex scientific predicates like mass, temperature and energy. This could perhaps be corrected with a more sophisticated inductive logic using a more holistic view of theories. Any move towards a more holistic view, however, undermines the use of basic sentences to determine a priori probabilities. The probabilities of individual sentences of a theory become dependent on the overall a priori probability of the theory. At the least, this will greatly complicate calculations of information content. The demise of the analytic/synthetic distinction and the implausibility of pure observation statements presents a similar but even more fundamental problem.

This problem arises from what has misleadingly been called the theory-ladenness of observation. Observation statements, if they are not purely demonstrative, assign predicates to evidence. This amounts to a classification of the evidence. Individual bits of evidence are particulars that don't carry with them the rules for their own classification. If the information content of a statement is information about the world, not an artifact of the way evidence is classified, the assignment of predicates to experiences must reflect some sort of regularity in the evidence itself. Any non-arbitrary classification of evidence under some predicate presumes its similarity (at least) to other evidence or possible evidence (see Collier 1987). Barring astonishingly good luck, this presumes some information about appropriate similarities. Thus knowledge is presupposed in determining the information content of a piece of evidence; the information is not determined purely by syntactic relations. Something about the evidence itself, the workings of the mind, or both, must determine this knowledge. Since none of these are purely formal, the formal top-down approach cannot work.

The informal top-down approach resembles the structuralist approach to scientific theories, which uses informal model theory and empirical constraints to determine acceptable interpretations of scientific theories (see Stegmüller 1979 for a review). The approach starts with meaningful representations and tries to specify their interpretation, making use of available empirical constraints. Barwise and Perry (1983) are engaged in this project. On their view, the interpretation of a representation is given in terms of the information it conveys. Unlike the formal top-down approach, in which information content is determined entirely by the structure of language (or other representational system), information in this approach is the content (or factual content) of a representation (or information-report) (Israel & Perry, this volume).

The goal of this approach is to connect meaningful representations to the concrete situations represented. Perry and Barwise base this connection on nomic regularities, which they call "constraints" (Israel & Perry, this volume). Information is conveyed to us by causal chains connecting situations in a lawful way. The information indicated by a situation is relative to the causal chains connecting the indicating situation both to our beliefs and to the situation the information is about. Thus, "The information a factual state of affairs carries is relative to a constraint." Complete determination of the reference of a representation (at least in cases involving indexical terms) also requires specific circumstances. The information content of a representation available to us is delimited by our knowledge of these constraints and circumstances. Further evidence could modify this available information, and it may be that complete knowledge would determine the absolute information content of a representation.

There is something "out there in the world" that can be transmitted to intelligent beings who can understand the information it contains, and pass it around among themselves. What the transmitted something is called at various stages along the route is not all that important. Either this something is pre-existing meaningful information or else it can be converted by cognition into meaningful information. If conversion is necessary, since the content of representations is characterized in intentional language, whereas the referents of veridical representations don't require intentional characterization, an account of reference should characterize referents in non-intentional terms commensurate with the characterization of representational content, otherwise there will be a gap. Barwise and Perry (1983:94) held that meaning pre-exists in the world, hoping thereby to avoid the gap. Their choice, though, is question-begging unless there is a characterization of things out in the world that is commensurate with intentional language.

Normally, things in the external world are not described in intentional terms. The Barwise/Perry approach needs an information-theoretic account of nomic regularities and causal interactions, and of the transmission of the information these nomic regularities and causal interactions contain. This account will be largely empirical. If nothing else, factual constraints are the only way to rule out clever but fanciful exceptions. The selection of an appropriate characterization, though, is a philosophical matter.

Superficially, Dretske's (1981) approach to information resembles the Carnap/Bar-Hillel approach. He also defines the information content of a piece of evidence in terms of the cases ruled out. A major difference is that Dretske does not try to specify representations purely syntactically. Rather than calculating the information content of statements, he uses states of affairs directly. His measure of information is similar to the inf definition (Dretske, 1981:52):

$$I(s) = -\log_2 p(s), \quad \text{in bits,} \qquad [3]$$

where p(s) is the probability of the state of affairs s.

From the examples and discussion of the first part of his book, Dretske appears to adopt a variety of the informal top-down approach. He describes perceived situations that rule out possibilities until only one relevant possibility is left (e.g., 1981:4-6, 48, 53, 78, 95). Dretske's treatment of intentionality in this part of the book supports this interpretation: "The ultimate source of intentionality inherent in the transmission and receipt of information is, of course, the *nomic regularities* on which the transmission of information depends" (1981:76). This is similar to Barwise and Perry's placement of meaning in the world. Although it is doubtful that the presumed intentionality of laws[2] satisfies Dretske's later characterization of intentionality (Dretske 1981:172-3), there is no need to pursue the matter here. Even if we accept that natural laws are intentional, the concrete situations that manifest them don't need intentional characterization, and they are not normally characterized intentionally. The gap in the Barwise/Perry treatment exists for Dretske as well.

Dretske's definition of information in terms of the cases ruled out might seem to fall afoul of the problem of background knowledge that plagues the Carnap/Bar-Hillel approach. Dretske admits that information received, as commonly understood, is relative to background knowledge (1981:80-1), but argues (1981:87) that the background knowledge can be treated in the same way as the received knowledge. This process is iterated until all background knowledge is accounted for. The problem with this solution is that prior knowledge of what possibilities need to be eliminated still seems to be necessary. Dretske meets this objection by turning in the second part of his book to the bottom-up approach.[3]

The idea behind the bottom-up approach is that concrete situations are sources of information that can be transmitted through channels that don't themselves generate new information (Dretske 1981:115; see also below, section on information transmission, pp. 00-00). Information is transmitted (perhaps indirectly) from structure to structure according to causal laws. If it is transmitted to a structure with the right order of intentionality[4] there is knowledge. The causal processes producing beliefs eliminate other possibilities from consideration; what is important is not that we have reasons for rejecting other possibilities, but that the correct causal connections are made. This approach is bottom-up because we start with information out in the world, and ask what makes it into the content of a belief.

There is a lot of information in the world. Much of it isn't very interesting to creatures like us. Some information is interesting and useful, though, and we are attuned to picking it up (see Israel & Perry, this volume). As long as there is a reliable causal connection (possibly indirect) between the source of the information and the higher order intentional structures Dretske describes, we have knowledge. We may have sceptical doubts about the efficacy of the channels involved, leading to doubts about the nature of the source, but these doubts are irrelevant to whether or not we really have knowledge. The bottom-up approach allows us to have knowledge without requiring justification. We do not need to eliminate all possibilities other than the correct one, or to know what the alternative possibilities are. The correct causal connections through reliable channels do all this work for us.

The bottom-up approach disarms sceptical doubts as an objection against the possibility of knowledge. This gain has its cost, though. Our pre-systematic intuitions about information are not suited to the bottom-up approach. We can no longer think of information in terms of the reduction of predetermined possibilities. The reduction is the *effect* of the receipt of information on us. The information source must have some intrinsic property that produces this effect. The whole approach is somewhat fanciful if the source does not really contain information (or some equivalent) capable of having this effect. The bottom-up approach needs a definition of intrinsic information content.

INTRINSIC INFORMATION

Physical things have properties that give them a definite structure and causal capabilities.[5] If information is an intrinsic property of physical objects, then it seems likely that it is contained in their physical structure. Brillouin (1962:152) distinguished two types of information: free and bound. The two forms of information differ in how they are regarded; Brillouin did not rule out the possibility that they might refer to the same thing. Free information depends on possible cases which are regarded as abstract and have no specific physical significance. Bound information depends on possible cases which are the complexions[6] of a physical system; it is a special case of free information. The information of a concrete situation, then, if it is either of these, is bound information.

On Brillouin's account, the bound information of a macroscopic system is inversely related to its entropy and directly related to its negentropy, the difference between the maximal entropy of the system and its actual entropy. Negentropy and information are reversibly interconvertible, and Brillouin may have considered them to be identical. A measurement, which produces information, must also produce negentropy, which must, according to the second law of thermodynamics, be produced at the expense of producing entropy someplace else. The duplication of information requires the conversion of available energy into entropy someplace in the system or its surroundings. Likewise, the transfer of information must be compensated for by the production of entropy in the system from which it is transferred. If a measured system is open, its entropy can remain constant (effectively duplicating the information transmitted), but the entropy of its surroundings must increase.

Landsberg (1984), following Layzer (1975), defined the order of a system in terms of the difference between its actual entropy and its maximal entropy:

$$\text{Order} = S_{max} - S_{act} \qquad [4]$$

where

$$S_{max} = k \log P, \qquad [5]$$

and

$$S_{act} = -k \sum_{i} \log[p(m_i)] \qquad [6],$$

where $p(m_i)$ is the probability of the ith microstate, and P is the number of elementary complexions of the system. The constant k can be eliminated by choosing base 2 for the logarithms. This gives the entropy in dimensionless entropy units (Brillouin 1962:118). Order, in this case, has the same form as the inf definition of information content. S_{max} is reached when all complexions are equally likely. This is the equilibrium condition. The order is non-zero only if the system is not in thermodynamic equilibrium, i.e., if some elementary complexions are more probable than others.

The order is sometimes called the "intropy" of the system (in contrast to its entropy). It is well defined only if the actual entropy S_{act} and the equilibrium entropy S_{max} are well defined. Both entropies are well defined only if the elementary complexions (the possible microstates) of the system are well defined. This requires that the macrostate be well-defined, or else it is not clear which microstates should be included as possible. As in the top-down approach, information in the statistical version of the bottom-up approach requires reference to possibilities. In this case, however, we do not need to know what the possibilities are; they are determined by the physical circumstances that prevail.

Although many writers have assumed that macrostates are subjective, I doubt they are right. Irreversibility is a consequence of the second law of thermodynamics under non-equilibrium conditions. The explanation of the second law is in terms of statistical mechanics, which assumes the objectivity of microstates. Thus irreversible chemical reactions depend (factually) on the objectivity of macroscopic phenomena. The existence of our bodies depends on such irreversible reactions; its maintenance in the living state depends on its not being at thermodynamic equilibrium. The modest assumption that the mind depends on the body, and that subjectivity depends on the mind, leads immediately to the conclusion that if macrostates are subjective, then a precondition for the mind's existence is its own existence. This is patently absurd, as Prigogine (Prigogine & Stengers 1984) has pointed out. One of these assumptions must go.

I maintain that macrostates have a mind-independent existence. I have argued elsewhere (Collier 1988b) that the basis of the objectivity of macrostates is *cohesion*.[7] Cohesion is produced by the causal interactions among the parts of a system that make it insensitive to fluctuations in its microstates. This insensitivity is not absolute; large enough perturbations can destroy a cohesive macroscopic system. Nonetheless, if cohesion is strong enough, there is a causal basis for macroscopic entities and their macroscopic properties. Cohesion keeps the macrostate stable while allowing minor external influences to change the microstate unpredictably, given the macrostate. These external influences are empirically likely, and justify a statistical treatment of microstates (Prigogine 1962:265-9), in turn justifying the use of probabilities in equation [6] above.

Cohesion acts as a constraint on the range of possible states of the system, determining not only the objectivity of the macrostate, but also the value of S_{max} for the system. Thus the basis of the objectivity of entropy is also the basis of the objectivity of intropy. Cohesive entities at one level can serve as the basis for the complexions at higher levels, allowing application of the statistical methods of modern thermodynamics to very large scale entities (e.g., Ulanowicz 1986; Brooks & Wiley, 1986, 1988; Collier 1986, 1988b). This allows the notion of intropy and its related information to be applied to a wide range of systems. Since cohesion guarantees the irrelevance of lower level fluctuations to the macrostate (Collier 1988b), the intropy of a cohesive system depends only on the states of the next lower cohesive level. Fluctuations at still lower levels are screened off by the cohesion of the elements at this level. This dramatically simplifies the calculation of intropy. For example, to determine the information in a gas we don't need to consider the nuclear microstates of its atoms.

Many situations are not cohesive, since they lack the necessary internal causal connections. These non-cohesive situations are epiphenomenal.[8] Epiphenomenal situations are either part of a larger cohesive situation that provides their causal basis, or are non-cohesive combinations of cohesive parts. In the first case, the intropy of the situation is determined by how it alters the probability of the microstates of the inclusive situation from the equilibrium expectation, i.e., according to equation [6]. Equation [4] then gives the intropy of the situation. The intropy of the partial situation will generally be lower than the intropy of the inclusive cohesive situation. In the second case, the intropy of the situation is the sum of the intropies of its cohesive parts. (Any apparent order of the parts has no macroscopic effect, since by assumption there is no cohesion among the parts, therefore this order can have no physical effect.) In both cases, the intropy of the situation is due entirely to the factual conditions that make it up. This is as required for a physical basis for the information of concrete situations.

Unfortunately, things are not quite this simple. The intropy is not the only part of the system that bears information. The causal relations producing cohesion also contain information. (This information is called *constraints* in information theory. It is not internal to the system in a special technical sense. See Collier 1986 for a discussion of internal information and constraints). For many purposes, like analysis of the self-organization and evolution of systems, the internal information is all that need be considered in order to understand system behaviour. For perceptual and semantic purposes, however, we must presume an external perspective. The constraints on the system are part of its complete description.

Some examples will be helpful here. The statistical measure of information (intropy) applies best to the idealized case of an ideal gas made up of non-interacting point sized molecules. The intropy ranges from zero to some maximal value depending solely on the position and momentum distribution of the molecules. The intropy is equivalent to the ability of the system to do work internally. It is zero at equilibrium, and non-zero under non-equilibrium conditions. Another extreme type of system is a crystal. A perfect crystal has only one possible microstate for its macrostate. It is completely ordered. Its information content is very low, as is its entropy. There can be no difference between S_{act} and S_{max}. The order of the crystal comes from causal interactions between its molecules, which confine it to an entirely rigid structure, so that only its size and location remain undetermined. Of course there are no perfect crystals, but actual systems can approximate this state. A more interesting case is a configuration which has a great deal of variety in its structure, but is causally constrained to take on only one possible state. An example is a state of the physical embodiment of a message with low redundancy, such as the signal in a telephone cable. In this case, as in a crystal, S_{act} and S_{max} are numerically close, but the information content of the state is intuitively very high.

The low intropy of both a low information crystal and a high information message shows that the information due to constraints cannot be intropy. Because it is based in the form of the system, I call it "enformation." Production and loss of enformation also involves loss and production of entropy, just as for intropy. The entropy increase, however, must be external to the system, whereas for intropy the changes can be internal as well.[9] The intrinsic information of a system, relative to its causal constraints, is the sum of its intropy and enformation:

$$I_{int} = \text{Intropy} + \text{Enformation} \quad [7]$$

Changes in intrinsic information depend on changes in entropy, but I_{int} depends on whatever determines the quantity of enformation as well as on S_{act} and S_{max}.

The concept of enformation could be avoided by enlarging the system to include the observer and the processes by which information is conveyed from the object to the observer. All information would be internal to the system, so that it could be treated as intropy. This approach has merit, but it runs into both practical and methodological problems. The approach is complicated and requires knowledge of physical and physiological processes we don't yet understand very well. Furthermore, information about an object would be contained entirely in the perceptual processes; the object would create only external constraints on this information. This goes against the idea that information contained in the object is transmitted to the observer.

Another way to avoid enformation is to use an absolute entropy not relative to any constraints. This approach also runs into trouble. In order to define an absolute entropy for a system, we need to know its state when all physical constraints have fully relaxed (i.e., the system has reached equilibrium). This state cannot be determined empirically for two reasons. First, it isn't known if all physical structures fully relax; for example, it is not known whether protons decay. It might be possible to use the equilibrium state of the energy equivalent to the matter in a system to define its S_{max}, but there is no guarantee that this is the highest entropy state. Furthermore, it is unclear in what form the energy should be assumed to be. Second, the system must be assumed to be closed. This assumption is unrealistic for anything less than the whole universe. The universe is expanding, so even the whole universe does not have an absolute S_{max}; S_{max} is always relative to the size of the universe at some time. We don't know enough to specify absolute entropies, and even if we did, there is reason to think they do not exist. These considerations are somewhat esoteric. I include them to indicate the difficulties in giving a purely statistical account of intrinsic information.

Even if we could define absolute entropies, the intropy obtained would not be very useful since we are usually concerned with small parts of the universe that are heavily constrained (in the short run, at least). It would be much more practical to find a measure of enformation. Standard approaches to information theory (e.g., Shannon & Weaver 1949; Brillouin 1962) use combinatorial or probabalistic definitions of the amount of information (Kolmogorov 1965). These definitions can be applied to intropy, but not to enformation. The standard approaches apply only to ensembles of systems; we need a conception of information which is applicable to individual cases. Algorithmic information theory (Kolmogorov 1965, 1968; Chaitin 1975) has this property.

The fundamental hypothesis of algorithmic information theory is that the information content of something is the length of the shortest program in binary form that can produce it. The idea behind this hypothesis is that a thing can be specified by making a series of binary distinctions (Spencer Brown 1972). The minimum number of distinctions required is its information content in bits.[10] The algorithmic definition of information content is equivalent to the combinatorial and probabalistic information contents except for an additive constant representing computational overhead (Kolmogorov 1968) that can be made arbitrarily small (Chaitin 1975). The basic concepts of information theory can be defined without recourse to probability theory, and are applicable to individual cases (Kolmogorov 1968). Furthermore, the relations between information and probability allow probability theory to be based on algorithmic information theory.

Algorithmic information theory is very abstract in its formulation. By analogy to Brillouin's (1962) definitions, we can distinguish between free algorithmic information and bound algorithmic information. The latter depends on computations which are physically possible. Computations, in this interpretation, are causal processes obeying natural laws. The information content of a state is determined by the most parsimonious causal process that can produce it from disorganized constituents. This is equal to its causal power, which is the measure of the number of distinctions the state can causally produce. The enformation is the bound algorithmic information of the constraints.

If we consider the total initial and total final states of any causal process, the information content of the final state must be less than or equal to the information content of the initial state. This is just a restatement of the second law of thermodynamics. If the process is 100 per cent efficient (i.e., there is no dissipation of available energy into lower grade forms), the total information content remains constant. If a state is causally effective, but is itself unchanged in a causal process, either the resulting final state contains only components of unchanged information content, or else information must be dissipated in one part of the total system in order to produce it elsewhere. This is the computational analogue to Brillouin's discussion of the negentropy budget. A system can causally maintain and change its form without dissipation of information, but this requires total efficiency of the processes involved. This will occur only if there is no friction involved, e.g., for state conditions, or if the system does not deviate from equilibrium.

The enformation and intropy of a system are independent if there is no dissipation. Enformation is converted into intropy if the rate of equilibration (relaxation time) of the system is greater than the rate of loss of enformation, i.e., under non- equilibrium conditions. Equilibrium and/or frictionless conditions are rarely if ever encountered in nature. Alternatively, intropy can do work on the system, sometimes producing enformation. I mention the close theoretical relations between enformation and intropy in order to strengthen the intuition that there is a common property of which they are different forms.

To summarize, the causal power of a state of a system has two components, its enformation and its intropy. Together, these constitute the information content of the state. Intropy is the ability to do work within the system; it represents the energetic aspect of causal power. Enformation, on the other hand, is the ability to alter the form of things; it represents the organizational aspect of causation, the ability to guide energy. Both are required for useful work. This simple distinction is complicated by the inter-convertibility of intropy and enformation. Work produces new enformation. Enformation can also produce intropy at expense of its own destruction if it is dissipated under non-equilibrium conditions. If friction is present in any causal process some information is lost; it becomes equivocation, or physical entropy. The general idea should be clear by now: *intrinsic information is a measure of causal power*. The less information, the less causal power.

It is now possible to define the intrinsic information of a situation. Situations contain both enformation and intropy. Many situations are not cohesive; they do not represent independent objects or systems, what might be termed individuals. Instead, they are epiphenomenal. The intropy and enformation of cohesive situations is not problematic. The intropy of epiphenomenal situations can be defined with respect to the cohesive situations (see above). Similarly, the enformation of an epiphenomenal situation is a part of the enformation of any inclusive cohesive system. Its enformation is the sum of the enformation of the cohesive parts that compose it. This enformation is equal to the sum of the algorithmic information of the most efficient causal processes that can make the parts. In other words, the enformation of an epiphenomenal situation is derivative from the enformation of the parts that make it up.

The direct relationship between information and causal power may seem paradoxical from the top-down view of information as a reduction of possibilities. It would seem that a situation with less information would allow more possibilities than one with more information, and consequently would have more causal capabilities. This, however, conflicts with our intuition that more complex things are capable of doing more than simpler things. The bottom-up approach is more in accord with this intuition. A situation with less intropy is capable of doing less work than one with more. Likewise, a situation with less enformation is capable of producing fewer distinctions than one with more enformation. It isn't the range of different abstract possibilities which determine intrinsic information content, but the productive power of a situation in optimal cases.

Information Transmission

The basics of information transmission are fairly easy to understand, given the above account of intrinsic information. Since intrinsic information is a measure of causal capacity, it can be detected through its effects. An information channel is affected by a source, according to its ability to do so, but is not significantly affected by other sources whose effects on the channel might interfere with those of the information source (see Dretske 1981:115). If there is extraneous interference, it produces equivocation. Equivocation results in lost information, because the effect on the channel is the net result of the various causal influences acting on it. Since there is only one effect for various causes, the individual causes cannot be discriminated on the basis of the intrinsic information in the channel alone. The desired information from a source can sometimes be detected by *tuning* the receiver correctly. This involves finding a channel with minimal interference from other sources. Such a channel can be termed *reliable*.

Our senses act as filters that select generally reliable channels. Presumably evolution has selected our particular sensory mechanisms because, among other things, they are reliable. At higher levels of cognition we learn to discriminate reliable channels from unreliable ones, ignoring the unreliable ones by using classifications that don't suit them. This sort of tuning allows us to detect abstract and general properties. This selection process allows us to receive information about particular situations (within the limits of the available reliable channels). Tuning for abstract and general properties creates classification systems that reflect natural classifications.

A consequence of this view of sensation and classification is that the selection of reliable channels will maximize transmitted information. The meaning conveyed from the world will thus be greatest. If unreliable (i.e., noisy or equivocal) channels are selected, the meaning conveyed will be diffuse and equivocal, even though it might give the illusion of clarity.

A channel cannot transmit more information than the intrinsic information of its states. However, a situation distant from, but causally connected to an information source can indicate more information about the source than it contains itself (Dretske 1981; Barwise & Perry 1983). For example, the decay of tritium might be the only thing which produces a gamma ray of a certain energy, so a photon with the correct frequency would carry the information that a tritium atom had decayed, even though the enformation of the photon is less than the enformation of either a tritium atom or its other decay products. The extra information is contained in the natural laws of radioactive decay and gamma ray production. How is this information transmitted?

Dretske (1981), as I mentioned above (p. 396), tried first to deal with background information by assuming it is obtained in the same way as the information that is directly present. This leaves open the objection that background information about the field of possibilities is still required. Dretske turned to the bottom-up approach to allow the causal conditions involved in the information transmitted to have the reduction of alternative possibilities as an effect. We don't, therefore, need information about the eliminated possibilities. The case of tritium decay, though, is not obviously amenable to this sort of treatment. Dretske's treatment requires that the information transmitted from the indicating situation (the gamma rays) together with previous transmitted information that we have retained determines the presence of a tritium decay. But if natural laws are the only thing that ensures that there was a tritium decay, then part of the information required is information transmitted to us from natural laws. It isn't clear how the information in a natural law can be transmitted. The universality of natural laws implies that information transmitted by any particular cases is always deficient. It seems that an inference beyond the available information is always required. This seems to open a gap into which the wedge if scepticism can be driven.

If we consider theories to be systems for classifying and organizing empirical evidence, then a true theory is one that selects reliable channels for all cases within its scope. If so, every instance of a natural law within this scope will be transmitted without corruption. This eliminates the need for inference. Usually, though, selected channels are less than completely reliable. This doesn't mean that we can't be right; some information will probably get through uncorrupted, except for very bad choices. It is true that choice of theory, and thus channel selection, involves inferences beyond immediate evidence, but if reliable channels are selected, knowledge is the result. On this approach inference plays a role in the production and identification of knowledge, but is not a constituent of knowledge.

FURTHER DEVELOPMENT OF THE BOTTOM-UP APPROACH

As I mentioned above in the section on top-down and bottom-up information, the latter approach requires an account of intrinsic information, an account of information transmission, and an account of the conversion of information into a cognitive form. In this paper I have focused on intrinsic information in particular concrete situations. I have also sketched an account of information transmission, but have had little to say about its conversion into cognitive form.

Further development of the account of transmission would involve a better characterization of noise and its avoidance, and of the selection process called tuning. Israel and Perry (this volume) discuss tuning from a biological perspective. Natural selection no doubt plays a central role in determining our basic repertoire of channels. The extension of biological concepts to cognition is most naturally carried out in the context of developmental psychology, along the general lines proposed by Piaget (1952). Piaget sees cognitive development as continuous with both ontogenetic and phylogenetic development. As Matthen (1988) has pointed out, ways of representing the world that are innate for the individual organism must be acquired by the lineage of which it is a member. Biological conditions provide not only an analogy for cognitive development, but are its foundation. Top-down philosophical and empirical studies of cognition, intentionality, and inductive logic place further constraints on acceptable accounts.

The transformation of raw information into cognitive content remains fairly mysterious. The continuity between biological selection and cognitive selection suggests that the roots of the transformation lie in evolution. Matthen (1988), for example, discusses biological functions from an evolutionary perspective, and sketches how it might shed light on how cognitive content appears. Adaptations are a form of tuning selected by the environment. On the present account, learning is also a process of tuning, suggesting that it is analogous to adaptation. This is in line with the general trends in Piaget's thought. Problems arise, though, from the fact that what is useful is not necessarily true. For example, Grandy (1987) used an ecological context to discuss perceptual content as mutual information of a representation and an organism's environment. He observed that optimal behaviour in conditions with less than complete information may favour having mistaken beliefs. This suggests that the selection of reliable channels is neither direct nor inevitable.

A full information-theoretic account from the biological perspective is required to make connections between raw physical information and cognitive content. Initial studies of this project have been done by Brooks and Wiley (1986, 1988), Collier (1986), and Ulanowicz (1986). These studies are still in their infancy, and there is much to be clarified about the interactions of biological organization, ontogeny, ecology, and evolution. The approach seems promising, since the basis of contemporary systematic biology, the genetic code, has both physical and representational properties that are amenable to information theoretic analysis (Collier 1988a). Ulanowicz (1986) has used the connections between information and entropy to do information theoretic analyses of ecology and fitness. This work needs to be connected to information theoretic studies of ontogeny and evolution before connections can be made between intrinsic information, adaptation, and eventually learning.

The current state of information-theoretic approaches to knowledge and semantics is very rudimentary. The applicability of related concepts at a variety of levels and to a number of apparently disparate applications suggests a large consilience. The major problem is to find general concepts that are broad enough to apply to a variety of applications, but concrete enough to avoid being vacuous. I believe that keeping close to the physical and causal basis of information will avoid vacuity. Generality will come from resolving problems that arise at the interface of different applications.

NOTES

I am grateful to the History and Philosophy of Science Department at Indiana University for providing me with office space and research facilities during the preparation of this paper. Final revisions were done during the tenure of a SSHRC Canada Research fellowship at the University of Calgary.

1 Inductive independence means that the conditional probability of each statement given the other is the same as its initial probability.

2 Dretske argues from the modal quality of laws.

3 Remnants of the top-down approach of the first part continue to appear throughout his book.

4 See (Dretske 1981:171-89) for a discussion of how higher orders of intentionality are formed from structures of lower order.

5 I will leave the question of whether it is possible for the same thing to have different causal capabilities under different natural laws open, since this question is irrelevant to intrinsic information content.

6 The complexions of a system are the possible microscopic states it can be in, given its macrostate.

7 It is possible to imagine non-cohesive macrostates. These are artificial, and have

no macroscopic causal properties. An example would be the state of the gas molecules in an arbitrary unbounded one metre sphere.

8 Often our interactions with situations will create a larger system with closed loops which can generate cohesion, as when we sort a deck of cards into suits, but this is certainly not necessary for something to be a situation.

9 Although he does not distinguish explicitly between intropy and enformation, Holzmüller (1984) discusses the relations of enformation and entropy in macromolecules.

10 The logic of distinctions can be shown to be equivalent to Boolean algebra (Banaschewski 1977).

References

Banaschewski, B. (1977). On G. Spencer Brown's laws of form. *Notre Dame Journal of Formal Logic* 18:507–9

Bar-Hillel, Y. (1964). *Language and Information*, esp. Chs. 15–17. Reading, MA: Addison-Wesley

Barwise, J. and J. Perry (1983). *Situations and Attitudes*. Cambridge, MA: MIT Press

Bennett, C.H. (1982). The Thermodynamics of Computation: A Review. *International Journal of Theoretical Physics* 21:905–10

— (1987). Demons, Engines and the Second Law. *Scientific American* 257:108–16

Bennett, C.H. and R. Landauer (1985). The fundamental physical Limits of Computation. *Scientific American* 256:48–56

Brillouin, L. (1962). *Science and Information Theory*. New York: Academic Press

Brooks, D.R. and E.O. Wiley (1986). *Evolution as Entropy*. Chicago: University of Chicago Press

Brooks, D.R. and E.O. Wiley (1988). *Evolution as Entropy*, 2nd ed. Chicago: University of Chicago Press

Chaitin, G.J. (1975). A theory of program size formally identical to information theory. *J. ACM* 22:329–40

Collier, J.D. (1986). Entropy in Evolution. *Biology and Philosophy*, 1:5–24

— (1987). "Theory ladenness," in P. Weingartner and G. Schurz (eds.), *Logic, Epistemology and Philosophy of Science: Proceedings of the 11th Annual Wittgenstein Congress*, 116–18

— (1988a). The Dynamics of biological order, in Depew, Weber, and Smith (1988)

— (1988b). Supervenience and reduction in biological hierarchichs. In M. Matthen and B. Linsky (eds.), *Biology and Philosophy, Canadian Journal of Philosophy Supplementary*, Vol. 14

Depew, D.J., B.H., Weber, and J.D. Smith (1988). *Entropy, Information and Evolution*. Cambridge, MA: MIT Press

Dretske, F. (1981). *Knowledge and the Flow of Information*. Cambridge, MA: MIT Press

Grandy, R. (1987). Information-based epistemology, ecological epistemology and epistemology naturalized. *Synthese* 70:171–204

Holzmüller, W. (1984). *Information in Biological Systems: The Role of Macromolecules*. Cambridge, Eng.: Cambridge University Press

Kolmogorov, A.N. (1965). "Three approaches to the quantitative definition of information." *Problems of Information Transmission* 1:1–7

— (1968). "Logical basis for information theory and probability theory." *IEEE Transactions on Information Theory* 14:662–4

Landsberg, P.T. (1984). Can entropy and order increase together? *Physics Letters* 102A:171–3

Layzer, D. (1975). "The Arrow of Time." *Scientific American* 233:56–69

Matthen, M. (1988). Biological functions and perceptual content. *Journal of Philosophy* 85:5–27

Piaget, J. (1963). *The Origins of Intelligence in Children*. New York: Norton

Prigogine, I. (1962). *Non-Equilibrium Statistical Mechanics*. New York: John Wiley and Sons

— and Stengers, I. (1984). *Order Out of Chaos*. New York: Bantam

Shannon, C.E. and Weaver, W. (1949). *The Mathematical Theory of Communication*. Urbana: University of Illinois Press

Solomonoff, R.J. (1964). A formal theory of inductive inference, *Inform. and Contr*. 7:1–22

Spencer Brown, G. (1972). *Laws of Form*. New York: Bantam

Stegmüller, W. (1979). *The Structuralist View of Theories*. New York: Springer-Verlag

Ulanowicz, R.E. (1986). *Growth and Development*. New York: Springer-Verlag

Contributors

Nicholas Asher, Department of Philosophy and Center for Cognitive Science, University of Texas
Lee R. Brooks, Department of Psychology, McMaster University
John D. Collier, Department of Philosophy, University of Calgary
Fred Dretske, Department of Philosophy, University of Wisconsin
Jerry Fodor, Department of Philosophy, Rutgers University, and Graduate Center, City University of New York
Robert F. Hadley, School of Computing Science, Simon Fraser University
John W. Heintz, Department of Philosophy, University of Calgary
David Israel, Artificial Intelligence Center, Stanford Research Institute International, and Center for the Study of Language and Information, Stanford University
Ali Akhtar Kazmi, Department of Philosophy, University of Calgary
David Kirsh, Cognitive Science Program, University of California at San Diego
Fred Landman, Department of Modern Languages and Linguistics, Cornell University
M. Andrew Moshier, Program in Computing, University of California at Los Angeles
John Perry, Department of Philosophy and Center for the Study of Language and Information, Stanford University
Carl J. Pollard, Department of Philosophy, Carnegie Mellon University
Ian Pratt, Department of Computer Science, University of Manchester
Zenon W. Pylyshyn, Department of Psychology and Center for Cognitive Science, University of Western Ontario

Brian Cantwell Smith, System Sciences Lab, Xerox Palo Alto Research Center, and Center for the Study of Language and Information, Stanford University
Scott Soames, Department of Philosophy, Princeton University
Edward P. Stabler Jr., Department of Linguistics, University of California at Los Angeles
Kim Sterelny, Department of Philosophy, Victoria University of Wellington, NZ
Paul Thagard, Cognitive Science Lab, Princeton University